Endocrine Disrupters in Wastewater and Sludge Treatment Processes

Endocrine Disrupters in Wastewater and Sludge Treatment Processes

Jason W. Birkett
John N. Lester

CRC Press
Taylor & Francis Group
Boca Raton London New York

CRC Press is an imprint of the
Taylor & Francis Group, an **informa** business

First published 2003 by Liwes Publishers

Published 2019 by CRC Press
Taylor & Francis Group
6000 Broken Sound Parkway NW, Suite 300
Boca Raton, FL 33487-2742

© 2003 by Taylor & Francis Group, LLC
CRC Press is an imprint of Taylor & Francis Group, an Informa business

First issued in paperback 2019

No claim to original U.S. Government works

ISBN 13: 978-0-367-45467-8 (pbk)
ISBN 13: 978-1-56670-601-8 (hbk)

Visit the Taylor & Francis Web site at
http://www.taylorandfrancis.com

and the CRC Press Web site at
http://www.crcpress.com

Library of Congress Card Number 2002016144

Library of Congress Cataloging-in-Publication Data

Endocrine disrupters in wastewater and sludge treatment processes / edited by J.W. Birkett and J.N. Lester.
p. cm.
Includes bibliographical references and index.
   ISBN 1-56670-601-7 (alk. paper)
   1. Reproductive toxicology. 2. Endocrine toxicology. 3. Sewage — purification — Organic compounds removal. 4. Sewage sludge — health aspects. I. Birkett, J.W. (Jason W.), 1970–. II. Lester, J.N. (John Norman), 1949–.
RA1224.2.E62 2002
571.9'.51—dc21   2002016144
                                                                                      CIP

Dedication

To Dr. Ka-man Lai, whose work at Imperial College in 1998 on natural and synthetic estrogens and their environmental impact inspired the editors and contributors to write this book.

Preface

Although the hypothesis that environmental chemicals may exhibit endocrine disrupting effects is not new, being raised by Allen and Doisy in 1924, again by Dodds et al. in 1938, and in the 1950s by Burlington and Lindeman, the issue has seen a growing level of concern due to reports of increased incidences of endocrine-related diseases in humans, including declining male fertility, and more significantly, to adverse physiological effects observed in wildlife where cause and effect relationships are more evident. In fact, the evidence from these effects in wildlife populations has suggested that the changes in the reproductive health of humans, including breast and testicular cancer, birth defects, and declining sperm counts, could be linked to exposure to endocrine disrupting chemicals. However, no definitive cause and effect data have yet been established.

While society has released large amounts of man-made chemicals to the environment since the 1940s, people born between 1950 and 1960 were the first generation to suffer exposure to these pollutants (from stores in maternal fat tissue) while they were growing in the womb. During the 1950s and 1960s, the pesticide DDT, which was used worldwide in vast quantities after 1945, was shown to be estrogenic and to affect the reproductive systems of mammals and birds. These findings, together with the publication in 1962 of *Silent Spring* by Rachel Carson, further highlighted the health problems afflicting some wildlife (e.g., egg shell thinning, deformities, and population declines) that were exposed to pesticides and other synthetic chemicals. The link to pollution was also thought to be a possible cause of the effects seen in the human reproductive system, with papers such as "Are estrogens involved in falling sperm counts and disorders of the male reproductive tract?", published in *The Lancet* in May 1993 provoking wider debate within the scientific community. In 1996, the publication of *Our Stolen Future* brought this complex scientific issue to the attention of the general public by providing a readable account of how man-made chemicals can disrupt hormonal systems. Society and politicians are now asking scientists if there is a link between exposure to pollutants and effects on the endocrine system.

Human health concerns are related to the effects on individuals, and the use of a synthetic steroid estrogen, diethylstilbestrol (DES), in the 1970s was demonstrated to cause abnormalities in the offspring of women treated with the drug. However, within the environment, the individual is of less immediate concern, and attention is focused on effects on the population level. The use of tributyl tin (TBT) as an antifouling agent in paints applied to the hulls of boats was observed to cause deformities in oysters and to be responsible for alarming declines in the populations of dog whelks. As a result, the use of this compound was restricted, with its use being phased out even further through international cooperation. Chemicals that can interfere with the natural functioning of the endocrine system have been termed endocrine disrupting chemicals (EDCs). The list of EDCs includes a range of anthropogenic compounds, for example, biocides, polychlorinated dioxins, furans and biphenyls, bisphenol-A, phthalates, and alkylphenolic compounds, among others. Naturally occurring steroid estrogens (estrone, estradiol) and synthetic estrogens (DES, ethynyl estradiol) are also implicated. While many EDCs identified to date have weak activity when compared with their endogenous counterparts, they can be present in significant concentrations in the environment, and furthermore, many bioaccumulate in fatty tissues of the body due to their lipophilic nature, increasing their concentration and bioavailability.

Within the aquatic environment, the presence of EDCs has concerned many scientists and water quality regulators. A comprehensive survey of estrogenic responses in fish undertaken throughout the U.K. revealed that chemicals in the water — generally natural steroid estrogens and alkyphenolic compounds — were responsible for the induction of abnormalities in male fish. These intersex fish were found downstream of many domestic sewage effluent outfalls in the U.K. As a result, an increasing number of scientists are investigating the role of sewage treatment works in preventing the release of such endocrine active compounds to the environment. This monograph outlines the current state of knowledge on endocrine disrupters in wastewater and sludge treatment processes. Discharge of effluents from treatment facilities is likely to be a significant source of input of contaminants to many systems, and the potential for concentration of hydrophilic compounds and transformation products within sludges has implications for their disposal. As such, understanding the processes and the fate of EDCs during treatment may facilitate controlling or limiting exposure of both humans and the environment to these compounds.

The Editors

Jason W. Birkett, Ph.D., is a senior lecturer in forensic science, Department of Biological Sciences, Faculty of Health and Life Sciences, University of Lincoln, U.K.

He graduated from Staffordshire University with a B.Sc. (Hons) in applied and analytical chemistry in 1991; he received a Ph.D. from the University of Manchester in 1996 for research into the characterization of natural organic matter from soils and natural waters. In 1997, he was appointed as a lecturer in environmental analytical chemistry at Manchester Metropolitan University. A year later, he joined Imperial College as a senior research fellow studying trace substances, such as mercury, methylmercury, and, more recently, estrogens and other endocrine disrupters, in lowland river systems. This work has expanded to include the study of the behavior of endocrine disrupters in biological wastewater treatment at laboratory and full-scale. In 1999, he was appointed lecturer in environmental chemistry and pollution at Imperial College. In 2002, Dr. Birkett joined the University of Lincoln.

Dr. Birkett, with Prof. John N. Lester, co-authored the book, *Microbiology and Chemistry for Environmental Scientists and Engineers*, a successful undergraduate and postgraduate textbook.

John N. Lester, Ph.D., D.Sc., is Professor of Water Technology, Environmental Processes and Water Technology Group, Imperial College of Science, Technology, and Medicine, London.

Prof. Lester graduated from the University of Bradford with a B.Sc. degree in applied biology in 1971; he was awarded the M.Sc. at the University of London and the Diploma of Imperial College (DIC) in 1972 for studies in biochemistry and plant physiology. In 1975, he received his Ph.D. from the University of London for research in the field of public health engineering and microbiology. After a postdoctoral fellowship in the Public Health Engineering Laboratory of the Civil Engineering Department at Imperial College, he joined the academic staff as a lecturer in 1977. In 1991, he was appointed Professor of Water Technology.

Prof. Lester is a fellow of the Institution of Water Engineers and Scientists, the Royal Society of Chemistry, and the Institute of Biology. In 1995, he was awarded a D.Sc. in environmental technology. In 1988 and again in 1997, he was technical advisor to the National Science Foundation in the United States on issues regarding research needs in water engineering and microbiology.

He has published over 350 research papers, presented numerous lectures at international and national conferences, and written or contributed to several books. His work in the environmental field has recently earned him the title of "the U.K.'s most highly cited environmental scientist" by the Institute of Scientific Information and the European Science Foundation.

Prof. Lester is currently continuing research into endocrine disrupting chemicals and their behavior in wastewater and sludge treatment processes in addition to their presence and behavior in environmental media.

Contributors

Jason W. Birkett, Ph.D.
Department of Biological Sciences
Faculty of Human and Life Sciences
University of Lincoln
Lincoln, England

Mark R. Gaterell, Ph.D.
Centre for Sustainable Construction
BRE
Garston
Watford, England

Rachel L. Gomes
Environmental Processes and Water
  Technology Group
Department of Environmental Science
  and Technology
Faculty of Life Sciences
Imperial College of Science,
  Technology, and Medicine
London, England

Katherine H. Langford
Environmental Processes and Water
  Technology Group
Department of Environmental Science
  and Technology
Faculty of Life Sciences
Imperial College of Science,
  Technology, and Medicine
London, England

John N. Lester, Ph.D.
Environmental Processes and Water
  Technology Group
Department of Environmental Science
  and Technology
Faculty of Life Sciences
Imperial College of Science,
  Technology, and Medicine
London, England

Mark D. Scrimshaw, Ph.D.
Environmental Processes and Water
  Technology Group
Department of Environmental Science
  and Technology
Faculty of Life Sciences
Imperial College of Science,
  Technology, and Medicine
London, England

Nick Voulvoulis, Ph.D.
Environmental Processes and Water
  Technology Group
Department of Environmental Science
  and Technology
Faculty of Life Sciences
Imperial College of Science,
  Technology, and Medicine
London, England

Table of Contents

1 Scope of the Problem

J.W. Birkett

CONTENTS

1.1 INTRODUCTION

Animals and plants have a system of chemical messengers that control various basic functions such as reproduction, growth, and maintenance. In animals, this system utilizes several glands that produce these chemical messengers (hormones) which are then transported to target organs. Although system similarities between species

1-56670-601-7/03/$0.00+$1.50
© 2003 by CRC Press LLC

have been noted (e.g., in the use of similar hormones), different evolutionary pathways can produce complex variations, resulting in a multifaceted regulatory system.

It is also known that certain chemicals can interfere with the endocrine system in several ways to produce an undesired response or disruption, which in turn may affect the health, growth, and reproduction of a wide range of organisms. These substances are collectively referred to as endocrine disrupting chemicals (EDCs).

The problem of endocrine disruption (ED) has been evident since the early 1900s,[1] but recently this phenomenon has emerged as a major environmental and human health issue, generating a vast amount of attention among scientific communities worldwide and considerable media interest.

While society has released large amounts of man-made chemicals into the environment since the 1940s, the generation born between 1950 and 1960 was the first to suffer exposure to these pollutants (from stores in maternal fat tissue) while they were growing in the womb. Since the oldest in this group did not reach childbearing age until the late 1970s, scientists are only now beginning to build a picture of the long-term impacts of this exposure on human health. Moreover, we must wait until nearly 2020 to really see how these chemicals may affect the adult well-being of children being conceived today.

During the 1950s and 1960s, the pesticide DDT, which was used worldwide in vast quantities after the second world war, was shown to be estrogenic and to affect the reproductive system in mammals and birds.[2,3] This compound is very persistent in the environment and is still used in some parts of the world. These findings, together with the publication in 1962 of *Silent Spring* by Rachel Carson,[4] further highlighted wildlife health problems (e.g., egg shell thinning, deformities, and population declines) which were linked to the exposure to pesticides and other synthetic chemicals. At that time, research also indicated that exposure in the early stages of life to naturally occurring hormones could produce harmful health effects and possibly cause cancer in young adult human populations.[5–7]

An example of the devastating consequences of exposure to endocrine disrupters is the use of the potent drug diethylstilbestrol (DES), a synthetic estrogen. Before it was banned in the early 1970s, medical practitioners prescribed DES to as many as five million pregnant women to block spontaneous abortion. DES was prescribed in the mistaken belief that it would prevent miscarriages and promote fetal growth. After children born to mothers who had taken DES went through puberty, it was discovered that DES affected the development of the reproductive system and caused vaginal cancer.[8–10] These findings were reinforced by laboratory studies which confirmed that DES causes developmental abnormalities in male and female mice.[11,12] The consequences of DES exposure are well known today.[13] A recent study by Kaufman et al.[14] looked at the births of the so-called DES daughters — women exposed to DES in their mother's womb. They concluded that DES-exposed women are more likely to have premature births, spontaneous abortions, and ectopic pregnancies than unexposed women. This suggests that potent estrogenic chemicals like DES can have long-lasting reproductive health effects in humans exposed *in utero*.

Perhaps the most widely used example of endocrine disruption in wildlife comes from the use of tributyl tin (TBT) as a component of antifouling paint on ships' hulls.[15] This compound was proven to have androgenic (masculinizing) properties resulting in the masculinization (imposex) of female mollusks, [16] which almost wiped out entire populations. This condition of imposex has been noted in over 110 marine species.[17] The effect on oyster populations in French waters was also severe (due to shell thickening) and prompted legislation in 1982 restricting TBT formulations to the application of larger vessels only (> 25m).[18] In 1987, regulation followed in the United States, the United Kingdom, and other European countries. Since then, the distribution, fate, and effects of organotin compounds in the marine[19,20] and freshwater environments[21,22] have been extensively studied. Alternative antifouling treatments based on copper and containing organic booster biocides to improve the efficacy of the formulation are now being used on small craft.[23] Only limited environmental monitoring of the biocides has been undertaken, so their effect on the environment has yet to be fully elucidated.

Numerous wildlife studies on a range of species have indicated possible endocrine disruption effects, the majority of which involve reproductive and developmental abnormalities, possibly resulting in population decreases. Some examples of these studies are given in Table 1.1.

One of the most comprehensive studies of endocrine disruption in wildlife is that on the impact of steroid estrogens and estrogenic chemicals (particularly alkylphenols) on British fish. This research has shown that a large number of sewage treatment work (STW) effluents in the United Kingdom are estrogenic for fish,[45,60,61] and that several of the receiving surface waters are also estrogenic.[56] Nationwide surveys in the United Kingdom have shown increased plasma levels of vitellogenin (a female-specific, estrogen-dependent plasma protein) in wild populations of a freshwater fish, roach (*Rutilus rutilus*), in rivers that receive effluent from sewage treatment plants. A large number of these fish also showed a high prevalence (locally up to 100%) of intersex (ovotestis).[47] Eight rivers were sampled upstream and downstream from STWs in the United Kingdom, and intersex roach were found at all sites, with the higher frequency downstream. However, intersex in roach can be found at low levels across the United Kingdom, and it is not known whether this is due to natural factors or a lack of pristine water habitats.

Despite early evidence, the phenomenon of ED has only become an overtly topical environmental issue since the early 1990s. The reason it came to the fore was that studies have revealed potential problems with human male reproductive health, in the form of reduced sperm quality/counts[62–64] as well as worldwide increases in testicular cancer,[64,65] with EDCs cited as a possible cause. Evidence was also emerging that certain wildlife were experiencing endocrine disruption to their reproductive systems,[34,39,43,44,47] with exposure to environmental pollutants cited as the cause (e.g., egg shelling thinning due to pesticides, estrogens and surfactants feminizing fish). The link to pollution was also thought to be a possible cause of the effects seen in the human reproductive system, although it has not been fully resolved as to whether ED is actually occurring in the human population.[66] The publication of *Our Stolen Future* in 1996[67] brought this complex scientific issue to the attention of the general public by providing a readable account of how man-made chemicals can disrupt hormonal systems.

TABLE 1.1
Endocrine-Disrupting Effects in Wildlife

Species	Contaminant/Effect	Reference
Mammals		
Panther	Hg, DDE, PCBs/cryptorchidism	24
Baltic seals	PCBs/sterility, adrenocortical hyperplasia	25–27
Beluga whales	PCBs, Dieldrin, 2,3,7,8-TCDD/hermaphroditism	28, 29
European otter	PCBs/reproductive impairment	30, 31
Dall's porpoises	PCBs, DDE/reduced testosterone levels	32
Birds		
Western gull	DDT compounds, methoxychlor/feminization, female–female pairing	33, 34
Peregrine falcon	DDE/egg shell thinning	35
Fish-eating birds (U.S., Great Lakes)	PCDD, PCDF/reproductive failure, deformities	36, 37
Common tern	PHAHs/reduced hatching, morphological abnormalities	38–40
Reptiles		
Snapping turtles	Organochlorine compounds/developmental abnormalities, feminization	41, 42
American alligator	DDE/low hatching rates, abnormalities in males and females	43, 44
Fish		
Roach	Steroid estrogens/increased vitellogenin in males, intersex	45–49
Flounder	Nonylphenol, octylphenol/vitellogenin in male fish	50–52
Flounder	Estrogens/vitellogenin in male fish	53–55
Rainbow trout	Estrogens, nonylphenol/vitellogenin in male fish	46, 56–59

1.2   THE ENDOCRINE SYSTEM

Within multicellular organisms it is necessary to regulate and integrate the functionality of different cells. The two systems employed to do this are the nervous system and the endocrine system. The latter system is crucial to both plants and animals because it is responsible for growth, reproduction, maintenance, homeostasis, and metabolism.[68]

The endocrine system consists of several glands in different areas of the body that produce hormones with different functions.[69] Endocrine glands are ductless and consist of the hypothalamus, pituitary, thyroid, parathyroid, adrenal glands, the pineal body, and the gonads. These synthesize hormones, which are then transported via the bloodstream to the target organs where they are used to invoke a natural response. These target cells are comprised of a binding site (receptor) and an effector site.[70] When hormones attach to the receptor, the effector site is altered, which, in turn, produces the desired response (Figure 1.1a). Some "free" hormone molecules will never reach the receptors and are inactivated prior to excretion, primarily by the liver and kidneys in a process called *metabolic clearance*. This process varies with the type of hormones, but the effective "life span" of a hormone in the body is from a few minutes to several hours. Thus, if the metabolic clearance rate is low, the hormone stays in the body longer and thus its availability to interact with receptors increases, resulting in more responses.

Hormone molecules are generally short lived in the body due to metabolic clearance mechanisms. However, when EDCs are present, these mechanisms may not apply, leading to the persistence and bioaccumulation of these chemicals in the body. The EDCs may ultimately interact with the endocrine system (such interaction tends to affect systems at certain stages of sexual development; e.g., juveniles are most susceptible).

The receptor sites have a very high affinity for a specific hormone, so only very low concentrations are required to achieve a response. Thus, the hormones utilized by the cell can be said to have a high potency, which can be defined as *the quantity of a substance required to produce a given effect*; i.e., the greater the potency, the less hormone that is required. Despite their high affinity for hormones, these receptors are also capable of binding other chemical compounds. This means that any EDCs present in low concentrations may cause an effect and elicit a response.

Endocrine disruption occurs when EDCs interact with the hormone receptors, altering the natural response patterns of the endocrine system. The types of processes involved are shown in Figures 1.1b and 1.1c. The chemical may bind to the receptor and activate a response, thus acting as a hormone mimic. This is defined as an *agonistic effect* (Figure 1.1b). If the chemical (a hormone blocker) binds to a receptor, but no response is produced, this prevents the natural hormone from interacting and is termed an *antagonistic effect* (Figure 1.1c). As agonists and antagonists bind to the same receptor, subtle changes in the receptor conformation are made to account

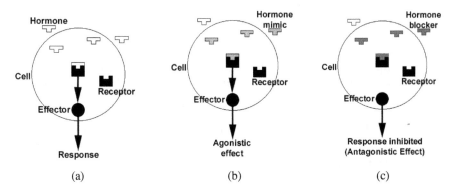

FIGURE 1.1 Endocrine disruption processes. (a) Natural response; (b) agonistic effect; (c) antagonistic effect.

for the distinct biological activities of the chemicals concerned. Other effects that can occur in the endocrine system are the disruption of the synthesis and removal of hormones and their receptors, and interaction with multiple hormone systems. Hence, the processes of endocrine disruption are complex.

Most EDCs are small molecules and therefore mimic or antagonize small hormones such as steroid or thyroid hormones.[71] They may eliminate the natural hormones responsible for regulating homeostasis and the development and reproductive processes.[72] The various mechanisms of action that EDCs can perform are summarized in Table 1.2.

1.3 DEFINITIONS

There have been many definitions proposed for endocrine disrupters, and they have been subject to some scrutiny. The term is taken to include any adverse health effects that are thought to be a result of alterations to any part of the endocrine system. The precise definition of an endocrine disrupter was proposed at the European Workshop on the impact of endocrine disrupters on human health and wildlife[73] and is commonly referred to as the "Weybridge" definition:

> An endocrine disrupter is an exogenous substance that causes adverse health effects in an intact organism, or its progeny, subsequent to changes in endocrine function.

Many chemicals have demonstrated the potential to be endocrine disrupting, but more research is needed. For example, *in vitro* studies, such as hormone binding assays,[74] have shown chemicals to possess endocrine disrupting effects, but activity has not yet been established *in vivo*. Thus, a separate definition of these potential endocrine disrupters was required:

> A potential endocrine disrupter possesses properties that might be expected to lead to endocrine disruption in an intact organism.

The U.S. Environmental Protection Agency (EPA) has proposed a more detailed definition of an endocrine disrupter:[68]

> An endocrine disrupter is an exogenous agent that interferes with the synthesis, secretion, transport, binding, action, or elimination of natural hormones in the body that are responsible for the maintenance of homeostasis, reproduction, development, and/or behavior.

The EPA definition reflects the wide diversity of mechanisms that are thought to be involved in the disruption of the endocrine system.

Other definitions have been proposed by the USEPA Endocrine Disrupter Screening and Testing Advisory Committee (EDSTAC),[75] and the International Programme on Chemical Safety (IPCS),[76] but these are effectively reworded versions of the definitions above.

TABLE 1.2

Different Mechanisms of Action Used by Endocrine Disrupters

Mechanism of action	Definition
Mimics	By mimicking a natural hormone, an endocrine disrupter is able to fit precisely into the hormone receptor. By occupying the receptor site, messages can be sent to receiving genes. Messages sent at the wrong time, or overproduction of messages has adverse effects on biological functions. The biological functions affected, depend on the hormone being mimicked.
Stimulators	Some endocrine disrupters are able to stimulate the formation of more hormone receptors on or within cells, causing hormone signals to multiply. This effect leads to the amplification of both natural and foreign hormones.
Blockers	By occupying the receptor site in the cell, some endocrine disrupters are able to block the natural hormone. This can have an increased or decreased effect on the gene depending on whether the blocker is more or less potent than the hormone being blocked. Chemicals that block or agonize hormones are labelled anti-androgens or anti-estrogens.
Endocrine flushers	Accelerating a hormone's breakdown and elimination from the body leads to depletion of the hormone.
Enzyme flushers	Endocrine disrupters can interfere with the enzymes that are required to break down hormones in the system. By deactivating enzymes necessary for hormone elimination, more hormone than is necessary (or than is even healthy) remains active. Their continued presence within the body sends more signals than normal or signals at inappropriate times.
Destructors	Destructors can destroy the hormone or the hormone's ability to carry out its function by acting directly or indirectly to alter its structure, causing the hormone to no longer fit its receptor site. Additionally, destruction can be achieved by influencing and thereby altering the pattern of hormone synthesis. Exposure can also alter the natural hormonal balance within the organism. This causes both sexes to have higher than normal levels and can cause feminization of the male if estrogen levels are higher than testosterone levels.

It is important to be aware that agreement is still pending on the specific biological effects that would be required to classify a substance as a potential EDC. EDCs can exhibit estrogenic or androgenic behavior. Thus, it seems prudent to include a definition of these terms.

Estrogens are defined as *"any of a family of steroid hormones that regulate and sustain female sexual development and reproductive function."* However, some scientists have broadened the definition of estrogens to include any compound that stimulates tissue growth by:[77]

Promoting cell proliferation in female sex organs
Promoting hypertrophy or increasing a cell's size, such as occurs in female
    breast and male muscle tissue during puberty
Initiating synthesis of specific proteins

According to this definition, many of the chemicals identified by the U.K. Environment Agency (UKEA) as endocrine disrupters would be classified as estrogens.[78] Other workers have therefore argued that the term *estrogen* should be limited to compounds capable of producing estrus, a phase in the sexual cycle characterized by the willingness of the female to accept the male.[79] Under this definition, only sex steroid hormones (i.e., natural hormones produced via biological processes) should be classified as estrogens, while other substances mimicking their estrogenic effect are classified as *environmental estrogens.*

An androgen is defined as *"a class of male sex hormones related to the steroid androstane and produced in the adrenal cortex and the testes; includes testosterone, androsterone, and androstenolone responsible for the development of secondary male characteristics, such as a deep voice and facial hair."*

Many EDCs compete with estradiol (female sex hormone) for the estrogen receptor, others compete with dihydrotestosterone (male sex hormone) for the androgen receptor. Hence, these substances exert a feminizing or masculinizing effect on the endocrine system. Substances that mimic these feminizing effects are known as "estrogenic;" those mimicking masculinizing effects are termed "androgenic." It therefore follows that an antiandrogenic substance, such as flutamide, inhibits the biological actions of androgens by binding to and thus inactivating the androgen receptor of target tissue. An antiestrogenic substance, such as tamoxifen, inhibits the biological actions of estrogens by binding to and thus inactivating the estrogen receptor of target tissue.

Endocrine disrupting chemicals can also be labeled as agonists and antagonists, which are more generic terms. A substance exhibiting agonistic behavior mimics a hormone (cf. estrogenic and androgenic), and conversely, a substance exhibiting antagonistic behavior inhibits a hormone (cf. antiestrogenic and antiandrogenic).

1.4 COMPOUNDS OF INTEREST (CHEMICALS WITH POTENTIALLY ENDOCRINE DISRUPTING PROPERTIES)

Many endocrine disrupting substances, or potential endocrine disrupters, were previously classified as organic micropollutants. These include such compounds as alkylphenols (APs), alkylphenol polyethoxylates (APEOs), polyaromatic hydrocarbons (PAHs), polychlorinated biphenyls (PCBs), phthalates, bisphenol-A, polybrominated flame retardants, dioxins, furans, herbicides, pesticides, and steroid hormones (e.g., estrogens). As our knowledge of endocrine disrupters increases, so does the list of chemicals that exhibit these endocrine disrupting properties. The two main approaches to determining any chemical endocrine disrupter function are *in vitro* and *in vivo* assays (see Chapter 3), although the majority of data available on endocrine disrupting chemicals have come from *in vitro* studies.

Table 1.3 includes a vast array of chemicals that have already been classed as endocrine disrupters, or potential endocrine disrupters, by organizations worldwide.

The European Union (EU) has also produced a report containing a range of substances suspected of interfering with the hormone systems of humans and wildlife.[83] The study identified 118 substances that were classed as endocrine disrupters or potential endocrine disrupters. Of these, 12 have been assigned priority for in-depth study. These are: carbon disulfide, *o*-phenylphenol, tetrabrominated diphenyl ether, 4-chloro-3-methylphenol, 2,4-dichlorophenol, resorcinol, 4-nitrotoluene, 2,2′-bis(4-(2,3-epoxypropoxy)phenyl)propane, 4-octylphenol, estrone, ethinyl estradiol, and estradiol.[84]

At present, the majority of known endocrine disrupters belong to the pesticide category. Several metals (e.g., cadmium, mercury) have also been shown to exhibit endocrine disrupter activity,[80] especially in their organoform (e.g., methylmercury). The fate and behavior of metals in wastewater treatment and sludge processes has been well documented[85,86] and will not be discussed in this book.

With such a vast number of compounds already determined to be endocrine disrupting, this book cannot attempt to describe the fate and behavior of all EDCs and potential EDCs. Discussions will therefore focus on specific groups of compounds known to be problematic. These include, but are not limited to:

Steroid compounds (e.g., estrogens)
Surfactants (e.g., nonylphenol and its ethoxylates)
Pesticides, herbicides, fungicides; hereafter referred to collectively as pesticides (e.g., DDT, dieldrin, 2, 4-D, tributyltin)
Polyaromatic compounds (e.g., PAHs, PCBs, brominated flame retardants)
Organic oxygen compounds (phthalates, bisphenol A)

1.5 PROPERTIES OF EDCs

1.5.1 Physicochemical Properties

The fate and behavior of an EDC are influenced by its physicochemical properties. Due to their physicochemical properties, the majority of EDCs tend to favor partitioning (adsorption) to solid surfaces or into biota. The partitioning of EDCs to the sediment phase or biota is defined by their solubility and partition coefficients (Table 1.4). To promote association with the sediment phase or biota, low solubility in water and a high octanol/water partition coefficient (K_{ow}) or high carbon/water

TABLE 1.3
List of Compounds Classified as EDCs by Various Organizations

Compound	UKEA[78]	USEPA[68]	OSPAR[80]		JEA [81]	WWF [82]
			vivo	vitro		
Steroids						
Ethinyl estradiol	X		X			
17β-estradiol	X		X			
Estrone	X		X			
Mestranol			X			
Diethylstilbestrol	X		X			
					X	X
Alkylphenols						
Nonylphenol	X	X	X		X	X
Nonylphenol ethoxylate	X			X		
Octylphenol	X	X	X		X	
Octylphenol ethoxylate	X					
Polyaromatic Compounds						
Polychlorinated biphenyls (PCBs)	X	X	X		X	X
Brominated flame retardants				X	X	X
Polyaromatic hydrocarbons (PAHs)		X		X		
Organic Oxygen Compounds						
Phthalates	X	X		X	X	X
Bisphenol A	X	X			X	X
Pesticides						
Atrazine	X	X		X	X	X
Simazine	X	X		X	X	X
Dichlorvos	X					
Endosulfan	X	X		X	X	X
Trifluralin	X	X				X
Demeton-S-methyl	X					
Dimethoate	X					X
Linuron						X
Permethrin	X	X			X	
Lindane	X	X	X			X
Chlordane	X			X	X	X
Dieldrin	X	X		X	X	X

(continued)

TABLE 1.3 (continued)
List of Compounds Classified as EDCs by Various Organizations

Compound	UKEA[78]	USEPA[68]	OSPAR[80]		JEA [81]	WWF [82]
			vivo	vitro		
Hexachlorobenzene	X			X	X	X
Pentachlorophenol (PCP)	X	X			X	X
Others						
Dioxins and furans	X		X		X	X
Tributyltin	X	X	X		X	

UKEA — United Kingdom Environment Agency

USEPA — United States Environmental Protection Agency

OSPAR — Oslo and Paris Commission

JEA — Japan Environment Agency

WWF — World Wildlife Fund

partition coefficient (K_{oc}) is preferred. For ease of transport to groundwater, high solubility in water is needed for greater mobility.

Environmental EDCs are not necessarily structurally related to the naturally occurring steroids, while others are structurally related to other types of compound (e.g., PAHs[88]). Therefore, many substances, such as foodstuffs, flavonoids, lignans, sterols, fungal metabolites, and synthetic chemicals of widely varying structural classes (e.g., phthalates, PCBs), can interact with hormone receptors and modulate the endocrine system.[89]

The following sections discuss the structure and endocrine activity of groups of compounds (both natural and anthropogenic) that behave as EDCs, with details on quantitative structure–activity relationship (QSAR) studies.

TABLE 1.4
Physicochemical Properties of EDCs Important for Fate and Behavior Processes

Physicochemical Property	Potential	Low	Moderate	High
Water solubility	Dissolving (mg l^{-1})	<1	—	1000
Henry's Law constant	Evaporation (atm m^3 mole $^{-1}$)	>10^{-2}	$10^{-2} - 10^{-7}$	<10^{-7}
Organic/carbon partition coefficient	Sorption (log K_{oc})	<3	—	>3
Log octanol/water partition coefficient[87]	Bioconcentration (log K_{ow})	<2.5	>2.5 – <4.0	>4

1.5.2 STEROID ESTROGENS (NATURAL AND SYNTHETIC)

Estrogens, like all steroids, share the same hydrocarbon ring nucleus as choles-
terol, their parent compound. Figure 1.2 illustrates this basic ring structure, consisting of three hexagonal rings, (A, B, C) and one pentagonal ring (D). Steroid estrogens
are characterized by their phenolic A-ring, which renders the 3-hydroxyl acidic[90]
and is essential for biological activity. Estrogens may be referred to as C_{18} steroids,[91]
because they have 18 carbon atoms within their structure.

The lipophilic cyclopentanophenanthrene nucleus shown in Figure 1.2 is modified by the addition of hydrophilic groups to form different steroids.[90] In the case
of the three natural free estrogens, hydroxyl and carbonyl groups are added, while
ethinyl groups are found in the structure of the synthetic free estrogen, 17β-ethi-
nylestradiol, a component of the contraceptive pill. Substituent groups above the
plane of the molecule are said to be in the "β" position, whereas those situated under
the plane of the molecule are said to be in the "α" position.

Log octanol/water coefficient (log K_{ow}) values for the free estrogens range from
2.81 to 4.15,[92,93] and thus, it is evident that these compounds are lipophilic and are
only sparingly soluble in water. When dissolved, these estrogens may be rapidly
removed from the aqueous phase as a result of binding to suspended solids.[93] Ester-
ification with glucuronic or sulfuric acid, however, dramatically alters the physical–chemical properties of free steroid estrogens. Sulfate and glucuronide conjugates
are far more hydrophilic than their unconjugated counterparts, although they are still
soluble to some extent in organic solvents.[90]

Environmental estrogens, sometimes referred to as *xenoestrogens*, are a diverse
group of substances that do not necessarily share any structural relationship with
natural estrogens such as 17β-estradiol, but still induce agonist or antagonist behavior
through shared mechanisms of action (e.g., pesticides, PCBs).

Several structural requirements for estrogenicity were identified based on
extensive studies of estradiol.[94] The relative positions of a phenolic hydroxy (OH)
group on ring A, lacking steric interference from alkyl substitutions in the ortho
positions, were considered to be crucial for high-affinity binding to the estrogen
receptor and *in vivo* estrogenicity. The alkyl substitution of the 3-phenolic OH
group of an estrogen reduces receptor binding, although this can be altered by
metabolic activation.

Structure–activity relationships (SAR) studies[95] suggest that when estradiol (E2)
is bound to the estrogen receptor (ER), there is only a close fit at the A-ring end of
the steroid. The most potent antagonists possess phenolic rings capable of mimicking
the E2 A-ring in achieving a high binding affinity to the ER. However, they do not
elicit a response because they are unable to stabilize the conformational change or
molecular interaction required for activation.

With the potent synthetic estrogen, DES, it has also been suggested[96] that, while
the A-ring performs the above role, the D-ring orientation modulates any subsequent
biological activity. This confirms the hypothesis that the ability of an EDC to produce
a hormonal response depends on its overall molecular configuration since planar
molecules are more active. Either the α or α′ rings (see DES structure-Figure 1.2)
can mimic the A-ring E2 binding to the ER.

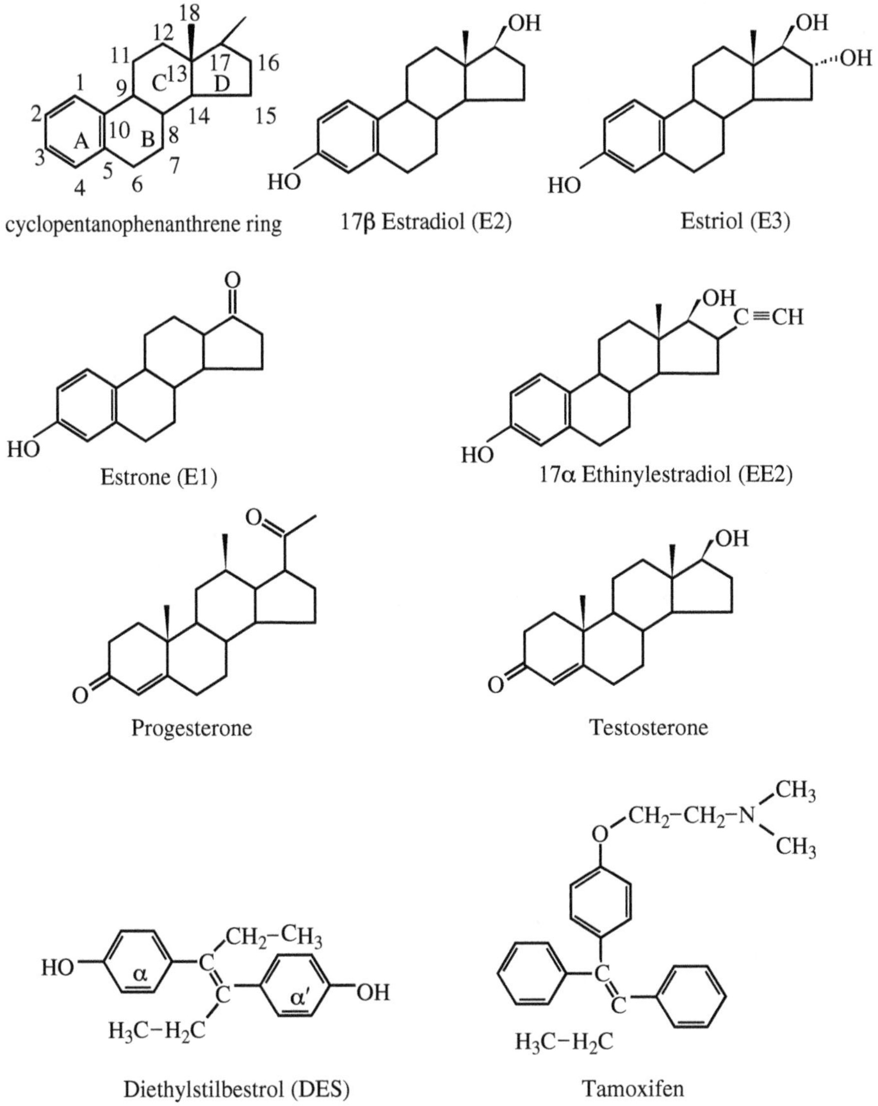

FIGURE 1.2 Structures of the steroid compounds.

Tamoxifen, an anticancer agent (Figure 1.2), has been shown to exhibit antiestrogenic properties in relation to breast tissue,[97] while it has an estrogenic effect on other parts of the endocrine system. The presence of the amino-ethoxy side chain of this compound is essential for its antiestrogenic activity, and triphenylethylene derivatives of tamoxifen lacking this side chain are estrogen agonists. It has also been suggested that side chain constituents that possess a lone pair of electrons (e.g., O and N atoms) are required to produce an antiestrogenic effect. Also, changes in tamoxifen from the *trans* to *cis* isomer alter the activity to that of a partial agonist.

Chemical interactions are further complicated because agonistic or antagonistic mechanisms depend on the target cell tissue type involved. This is known as tissue selective agonism/antagonism. For example, tamoxifen is estrogenic (agonistic) in uterine, liver, and osteoblastic cells, but antiestrogenic (antagonistic) in breast cells.[97]

1.5.3 PHYTOESTROGENS

Phytoestrogens are plant chemicals that can act as fungicides, regulate plant hormones, protect plants against UV radiation, and deter herbivores. Many different plants produce these compounds that may mimic or interact with estrogen hormones in animals. At least 20 compounds have been identified in at least 300 plants from more than 16 different plant families.[67] These natural compounds are weaker than the endogenous estrogens and are present in herbs, grains, vegetables, and fruits.

Phytoestrogens have a 2-phenylnaphthalene-type structure which resembles those of the endogenous estrogens. They have also been found to bind to estrogen receptors, and research suggests that they may function as agonists and antagonists on the endocrine system.[98]

Scientists have focused their attention on the two main groups of phytoestrogens: the isoflavones and the lignans. The isoflavones are found in soybeans and other legumes, while lignans are produced from the microbial breakdown in the gut of grains, fibers, and several fruits and vegetables. The structures of these types of compounds are given in Figure 1.3. Isoflavones have a diphenolic structure that resembles the structure of the potent synthetic estrogen, diethylstilbestrol. Daidzein and genistein are the two major isoflavone compounds found in humans (Figure 1.3). Lignans possess a 2,3-dibenzylbutane structure and are essentially the precursors of the formation of lignins, which are produced in plant cell walls.

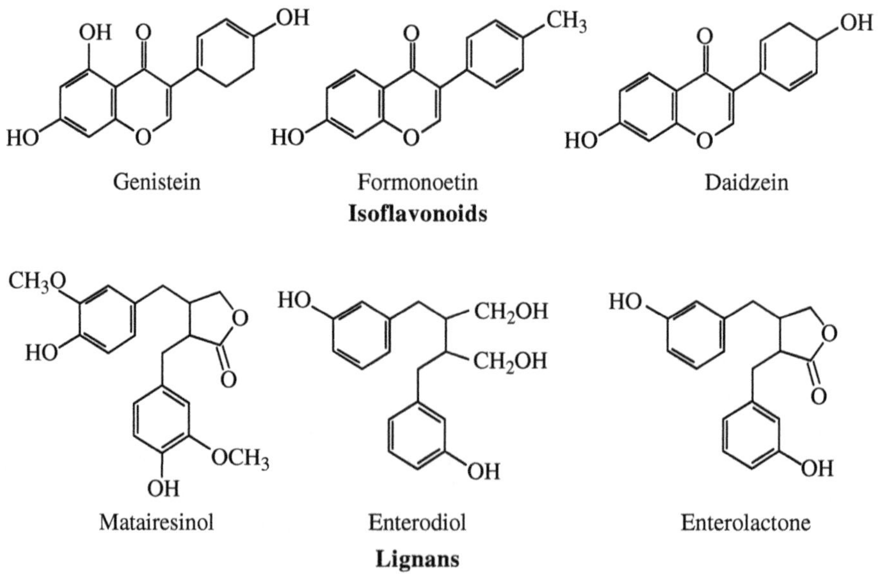

Genistein   Formonoetin   Daidzein

Isoflavonoids

Matairesinol   Enterodiol   Enterolactone

Lignans

FIGURE 1.3 Structures of phytoestrogens.

Both types of phytoestrogen have been found to produce weak estrogenic activity of the order of 10^{-2} to 10^{-3} when compared to 17β-estradiol.[99–101] This means that if a phytoestrogen binds to an estrogen receptor, it produces a response that is 100 to 1000 times less than that of an endogenous estrogen. However, the concentrations of these compounds in the body can be 100 times higher than the endogenous estrogens.[102–104] Despite this, it is generally accepted that phytoestrogens in the human diet can have a beneficial effect rather than a deleterious one.

1.5.4 Organic Oxygen Compounds

1.5.4.1 Bisphenols

The chemical structure of hydroxylated diphenylalkanes (bisphenols) consists of two phenolic rings joined together by a bridging carbon atom (Figure 1.4). Bisphenols with OH groups in the *para* position (i.e., bisphenol A) and an angular configuration are suitable for hydrogen bonding to the acceptor site of the estrogen receptor. The log K_{ow} value for bisphenol A is 3.4, indicating its lipophilicity and tendency to bind to solid phases in the aquatic environment. The estrogenic potency of bisphenols is influenced by the length and chemical nature of the substituents at the bridging central carbon atom,[105] with the most active compound containing two propyl chains at the bridging carbon. Other studies[106,107] have reported the estrogenic activity of bisphenol A, with a potency of four to six orders of magnitude less than 17β-estradiol. A recent study by Chen et al.[108] determined that other bisphenols used in industrial applications are also weakly estrogenic. In addition to its weakly estrogenic behavior, bisphenol A has been shown to possess some antiandrogenic activity.[109]

1.5.4.2 Dioxins

This group of chemicals includes seven of the polychlorinated dibenzodioxins (PCDDs) and 10 of the polychlorinated dibenzofurans (PCDFs). Structural examples of these compounds are given in Figure 1.4. Tetrachlorodibenzo-p-dioxin (TCDD)

Bisphenol A

Dibenzo-p-dioxin

2,3,7,8-tetrachlorodibenzo-p-dioxin
(TCDD)

2,3,7,8-tetrachlorodibenzofuran
(TCDF)

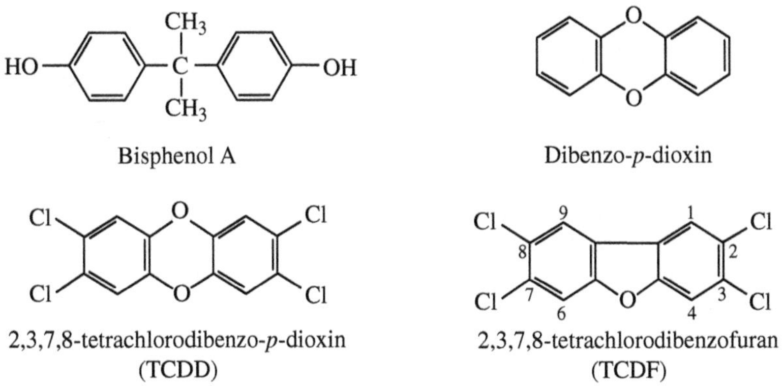

FIGURE 1.4 Structures of organic oxygen compounds.

is the most biologically active and toxic member of this group. Animal studies have shown that TCDD has a deleterious effect on reproductive functions, and this compound can produce both antiandrogenic and antiestrogenic effects.[110] Dioxins are persistent in the environment and have the potential for bioaccumulation, which is cause for concern because the EPA has characterized dioxins as likely human carcinogens as well as having the capacity for endocrine disruption.

1.5.4.3 Phthalates

Phthalate esters represent a group of chemicals that are widely used as plasticizers and, as such, they are not chemically bound to the end product. Therefore, they have the potential to leach into their surrounding environment. The structures of several of these are given in Figure 1.5. Certain phthalates have also shown estrogenic behavior.[111,112] The log K_{ow} values for this group of chemicals range from 1.46 to 13.1,[113] indicating greater lipophilicity with increasing alkyl chain length. The relative estrogenic potencies of several phthalates were in the following order:[111]

Butyl benzyl phthalate (BBP) > dibutyl phthalate (DBP) > diisobutyl phthalate (DIBP) > diethyl phthalate (DEP) > diisononyl phthalate (DINP).

Actual potencies compared to 17β-estradiol were six to seven orders of magnitude lower, indicating that these compounds are very weakly estrogenic. The most widely used phthalate, di (2-ethylhexyl) phthalate (DEHP), showed no estrogenic activity in *in vitro* assays.[111] The activities of simple mixtures of BBP, DBP, and 17β-estradiol have been tested for synergistic effects, with no synergism being observed although the activities of the mixtures were proven to be approximately additive. No structure activity relationships have been reported for these group of chemicals.

1.5.5 SURFACTANTS

1.5.5.1 Alkylphenolic Compounds

Certain alkylphenols (AP), their ethoxylates (APnEO), and carboxylates (APEC) have been shown to exhibit varying degrees of estrogenicity (where n = the number of ethylene oxide groups).[58,114–116] Alkylphenols are basically an alkyl group, which can vary in size and position, attached to a phenolic ring. Figure 1.6 illustrates some structures of APs, APEOs, and APECs. With log K_{ow} values ranging from 4.17 to 4.48, these compounds are lipophilic and will tend to partition to the solid phases in the environment.

Compounds such as octylphenol (OP) and nonylphenol (NP) are estrogenic, and their activities have been shown to depend on the nature of the alkyl substituent. White et al.[117] found the estrogenic potencies of these compounds, together with nonylphenol carboxylic acid (NP$_1$EC) and nonylphenol ethoxylate (NP$_1$EO) to be in the order:

OP > NP$_1$EC > NP > NP$_2$EO

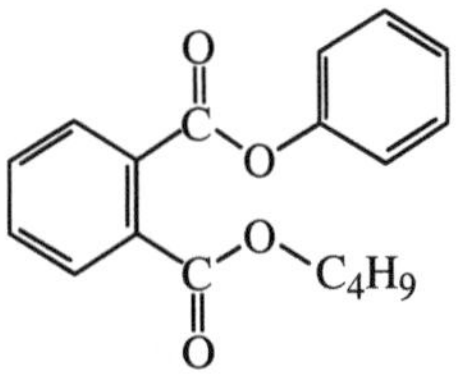

Butylbenzylphthalate

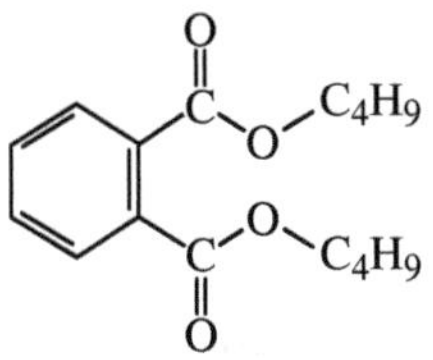

Dibutylphthalate

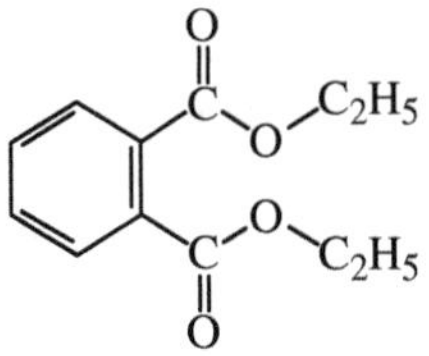

Diethylphthalate

FIGURE 1.5 Structures of several phthalates.

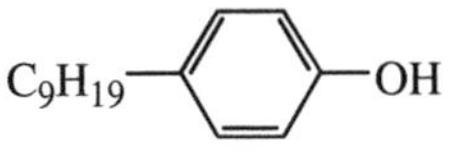

Nonylphenol

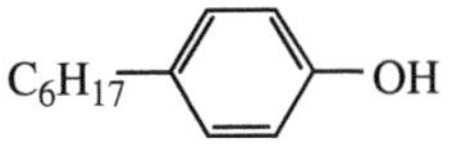

Octylphenol

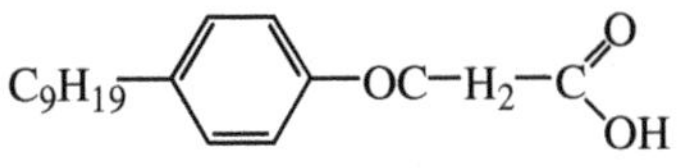

Nonylphenolethoxycarboxylate (NPEC)

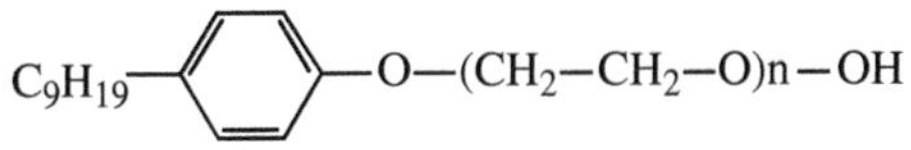

Nonylphenol ethoxylates (NPnEO)

FIGURE 1.6 Structures of alkylphenols, their ethoxylates and carboxylates.

irrespective of the assay used. It was also concluded that the estrogenicity of APEOs depends on chain length. The estrogenic potencies of these compounds are in the order of five to six times lower than 17β-estradiol. Structure activity relationships on several alkylphenols[118] have revealed that binding to the estrogen receptor results from the covalent bonding of two constituents of the phenol and alkyl groups, which respectively correspond to the A-ring and hydrophobic moieties of 17β-estradiol. Another study on structural features[114] indicates that both the position (para > meta > ortho) and branching (tertiary > secondary = normal) of the alkyl group affect the estrogenicity. The alkyl group also requires at least three carbon atoms to exhibit estrogenic activity.

1.5.6 POLYCHLORINATED COMPOUNDS

1.5.6.1 Polychlorinated Biphenyls (PCBs)

The general structure of PCBs is illustrated in Figure 1.7. If all the potential isomers are taken into consideration, there are a total of 209 PCB congeners, all exhibiting varying degrees of toxicity. These compounds are ubiquitous in the environment and have the potential for bioaccumulation and biomagnification. Log K_{ow} values range from 4.6 to 8.4, so these compounds are rapidly adsorbed to sediments and other particulate matter. In the United States and the European Union (EU), the use and disposal of PCBs has come under strict control since 1976, and many of these compounds have now been banned, although large amounts may still be in use.

PCBs and their metabolites, the hydroxylated PCBs (OH–PCBs) can exhibit agonistic and antagonistic behavior in estrogenic systems,[119,120] although the exact mechanism by which OH–PCBs produce their effects has not been established.[121] For PCBs, it has been shown that di-*ortho* and multiple chloro substituted biphenyls can compete for binding to the estrogen receptor.[122] In the case of OH–PCBs, the estrogenic potencies are largely dependent on the *ortho*-Cl and *para*-OH substitutions on the rings, the most potent being 2,5-dichloro-4, hydroxybiphenyl (see Figure 1.5). Other studies,[121,123–126] using a variety of animal and human cell bioassays, have shown that PCBs and OH–PCBs possess estrogenic activity.

1.5.6.2 Brominated Flame Retardants

These compounds, used in conjunction with other materials to prevent fires, have been shown to produce estrogenic responses *in vitro*.[127] This group of compounds is ubiquitous in the environment and includes polybrominated biphenyls (PBBs),

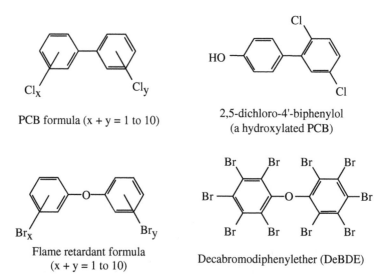

PCB formula (x + y = 1 to 10)

2,5-dichloro-4'-biphenylol
(a hydroxylated PCB)

Flame retardant formula
(x + y = 1 to 10)

Decabromodiphenylether (DeBDE)

FIGURE 1.7 General structures of PCBs and brominated flame retardants.

tetrabromobisphenol A (TBBPA), hexabromocyclododecane (HBCD), and poly-brominated diphenyl ethers (PBDEs), the latter receiving the most scientific attention.[128]

The PBDEs have low vapor pressures and are very lipophilic (log K_{ow} = 5.6 to 10). Therefore, these compounds are persistent, have a low water solubility, and a high binding affinity for particulates. As a result, they tend to accumulate in the sediment compartment. Similar to PCBs, there are theoretically 209 PBDE congeners. However, PBDEs are more likely to be susceptible to degradation in the environment than PCBs because the C–Br bond is weaker than the C–Cl bond.

PBDE agonists have been shown to be 250,000 to 390,000 times less potent than the endogenous estrogens,[127] (potencies similar to bisphenol A). The congeners with the highest estrogenic activity were: 2,2′,4,4′,6-pentaBDE, 2,4,4′,6-tetraBDE, and 2,2′,4,6′-tetraBDE. The congener 2,2′,4,4′,6-pentaBDE has also been reported among the PBDEs found in humans and other mammals.[129–131]

PBDEs are structurally similar to PCBs and DDT and, therefore, their chemical properties, persistence, and distribution in the environment follow similar patterns. Moreover, the concentrations of PBDEs in environmental samples are now higher than those of PCBs. The general structure of PBDEs is given in Figure 1.7. For estrogenic activity, PBDEs require two ortho (2,6) bromine atoms on one phenyl ring, at least one para atom (preferably on the same ring as the ortho bromines), and a nonbrominated ortho –meta or meta carbons on the other phenyl ring. This structure–activity relationship is similar to that of hydroxylated PCBs.

1.5.6.3 Polycyclic Aromatic Hydrocarbons (PAHs)

These compounds generally consist of benzene rings that are fused together (Figure 1.8). PAHs also have the potential to bioaccumulate and have log K_{ow} values in the region of 6, indicating a high degree of lipophilicity. Hence, partitioning to solid phases will predominate. Due to the structural similarity of PAHs and steroids,

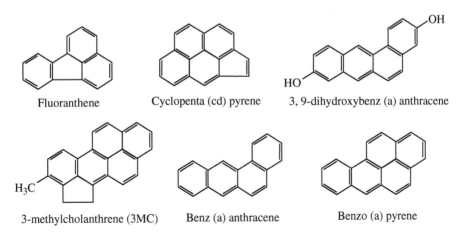

FIGURE 1.8 Structures of polyaromatic hydrocarbons (PAHs).

it was hypothesized in the 1960s that these compounds would be able to act upon steroid hormone receptor sites.[132] Moreover, these compounds have since been found to exhibit either weakly estrogenic or antiestrogenic (antagonistic) responses in *in vivo* and *in vitro* assays.[88,133–135] Heterocyclic PAHs, which contain an O, S, or N atom in one of the rings, have also been reported to possess estrogenicity.[136] The hydroxylated metabolites of PAHs, specifically benzo[a]pyrene, are considered to produce the estrogenic activity observed for this compound.[137] Both laboratory and field studies in the aquatic environment have revealed that PAH exposure affects steroid levels, gonadal development, fertilization, and hatching success.[138] Santodonato[88] has produced a comprehensive review concerning the estrogenic activity of PAHs.

1.5.7 PESTICIDES

This is by far the largest group of endocrine and potential endocrine disrupters. The chlorinated pesticides (examples in Figure 1.9), are particularly known to alter the reproductive capacity of animals by interacting with estrogen target tissues.[139–141] The o',p' isomers of DDT[142] and methoxychlor (MC)[143] have shown estrogenic activity in rat uterine weight assays. The p',p' isomer of DDT (shown in Figure 1.9) has an extremely low affinity for the ER, but o',p' DDT will inhibit binding of estradiol *in vivo* and *in vitro*.

The *in vivo* metabolism of EDCs is important since several hormonally active chemicals, including some naturally occurring EDCs, are known to be subject to metabolism *in vivo*, whereas others are not. An example of the importance of these

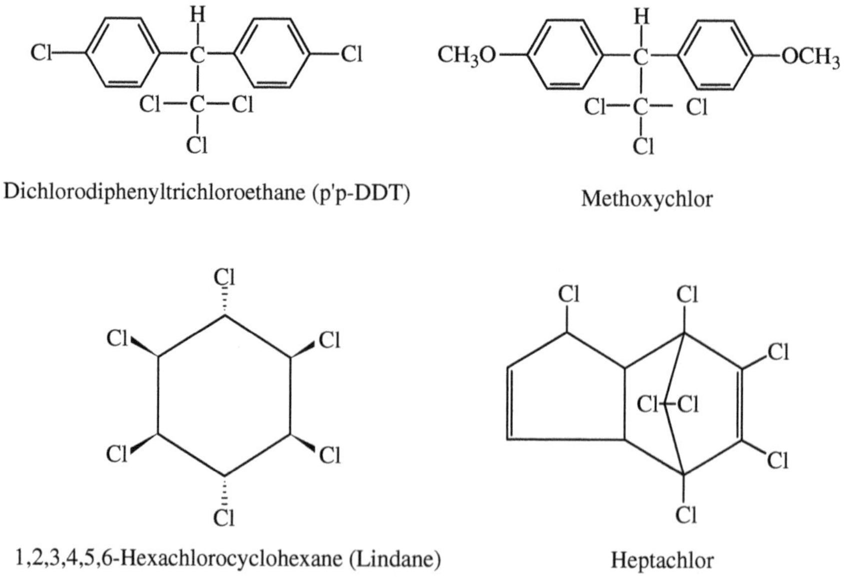

Dichlorodiphenyltrichloroethane (p'p-DDT)

Methoxychlor

1,2,3,4,5,6-Hexachlorocyclohexane (Lindane)

Heptachlor

FIGURE 1.9 Structures of selected pesticides.

metabolism processes for EDCs is MC, or bis[*p*-methoxydichlorophenyl] trichloro-ethane. This chemical is considerably less estrogenic, toxic, and persistent, with a shorter half life, than its parent compound, DDT.[144] This reduced toxicity of MC is attributed to its conversion of polar metabolites.

1.5.8 QUANTITATIVE STRUCTURE–ACTIVITY RELATIONSHIP (QSAR) STUDIES

This type of analysis correlates a chemical's structural characteristics with its bio-logical activity, and the models derived from these relationships can be used to predict the activity of untested environmental contaminants. Models are based on the biological activity of small groups of compounds with similar activity and structural features. Such models could yield information that would be useful in developing assay batteries for toxicity testing and possibly replace the high-through-put screening for chemical priority setting.

These studies have been used to establish some three-dimensional and physic-ochemical characteristics (biophore and toxicophore). These parameters can be used to identify characteristics of molecules that are responsible for ED activity. Cun-ningham et al.[145] used an artificial intelligence system to identify biophores associated with the activity of several estrogens and antiestrogens. It was found that chemicals possessed a common 6 Å biophore which modulates their activities. However, this biophore is not widespread among estrogens.[146] The precise role this biophore plays in receptor binding and its relationship between estrogenicity and carcinogenicity has yet to be evaluated.

Several chemicals containing the biophore, and thus suspected EDCs, have been identified and are illustrated in Figure 1.10. It is worthwhile noting that 2,5,-dichloro-4-hydroxylbiphenyl (Figure 1.7) also has the same biophore, analogous to 2-chloro-4-hydroxybiphenyl.

Such studies have revealed that almost all chemical estrogens contain one or more phenolic groups, which are required for binding to the ER. Many natural and synthetic estrogens have a phenolic OH group on a small lipophilic molecule of about 200 to 300 Daltons.[89] This chemical moiety is found in the body arising from the metabolism of endogenous and xenobiotic chemicals. For example, tamoxifen does not have the 6 Å biophore, but its 4-hydroxy derivative does (Figure 1.10). Metabolites of compounds are also suspected to be EDCs based on the presence of this 6 Å biophore. An example is 2,2-bis (4-hydroxyphenyl)-1,1,1, trichloroethane, which is a metabolite of MC, and 2-chloro-4 hydroxy biphenyl, which is a metabolite of chlorobiphenyl.[89]

1.6 ENVIRONMENTAL EDC PROGRAMS

The high level of concern about EDCs affecting humans and wildlife has prompted worldwide legislation and research programs to achieve a better understanding of the endocrine disrupter issue and its potential health effects. In response to these issues, the United States implemented the Food Quality Protection Act and the Safe Drinking Water Act amendments in 1996. These acts require the EPA to screen

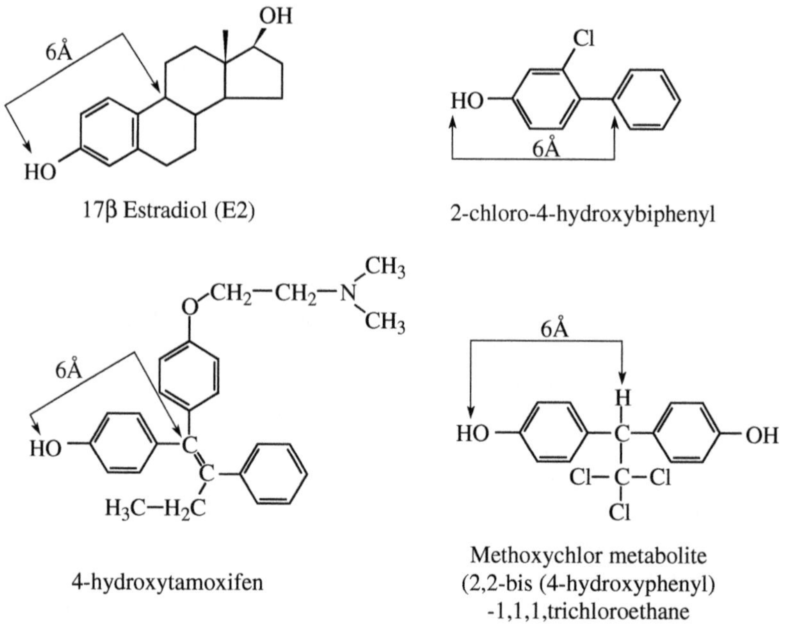

FIGURE 1.10 Biophore identification on estradiol and some potential EDCs. (Adapted from Combes R.D., Endocrine disruptors: a critical review of *in vitro* and *in vivo* testing strategies for assessing their toxic hazard to humans, *ATLA*, 28, 81, 2000. With permission.)

chemicals for "effects in humans similar to that produced by a naturally occurring estrogen or other endocrine disrupting effects." With this in mind, the EPA established the Endocrine Disrupter Screening and Testing Advisory Committee (EDSTAC) in 1996 to advise on the screening and testing of pesticides and other chemicals for their potential to act as EDCs. The aim of this program was to provide advice on the following:[147]

1. Developing a strategy for selection and prioritization of chemicals for screening and analysis
2. Defining a process for identifying new and existing screening assays
3. Assembling a set of available screens for early application
4. Determination of what definitive tests should be used and when, beyond screening
5. Identifying mechanisms for standardization and validation of screens and tests

The final report[75] contains over 70 recommendations, many being incorporated by the EPA into its proposed screening program. The program prioritizes chemicals to be screened based on existing information followed by a screening battery (Tier 1) and a testing battery (Tier 2). Tier 1 is designed to detect the potential for a substance to be endocrine disrupting and consists of a variety of *in vivo* and *in vitro* assays that are thought to include all known mechanisms of endocrine disruption in the three hormone systems: estrogen, androgen, and thyroid. Tier 2 is designed to characterize the endocrine effects of chemicals from Tier 1 testing that have shown to be endocrine disrupting. The initial number of chemicals identified for screening was around 87,000.[75]

Within the European Union, there are two Directorates General (DG) that are involved with the EDC issue: DG XII and DG XXIV. DG XII, Science, Research, and Development, has produced a number of research and monitoring recommendations covering modeling, methodology, human health, and wildlife studies.[76] DGXXIV, Consumer Policy and Consumer Health Protection, considers the impact of chemical and biological compounds, including EDCs, on human health and the environment.

In the United Kingdom, the Department of the Environment, Food and Rural Affairs (DEFRA, formerly the DETR), in conjunction with the Ministry of Agriculture, Fisheries and Food (MAFF), the UK Environment Agency (UKEA), Scotland and Northern Ireland Forum for Environmental Research (SNIFFER), and the European Chemical Industry Council, initiated the EDMAR program in 1998 (Endocrine Disruption in the Marine Environment). This program was to investigate whether there is evidence of changes associated with endocrine disruption in marine life and the possible causes and potential impacts. EDMAR has six main objectives:[148]

1. To develop biomarkers for detecting androgenic activity in marine fish and invertebrates and estrogenic activity in invertebrates
2. To conduct surveys of estrogenic and androgenic activity in key indicator species in the estuarine, coastal, and offshore environments
3. To observe the impact of this activity on a fish and invertebrate species
4. To conduct confirmatory experiments with suspect effluents and substances in laboratory test systems
5. To model the possible effects at the population level
6. To isolate the substances causing this marine endocrine disruption and identify the main sources of marine contamination

Much of this research has already been published[66,149–156] as have two interim reports[148,157] on the progress of this program.

Other work involving the U.K. Environment Agency has investigated the effects of EDCs on the fish population of U.K. rivers,[47–50,58,158] the estrogenicity of STW effluents,[45,52,61] and current work is focusing on the development of predicted-no-effect-concentrations (PNECs) for steroid estrogens[159] and other EDCs. Since 1993, there have been over 140 U.K. government research-related projects concerning endocrine disrupters, ranging from methodology to monitoring and risk assessment.

In 1998, the Japanese Environment Agency produced a Strategic Program on Environmental Endocrine Disrupters (SPEED).[81] Prior to this in 1997, they had set up an exogenous endocrine disrupting chemical task force. The 67 substances listed in the SPEED document as "chemicals suspected of causing endocrinological disruption," have not been conclusively proven to have, or not have, such a disruptive effect. They have only been identified as high-priority targets for future research, leading to frequent misunderstanding about "environmental hormones." The program is designed to promote research to first assign priorities to these suspected substances, and then clarify the mechanisms, existence, and strength of endocrine disrupters,[81] as well as promoting the strengthening of international networks.

The Organization for Economic Co-operation and Development (OECD) is an intergovernmental organization in which representatives of 30 industrialized countries in North America, Europe, and the Pacific, as well as the European Commission, meet to coordinate and harmonize policies, discuss issues of mutual concern, and work together to respond to international problems. The OECD endocrine disrupter activity program was initiated in 1996, at the request of OECD member countries and industry to ensure that testing and assessment approaches would not differ substantially among countries.[160] A task force on Endocrine Disrupter Testing and Assessment (EDTA) was set up to oversee this program. This has resulted in an initial framework for the testing of EDCs and consists of three tiers: initial assessment, screening, and testing.[160] The development of this framework was necessary for consolidating the existing work in the United States, Europe, and Japan, and to provide a platform for the discussion of testing needs and future work.

1.7 REASONS FOR CONCERN

One of the main reasons for concern is the possible effect EDCs may have on human health. Endocrine disrupting chemicals are now being linked as potentially responsible for various human health problems, so should we be concerned? Environmental chemicals with endocrine activity are thought to be responsible for the decrease in the quality and quantity of human sperm during the past 40 years. A study by Carlsen et al.[63] examined sperm quality over the last 50 years from 61 studies and found that there was a decrease in sperm quality and quantity over this period. However, according to Handelsman,[62] this meta analysis is marred by numerous flaws that invalidate its claims. Major defects include severe heterogeneity of component studies, rendering them unsuitable for aggregation, and defective analysis of data which showed no significant changes over time. Therefore, the data did not support the claim of falling sperm counts or any deterioration in male health. The question of whether sperm quality has actually declined is likely to remain unanswered until valid, representative, population-based studies of human sperm output can be produced. Other human health effects include increases in testicular and prostate cancer,[64,65,161] cryptorchidism (undescended testes),[65] hypospadias (penis malformation),[65,162] and female breast cancer.[163] However, none of these effects has been shown to have a definite cause–effect relationship with EDCs.

In the case of wildlife, it appears that, since the 1950s, many species have been affected by endocrine disrupting chemicals in the environment. More recently, STW effluents have been shown to have estrogenic activity which is causing feminization of river fish,[45,52,61] and this may ultimately affect their population densities. Hence, research into the fate and behavior of the natural and synthetic estrogens and other EDCs within an STW is paramount if we are to minimize this source of chemicals entering the environment.

The scientific community is divided over the endocrine disruption issue, and opinions can be placed into one of three categories:

1. Wildlife and laboratory evidence demonstrates that some chemicals behave as endocrine disrupters and have the potential to cause severe health problems.
2. More research is needed to clarify certain areas where there may be reason for concern over possible endocrine disruption.
3. The scientific data are inconclusive, and there is a lack of substantial cause and effect evidence.

At present, the majority of the scientific community falls into the second category, believing that, while endocrine disruption does exist, more research is required to identify mechanisms, the substances responsible, and the extent of the problem. Although it is now generally accepted that endocrine disruption is impacting wildlife, it remains unclear as to whether EDCs are actually affecting the human population or if the observed trends in human health are a result of other factors (e.g., diet and lifestyle). Only continued research in this complex field of study will ultimately reveal the answers.

REFERENCES

1. Dodds, E.C., Goldberg, L., Lawson, W., and Robinson, R., Oestrogenic activity of certain synthetic compounds, *Nature*, 141, 247, 1938.
2. Burlington, H. and Lindeman, V., Effect of DDT on testes and secondary sex characters of white leghorn cockerels, *Proc. Soc. Expt. Biol. Med.*, 74, 48, 1950.
3. Bitman, J., Cecil, H.C., Harris, S.J., and Fries, G.F., Estrogenic activity of o,p'-DDT in the mammalian uterus and avian oviduct, *Science*, 162, 371, 1968.
4. Carlson, R., *Silent Spring*, Penguin Books, London, 1962, pp. 301.
5. Dunn, T. and Green, A., Cysts of the epididymis, cancer of the cervix, granular cell myoblastoma and other lesions after estrogen injection in newborn mice, *J. Natl. Canc. Inst.*, 31, 425, 1963.
6. Takasugi, N. and Bern, H.A., Tissue changes in mice with persistent vaginal cornification induced by early postnatal treatment with estrogen, *J. Natl. Canc. Inst.*, 33, 855, 1964.
7. Foresbert, J. G., The development of a typical epithelium in the mouse uterine cervix and vaginal fornix after neonatal estradiol treatment, *Br. J. Exp. Pathol.*, 50, 187, 1969.
8. Herbst, A., Ulfelder, H., and Poskanzer, D., Adenocarcinoma of the vagina: Association of the maternal stilbestrol therapy with tumour appearance in young women, *New Eng. J. Med.*, 284, 878, 1971.
9. Herbst, A.L., Poskanzer, D.C., Robboy, S.J., Fiedlander, L., and Scully, R.E., Prenatal exposure to stilbestrol: a prospective comparison of exposed female offspring with unexposed controls, *New Eng. J. Med.*, 292, 334, 1975.
10. Gill, W.B., Schumacher, G.F.B., and Bibbo, M., Structural and functional abnormalities in the sex organs of male offspring of mothers treated with diethylstilbestrol (DES), *J. Reproduct. Med.*, 16, 147, 1976.

11. MacLachlan, J.A., Newbold, R., and Bullock, B., Reproductive tract lesions in male mice exposed prenatally to diethylstilbestrol, *Science*, 190, 991, 1975.

12. MacLachlan, J.A., Newbold, R., and Bullock, B., Long term effects on the female mouse genital tract associated with prenatal exposure to diethylstilbestrol, *Cancer Res.*, 40, 3988, 1980.

13. Melnick, S., Cole, P., Anderson, D., and Herbst, A., Rates and risks of diethylstilbestrol related clear-cell adenocarcinoma of the vagina and the cervix. An update, *New Eng. J. Med.*, 316, 514, 1987.

14. Kaufman, R.H., Adam, E., Hatch, E.E., Noller, K., Herbst, A.L., Palmer, J.R., and Hoover, R.N., Continued follow-up of pregnancy outcomes in diethylstilbestrol-exposed offspring, *Obst. Gynecol.*, 96, 483, 2000.

15. DeMora, S.J., *Tributyltin: Case Study of an Environmental Contaminant,* Cambridge University Press, Cambridge, 1996.

16. Gibbs, P.E., and Bryan, G.W., TBT-induced imposex in neogastropod snails: masculinization to mass extinction, in *Tributyltin: Case Study of an Environmental Contaminant*, DeMora, S.J., Ed., Cambridge University Press, Cambridge, 1996, pp. 212.

17. Fioni, P., Oehlmann, J., and Stroben, E., The pseudohermaphrodistism of prosobranchs; morphological aspects, *Zoo. Anz.*, 226, 1, 1991.

18. Alzeiu, C., Environmental problems caused by TBT in France: assessment, regulations, prospects, *Mar. Environ. Res.*, 32, 7, 1991.

19. Clark, E.A., Steritt, R.M., and Lester, J.N., The fate of tributyltin in the aquatic environment: an overview, *Environ. Sci. Technol.*, 22, 600, 1988.

20. Maguire, R.J., Environmental assessment of tributyltin in Canada, *Wat. Sci. Technol.*, 25, 125, 1992.

21. Dowson, P.H., Bubb, J.M., and Lester, J.N., Organotin distribution in sediments and waters of selected east coast estuaries in the UK, *Mar. Pollut. Bull.*, 24, 492, 1992.

22. Dowson, P.H., Pershke, D., Bubb, J.M., and Lester, J.N., Spatial distribution of organotin in sediments of lowland river catchments, *Environ. Pollut.*, 76, 259, 1992.

23. Voulvoulis, N., Scrimshaw, M.D., and Lester, J.N., Alternative antifouling biocides, *Appl. Organomet. Chem.*, 13, 135, 1999.

24. Facemire, C.F., Gross, T.S., and Guillette, L.L.J., Reproductive impairment in the Florida panther: nature or nurture?, *Environ. Health Perspect.*, 103, 79, 1995.

25. Bergmann, A. and Olsson, M., Pathology of Baltic ringed seals and grey seal females with special reference to adrenocortical hyperplasma: is environmental pollution the cause of a widely distributed disease syndrome?, *Finn. Game Res.*, 44, 47, 1985.

26. Helle, E., Lowered reproductive capacity in female ringed seals (*Phoca hispida*) in the Bothnian Bay, northern Baltic Sea, with special reference to uterine conditions, *Ann. Zool. Fenn.*, 17, 147, 1980.

27. Olsson, M., Karlsson, B., and Ahnland, E., Diseases and environmental contaminants in seals from the Baltic and the Swedish west coast, *Sci. Tot. Environ.*, 154, 217, 1994.

28. DeGuise, S., Lagace, A., and Beland, P., True hermaphroditism in a St. Lawrence beluga whale (*Delphinapterus leucas*), *J. Wildl. Dis.*, 30, 287, 1994.

29. Martineau, D., Lagace, A., Beland, P., Higgins, R., Armstrong, D., and Shugart, L.R., Pathology of Stranded Beluga Whales (Delphinapterus-Leucas) from the St-Lawrence Estuary, Quebec, Canada, *J. Comp. Pathol.*, 98, 287, 1988.

30. Roos, A., Greyerz, E., Olsson, M., and Sandegren, F., The otter (*Lutra lutra*) in Sweden — population trends in relation to Sigma DDT and total PCB concentrations during 1968–99, *Environ. Pollut.*, 111, 457, 2001.

31. Gutleb, A.C. and Kranz, A., Estimation of polychlorinated biphenyl (PCB) levels in livers of the otter (*Lutra lutra*) from concentrations in scats and fish, *Water Air Soil Pollut.*, 106, 481, 1998.

32. Subramanian, A.N., Tanabe, S., Tatuskawa, R., Saito, S., and Myazaki, N., Reductions in the testosterone levels by PCBs and DDE in Dall's porpoises of Northwestern North Pacific, *Mar. Pollut. Bull.*, 18, 643, 1987.

33. Fry, D.M. and Toone, C. K., DDT-induced feminization of gull embryos, *Science*, 213, 922, 1981.

34. Fry, D.M., Reproductive effects in birds exposed to pesticides and industrial chemicals, *Environ. Health Perspect.*, 103, 165, 1995.

35. Ratcliffe, D.A., Decrease in eggshell weight in certain birds of prey, *Nature*, 215, 208, 1967.

36. Gilberston, M., Kubiak, T., Ludwig, J., and Fox, G., Great Lakes embryo mortality, edema and deformities syndrome (GLEMEDS) in colonial fish-eating birds: similarity to chick-edema disease, *J. Toxicol. Environ. Health*, 33, 455, 1991.

37. Giesy, J.P., Ludwig, J., and Tiilitt, D., Deformities in birds of the Great Lakes region. Assigning causality, *Environ. Sci. Technol.*, 28, 128, 1994.

38. Murk, A.J., Bosveld, A.T.C., Van den Berg, M., and Brouwer, A., Effects of poly-halogenated aromatic hydrocarbons (PHAHs) on biochemical parameters in chicks of the common tern (*Sterna hirundo*), *Aquat. Toxicol.*, 30, 91, 1994.

39. Murk, A.J., Boudewijin, T.J., Meininger, P.L., Bosveld, A.T.C., Rossaert, G., Ysebaert, T., Meire, P., and Dirksen, S., Effects of polyhalogenated aromatic hydrocarbons and related contaminants on common tern reproduction: Integration of biological, biochemical and chemical data, *Arch. Environ. Contam. Toxicol.*, 31, 128, 1996.

40. Bosveld, A.T.C., Gradener, J., Murk, A., Brouwer, A., Van Kampen, M., Evers, E. H. G., and Van den Berg, M., Effects of PCDDs, PCDFs and PCBs in common tern (*Sterna hirundo*) breeding in estuarine and coastal colonies in the Netherlands and Belgium, *Environ. Toxicol. Chem.*, 1, 99, 1995.

41. Bishop, C.A., Brooks, R.J., and Carey, J.H., The case for a cause-effect linkage between environmental contamination and development in eggs of the common snapping turtle (*Chelydra s. serpentina*) from Ontario, Canada, *J. Toxicol. Environ. Health*, 33, 521, 1991.

42. De Solla, S.R., Bishop, C.A., Van Der Kraak, G., and Brooks, R.J., Impact of organochlorine contamination on levels of sex hormones and external morphology of common snapping turtles (*Chelydra s. serpentina*) in Ontario, Canada, *Environ. Health Perspect.*, 106, 253, 1998.

43. Guillette, L.L.J., Gross, T.S., Masson, G.R., Matter, J.M., Percival, H.F., and Woodward, A.R., Developmental abnormalities of the gonad and the abnormal sex hormone concentrations in juvenile alligators from contaminated and control lakes in Florida, *Environ. Health Perspect.*, 102, 680, 1994.

44. Guillette, L.L.J., Pickford, D.B., Crain, D.A., Rooney, A.A., and Percival, H.F., Reduction in penis size and plasma testosterone concentrations in juvenile alligators living in a contaminated environment, *Gen. Comp. Endocrinol.*, 101, 32, 1996.

45. Purdom, C.E., Hardiman, P.A., Bye, V.J., Eno, N.C., Tyler, C.R., and Sumpter, J.P., Estrogenic effects of effluents from sewage treatment works, *Chem. Ecol.*, 8, 275, 1994.

46. Routledge, E.J., Sheanan, D., Desbrow, C., Brighty, G.C., Waldock, M., and Sumpter, J.P., Identification of estrogenic chemicals in STW effluent. 2. *In vivo* responses in trout and roach, *Environ. Sci. Technol.*, 32, 1559, 1998.

47. Jobling, S., Nolan, M., Tyler, C.R., Brighty, G.C., and Sumpter, J.P., Widespread sexual disruption in wild fish, *Environ. Sci. Technol.*, 32, 2498, 1998.

48. Nolan, M., Jobling, S., Brighty, G., Sumpter, J.P., and Tyler, C.R., A histological description of intersexuality in the roach, *J. Fish Biol.*, 58, 160, 2001.

49. Jobling, S., Beresford, N., Nolan, M., Rodgers-Gray, T., Brighty, G.C., Sumpter, J.P., and Tyler, C.R., Altered sexual maturation and gamete production in wild roach (*Rutilus rutilus*) living in rivers that receive treated sewage effluents, *Biol. Reprod.*, 66, 272, 2002.

50. Lye, C.M., Frid, C.L.J., Gill, M.E., Cooper, D.W., and Jones, D.M., Estrogenic alkylphenols in fish tissues, sediments and waters from the UK Tyne and Tees estuaries, *Environ. Sci. Technol.*, 33, 1009, 1999.

51. Rodgers-Gray, T.P., Jobling, S., Kelly, C., Morris, S., Brighty, G., Waldock, M.J., Sumpter, J.P., and Tyler, C.R., Exposure of juvenile roach (*Rutilus rutilus*) to treated sewage effluent induces dose-dependent and persistent disruption in gonadal duct development, *Environ. Sci. Technol.*, 35, 462, 2001.

52. Rodgers-Gray, T.P., Jobling, S., Morris, S., Kelly, C., Kirby, S., Janbakhsh, A., Harries, J. E., Waldock, M. J., Sumpter, J.P., and Tyler, C.R., Long-term temporal changes in the estrogenic composition of treated sewage effluent and its biological effects on fish, *Environ. Sci. Technol.*, 34, 1521, 2000.

53. Allen, Y., Mathiessen, P., Haworth, S., Thain, J.E., and Feist, S., A survey of the estrogenic activity in the UK estuarine and coastal waters and its effects on the gonadal development of the flounder (*Platichthys flesus*), *Environ. Toxicol. Chem.*, 18, 1791, 1999.

54. Allen, Y., Mathiessen, P., Scott, A.P., Haworth, S., Feist, S., and Thain, J.E., The extent of estrogenic contamination in the UK estuarine and marine environments: Further survey of flounder, *Sci. Tot. Environ.*, 233, 5, 1999.

55. Matthiessen, P., Allen, Y., Allchin, C. R., Feist, S.W., Kirby, M.F., Law, R. J., Scott, A.P., Thain, J.E., and Thomas, K.V., Oestrogenic endocrine disruption in flounder (*Platichthys flesus*) from United Kingdom estuarine and marine waters, Scientific Services Technical Report (CEFAS), Lowestoft, 1998, pp. 48.

56. Harries, J.E., Sheanan, D.A., Jobling, S., Mathiessen, P., Neall, P., Routledge, E., Rycroft, R., Sumpter, J.P., and Tylor, T., A survey of estrogenic activity in United Kingdom inland waters, *Environ. Toxicol. Chem.*, 15, 1993, 1996.

57. Harries, J.E., Sheanan, D. A., Jobling, S., Mathiessen, P., Neall, P., Routledge, E., Rycroft, R., Sumpter, J. P., Tylor, T., and Zaman, N., Estrogenic activity in five United Kingdom rivers detected by measurement of vitellogenesis in caged male trout, *Environ. Toxicol. Chem. 16*, 534, 1997.

58. Jobling, S. and Sumpter, J.P., Detergent components in sewage effluent are weakly oestrogenic to fish: An *in vitro* study using rainbow trout (*Oncorhynchus mykiss*) hepatocytes, *Aquat. Toxicol.*, 27, 361, 1993.

59. Thorpe, K.L., Hutchinson, T.H., Hetheridge, M.J., Scholze, M., Sumpter, J.P., and Tyler, C.R., Assessing the biological potency of binary mixtures of environmental estrogens using vitellogenin induction in juvenile rainbow trout (*Oncorhynchus mykiss*), *Environ. Sci. Technol.*, 35, 2476, 2001.

60. Sumpter, J.P. and Jobling, S., Vitellogenesis as a biomarker for oestrogenic contamination of the aquatic environment, *Environ. Health Perspect.*, 103, 173, 1995.

61. Sheahan, D.A., Brighty, G.C., Daniel, M., Jobling, S., Harries, J.E., Hurst, M. R., Kennedy, J., Kirby, S.J., Morris, S., Routledge, E.J., Sumpter, J.P., and Waldock, M.J., Reduction in the estrogenic activity of a treated sewage effluent discharge to an English river as a result of a decrease in the concentration of industrially derived surfactants, *Environ. Toxicol. Chem.*, 21, 515, 2002.

62. Handelsman, D.J., Estrogens and falling sperm counts, *Reprod. Fertil. Develop.*, 13, 317, 2001.

63. Carlsen, E., Giwercman, A., Keiding, N., and Skakkebaek, N.E., Evidence for decreasing quality of semen during the past 50 years., *Br. Med. J.*, 305, 609, 1992.

64. Sharpe, R.M. and Skakkebaek, N.E., Are oestrogens involved in the falling sperm counts and disorders of the male reproductive tract?, *Lancet*, 341, 1392, 1993.

65. Toppari, J. and Skakkebaek, N.E., Endocrine disruption in male human reproduction, in *Environmental Endocrine Disrupters: An Evolutionary Perspective*, Guillette, L.J. and Crain, D.A., Eds., Taylor and Francis, London, 2000, pp. 269.

66. Matthiessen, P., Is endocrine disruption a significant ecological issue?, *Ecotoxicology*, 9, 21, 2000.

67. Colborn, T., Dumanoski, D., and Myers, J.P., *Our Stolen Future*, Abacus, London, 1996, pp. 294.

68. U.S. Environmental Protection Agency, Special Report on Environmental Endocrine Disruption: An Effects Assessment and Analysis, Report No. EPA/630/R-96/012, Washington D.C., 1997, pp. 111.

69. Griffin, J.E. and Ojeda, S.R., *Textbook of Endocrine Physiology*, 3rd ed., Oxford University Press, New York, 1996, pp. 396.

70. Colborn, T., Vom Saal, F.S., and Soto, A.M., Developmental effects of endocrine disrupting chemicals in wildlife and humans, *Environ. Health Perspect.*, 101, 378, 1993.

71. Sumpter, J.P., Xenoendocrine disrupters — environmental impacts, *Toxicol. Letts.*, 102–103, 337, 1998.

72. Sadik, O.A. and Witt, D.M., Monitoring endocrine disrupting chemicals, *Environ. Sci. Technol.*, 33, 368A, 1999.

73. European Commission, Report No. EUR 17549, 1996.

74. Zacharewski, T., *In vitro* bioassays for assessing estrogenic substances, *Environ. Sci. Technol.*, 31, 613, 1997.

75. Endocrine Disruptor Screening and Testing Advisory Committee, Final Report, Washington, D.C., 1998, pp. 239.

76. Phillips, B. and Harrison, P., Overview of the endocrine disrupters issue, in *Endocrine Disrupting Chemicals*, Hester, R.E. and Harrison, R.M., Eds., The Royal Society of Chemistry, Cambridge, 1999, p. 1.

77. Hertz, R., The estrogen problem: retrospect and prospect, in *Estrogens in the Environment II: Influences on Development*, MacLachlan, J.A., Ed., Elsevier, New York, 1985, p. 1.

78. United Kingdom Economics Account, Endocrine disrupting substances in the environment: The Environment Agency's strategy, *Environment Agency*, 2000, p. 23.

79. Hughes, C., Are the differences between estradiol and other estrogens merely semantical?, *J. Clin. Endocrinol. Metab.*, 81, 2405, 1996.

80. Oslo and Paris Commissions, OSPAR strategy with regard to hazardous substances, OSPAR convention for the protection of the marine environment of the North-East Atlantic, OSPAR 98/14/1, Annex 34, 1998, pp. 22.

81. Japan Environmental Agency, Strategic Programs on Environmental Endocrine Disrupters, SPEED 98, 1998, pp. 49.

82. World Wildlife Foundation, Chemicals in the environment reported to have reproductive and endocrine disrupting effects, http://wwf.acenetx.com/satellite/hormone-disruptors/science/edclist.html, accessed 5-16-2002, pp. 2.

83. Commission of the European Communities, The implementation of the Community strategy for endocrine disrupters: A range of substances suspected of interfering with the hormone systems of humans and wildlife (COM [1999] 706), 2001, pp. 45.

84. Commission lists priority hormone chemicals, *ENDS Environment Daily*, Jun. 20, 2001.

85. Lester, J.N., *Heavy Metals in Wastewater and Sludge Treatment Processes*, CRC Press, Boca Raton, FL, 1987.

86. Lester, J.N., *Heavy Metals in Wastewater and Sludge Treatment Processes*, CRC Press, Boca Raton, FL, 1987.

87. Rogers, H.R., Sources, behaviour and fate of organic contaminants during sewage treatment and in sewage sludges, *Sci. Tot. Environ.*, 185, 3, 1996.

88. Santodonato, J., Review of the estrogenic and antiestrogenic activity of polycyclic aromatic hydrocarbons: relationship to carcinogenicity, *Chemosphere*, 34, 835, 1997.

89. Combes, R. D., Endocrine disruptors: a critical review of *in vitro* and *in vivo* testing strategies for assessing their toxic hazard to humans, *ATLA*, 28, 81, 2000.

90. Makin, H.L.J., Honour, J.W., and Shackleton, C.H.L., General methods of steroid analysis: Extraction, purification and measurement of steroids by HPLC, GLC and MS, in *Steroid Analysis*, Makin, H.L.J., Gower, D.B., and Kirk, D.N., Eds., 1995, pp. 114.

91. Makin, H. L. and Heftmann, E., High-performance liquid chromatography of steroid hormones, *Mono. Endocrinol.*, 30, 1998.

92. Tabak, H.H., Bloomhuff, R.N., and Bunch, R.L., Steroid hormones as water pollutants. II. Studies on the persistence and stability of natural urinary and synthetic ovulation-inhibiting hormones in untreated and treated wastewaters, *Develop. Ind. Microbiol.*, 22, 497, 1981.

93. Lai, K.M., Johnson, K.L., Scrimshaw, M.D., and Lester, J.N., Binding of waterborne steroid estrogens to solid phases in river and estuarine systems, *Environ. Sci. Technol.*, 34, 3890, 2000.

94. Jordan, V.C., Mittal, S., Gosden, B., Koch, R., and Lieberman, M.E., Structure-activity relationships of estrogens, *Environ. Health Perspect.*, 61, 97, 1985.

95. Daux, W.L. and Griffin, J.F., Structural features which distinguish estrogen agonists from antagonists, *J. Steroid Biochem.*, 27, 271, 1987.

96. Daux, W.L., Sweenson, D.C., Strong, P.D., Korach, K.S., McLachlan, J., and Metzler, M., Molecular structures of metabolites and analogues of DES and their relationship to receptor binding and biological activity, *Mol. Pharm.*, 26, 520, 1984.

97. Murphy, C.S. and Jordan, V.C., The biological significance of the interaction of estrogen agonists and antagonists with the estrogen receptor, *Receptor*, 1, 65, 1990.

98. Tham, D.M., Potential health benefits of dietary phytoestrogens: a review of the clinical, epidemiological, and mechanistic evidence, *J. Clin. Endocrinol. Metab.*, 83, 2223, 1998.

99. Zava, D.T. and Duwe, G., Estrogenic and antiproliferative properties and other flavonoids in human breast cancer cells *in vivo*, *Nutr. Cancer*, 27, 31, 1997.

100. Miksicek, R. J., Interactions of naturally occurring nonsteroidal estrogens with expressed recombinant human estrogen receptor, *J. Steroid Biochem. Mol. Biol.*, 49, 153, 1994.

101. Santell, R.C., Cheng, Y.C., Nair, M.G., and Helferich, W.G., Dietary genistein exerts estrogenic effects upon the uterus, mammary gland and the hypothalamic/pituitary axis in rats, *J. Nutr.*, 127, 263, 1997.

102. Adlercreutz, H., Heikkinen, R., Woods, M., Fotsis, T., Dwyer, J.T., and Goldin, B.R., Excretion of the lignans enterolactone and enterodiol and of equol in omnivorous and vegetarian postmenopausal women and in women with breast cancer, *Lancet*, 2, 1295, 1982.

103. Adlercreutz, H., Fotsis, T., Bannwart, C., Hamalainen, E., Bloigu, S., and Ollus, A., Urinary estrogen profile determination in young Finnish vegetarian and omnivorous women, *J. Steroid Biochem. Mol. Biol.*, 24, 289, 1986.

104. Adlercreutz, H., Markkanen, H., and Watanabe, S., Plasma concentrations of phyto-oestrogens in Japanese men, *Lancet*, 342, 1209, 1993.

105. Perez, P., Pulgar, R., Olea-Serrano, F., Villalobos, M., Rivas, A., Metzler, M., Pedraza, V., and Olea, N., The estrogenicity of bisphenol A-related diphenylalkanes with various substituents at the central carbon and the hydroxy groups, *Environ. Health Perspect.*, 106, 167, 1998.

106. Bergeron, R. M., Thompson, T. B., Leonard, L. S., Pluta, L., and Gaido, K. W., Estrogenicity of bisphenol A in a human endometrial carcinoma cell line, *Mol. Cell. Endocrinol.*, 150, 179, 1999.

107. Schafer, T.E., Lapp, C.A., Hanes, C.M., Lewis, J.B., Wataha, J.C., and Schuster, G.S., Estrogenicity of bisphenol A and bisphenol A dimethacrylate *in vitro*, *J. Biomed. Mater. Res.*, 45, 192, 1999.

108. Chen, M.Y., Ike, M., and Fujita, M., Acute toxicity, mutagenicity, and estrogenicity of bisphenol-A and other bisphenols, *Environ. Toxicol.*, 17, 80, 2002.

109. Sohoni, P. and Sumpter, J.P., Several environmental oestrogens are also anti-andro-gens, *J. Endocrinol.*, 158, 327, 1998.

110. Massaad, C., Entezami, F., Massade, L., Benahmed, M., Olivennes, F., Barouki, R., and Hamamah, S., How can chemical compounds alter human fertility?, *Eur. J. Obstet. Gynecol. Reprod. Biol.*, 100, 127, 2002.

111. Harris, C.A., Henttu, P., Parker, M.G., and Sumpter, J.P., The estrogenic activity of phthalate esters *in vitro*, *Environ. Health Perspect.*, 105, 802, 1997.

112. Jobling, S., Reynolds, T., White, R., Parker, M.G., and Sumpter, J.P., A variety of environmentally persistent chemicals, including some phthalate plasticizers, are weakly estrogenic, *Environ. Health Perspect.*, 103, 582, 1995.

113. Staples, C. A., Peterson, D. R., Parkerton, T. F., and Adams, W. J., The environmental fate of phthalate esters: a literature review, *Chemosphere*, 35, 667, 1997.

114. Routledge, E.J. and Sumpter, J.P., Structural features of alkylphenolic chemicals associated with estrogenic activity, *J. Biol. Chem.*, 272, 3280, 1997.

115. Nimrod, A.C. and Benson, W.H., Environmental estrogenic effects of alkylphenol ethoxylates, *Crit. Rev. Toxicol.*, 26, 335, 1996.

116. Soto, A.M., Justicia, H., Wray, J.W., and Sonnenschein, C., p-Nonylphenol: an estro-genic xenobiotic released from "modified" polystyrene, *Environ. Health Perspect.*, 92, 167, 1991.

117. White, R., Jobling, S., Hoare, S.A., Sumpter, J.P., and Parker, M.G., Environmentally persistent alkylphenolic compounds are estrogenic, *Endocrinol.*, 135, 175, 1994.

118. Tabira, Y., Nakai, M., Asai, D., Yakabe, Y., Tahara, Y., Shinmyozu, T., Noguchi, M., Takatsuki, M., and Shimohigashi, Y., Structural requirements of para-alkylphenols to bind to estrogen receptor, *Eur. J. Biochem.*, 262, 240, 1999.

119. Safe, S.H., Polychlorinated biphenyls (PCBs): environmental impact, biochemical and toxic responses, and implications for risk assessment, *Crit. Rev. Toxicol.*, 24, 87, 1994.

120. Li, M.H. and Hansen, L.G., Consideration of enzyme and endocrine interactions in the risk assessment of PCBs, *Rev. Toxicol.*, 1, 71, 1997.

121. Kester, M.H.A., Bulduk, S., Tibboel, D., Meinl, W., Glatt, H., Falany, C.N., Coughtrie, M.W.H., Bergman, A., Safe, S.H., Kuiper, G., Schuur, A.G., Brouwer, A., and Visser, T.J., Potent inhibition of estrogen sulfotransferase by hydroxylated PCB metabolites: A novel pathway explaining the estrogenic activity of PCBs, *Endocrinol.*, 141, 1897, 2000.

122. Bonefeld-Jorgensen, E.C., Andersen, H.R., Rasmussen, T.H., and Vinggaard, A. M., Effect of highly bioaccumulated polychlorinated biphenyl congeners on estrogen and androgen receptor activity, *Toxicology*, 158, 141, 2001.

123. Vakharia, D. and Gierthy, J., Rapid assay for oestrogen receptor binding to PCB metabolites, *Toxicol. Vitro*, 13, 275, 1999.

124. Andersson, P.L., Blom, A., Johannisson, A., Pesonen, M., Tysklind, M., Berg, A.H., Olsson, P.E., and Norrgren, L., Assessment of PCBs and hydroxylated PCBs as potential xenoestrogens: *in vitro* studies based on MCF-7 cell proliferation and induction of vitellogenin in primary culture of rainbow trout hepatocytes, *Arch. Environ. Contam. Toxicol.*, 37, 145, 1999.

125. Matthews, J. and Zacharewski, T., Differential binding affinities of PCBs, HO-PCBs, and aroclors with recombinant human, rainbow trout (*Oncorhynchus mykiss*), and green anole (*Anolis carolinensis*) estrogen receptors, using a semi-high throughput competitive binding assay, *Toxicol. Sci.*, 53, 326, 2000.

126. Carlson, D.B. and Williams, D.E., 4-hydroxy-2',4',6'-trichlorobiphenyl and 4-hydroxy-2',3',4',5'-tetrachlorobiphenyl are estrogenic in rainbow trout, *Environ. Toxicol. Chem.*, 20, 351, 2001.

127. Meerts, I.A.T.M., Letcher, R.J., Hoving, S., Marsh, G., Bergmann, A., Lemmen, J.G., Van der Burg, B., and Brouwer, A., *In vitro* estrogenicity of polybrominated diphenyl ethers, hydroxylated PBDEs, and polybrominated bisphenol A, *Environ. Health Perspect.*, 109, 399, 2001.

128. Renner, R., What fate for brominated fire retardants?, *Environ. Sci. Technol.*, 34, 222A, 2000.

129. Sjodin, A., Hagmar, L., Klasson-Wehler, E., Kronholm-Diab, K., Jakobsson, E., and Bergman, A., Flame retardant exposure: polybrominated diphenyl ethers in blood from Swedish workers, *Environ. Health Perspect.*, 107, 643, 1999.

130. Lindstrom, G., Wingfors, H., Dam, M., and Van Bavel, B., Identification of 19 polybrominated diphenyl ethers (PBDEs) in long-finned pilot whale (*Globicephala melas*) from then Atlantic, *Arch. Environ. Contam. Toxicol.*, 36, 355, 1999.

131. De Boer, J., De Boer, K., and Boon, J.P., Polybrominated biphenyl and diphenyl ethers, in *The Handbook of Environmental Chemistry*, Paasivirta, J., Ed., Springer-Verlag, Heidelberg, 1999, pp. 61.

132. Yang, N.C., Castro, A.J., Lewis, M., and Wong, T.W., Polynuclear aromatic hydrocarbons, steroids and carcinogenesis, *Science*, 134, 386, 1961.

133. Morreal, C.E., Sinha, D.K., Schneider, S.L., Bronstein, R.E., and Dawidzik, J., Anti-estrogenic properties of substituted benz[a]anthracene-3,9-diols., *J. Med. Chem.*, 25, 323, 1982.

134. Tran, D.Q., Ide, C.F., McLachlan, J.A., and Arnold, S.F., The anti-estrogenic activity of selected polynuclear aromatic hydrocarbons in yeast expressing human estrogen receptor, *Biochem. Biophys. Res. Commun.*, 229, 101, 1996.

135. Clemons, J.H., Allan, L.M., Marvin, C.H., Wu, Z., McCarry, B.E., Bryant, D.W., and Zacharewski, T.R., Evidence of estrogen and TCDD-like activities in crude and fractionated extracts of PM_{10} air particulate material using *in vitro* gene expression assays, *Environ. Sci. Technol.*, 32, 1853, 1998.

136. Fertuck, K.C., Kumar, S., Sikka, H.C., Matthews, J.B., and Zacharewski, T.R., Interaction of PAH-related compounds with the alpha and beta isoforms of the estrogen receptor, *Toxicol. Lett.*, 121, 167, 2001.

137. Charles, G.D., Bartels, M.J., Zacharewski, T.R., Gollapudi, B.B., Freshour, N.L., and Carney, E.W., Activity of benzo a pyrene and its hydroxylated metabolites in an estrogen receptor-alpha reporter gene assay, *Toxicol. Sci.*, 55, 320, 2000.

138. Wright, D.A. and Welborn, P., *Environmental Toxicology*, Cambridge University Press, Cambridge, 2002, pp. 630.

139. Bulger, W.H. and Kupfer, D., Estrogenic action of DDT analogs, *Am. J. Ind. Med.*, 4, 163, 1983.

140. Kupfer, D., Effects of pesticides and related compounds on steroid metabolism and function, *Crit. Rev. Toxicol.*, 4, 83, 1975.

141. Nelson, J.A., Struck, R.F., and James, R.E., Estrogenic activities of chlorinated hydrocarbons, *J. Toxicol. Environ. Health*, 4, 325, 1978.

142. Ireland, J.S., Mukku, V.R., Robison, A.K., and Stancel, G.M., Stimulation of uterine deoxyribonucleic acid synthesis by 1,1,1,-trichloro-2-(p-chlorophenyl)-2-(*o*-chlorophenyl)ethane (*o,p'*-DDT), *Biochem. Pharmacol.*, 24, 1469, 1980.

143. Bulger, W.H., Muccitelli, R.M., and Kupfer, D., Studies on the *in vivo* and *in vitro* estrogenic activities of methoxychlor and its metabolites. Role of hepatic monooxygenase in methoxychlor activation, *Biochem. Pharmacol.*, 27, 2417, 1978.

144. Ratcliffe, J.M., McElhatton, P.R., and Sullivan, F.M., Reproductive toxicity, in *General and Applied Toxicology*, Ballantyne, B., Marrs, T., and Turner, P.E., Eds., Macmillan Press, Basingstoke, 1994, pp. 989.

145. Cunningham, A., Klopman, G., and Rosenkranz, H.S., The carcinogenicity of DES: structural evidence for a non-genotoxic mechanism, *Arch. Toxicol.*, 70, 356, 1996.

146. Cunningham, A.R., Rosenkranz, H.S., and Klopman, G., Structural analysis of a group of phytoestrogens for the presence of a 2-D geometric descriptor associated with nongenotoxic carcinogens and some estrogens, *Proc. Soc. Expt. Biol. Med.*, 217, 288, 1998.

147. Fenner-Crisp, P.A., Maciorowski, A.F., and Timm, G.E., The endocrine disruptor screening program developed by the U.S. environmental protection agency, *Ecotoxicology*, 9, 85, 2000.

148. Endocrine Disruption in the Marine Environment, Report of the 1st Annual Seminar, Central Science Laboratory, York, 1999, pp. 23.

149. Hutchinson, T.H. and Mathiessen, P., Endocrine disruption in wildlife: identification and ecological relevance, *Sci. Tot. Environ.*, 233, 1, 1999.

150. Hutchinson, T.H., Pounds, N.A., Hampel, M., and Williams, T.D., Impact of natural and synthetic steroids on the survival, development and reproduction of marine copepods *Tisbe battagliai*, *Sci. Tot. Environ.*, 233, 167, 1999.

151. Hutchinson, T.H., Pounds, N.A., Hampel, M., and Williams, T.D., Life cycle studies with marine copepods (*Tisbe battagliai*) exposed to 20-hydroxyecdysone and diethylstilbestrol, *Environ. Toxicol. Chem.*, 18, 2914, 1999.

152. Katsiadaki, I., Scott, A.P., and Matthiessen, P., The use of the three-spined stickleback as a potential biomarker for androgenic xenobiotics, in *Proceedings of the 6th International Symposium on the Reproductive Physiology of Fish*, Norberg, B., Kjesbu, O.S., Taranger, G.L., Andersson, E. and Stefansson, S.O., Eds., Institute of Marine Research and University of Bergen, 2000.

153. Matthiessen, P., Allen, Y., Bignell, J., Craft, J., Feist, S., Jones, G., Katsiadaki, I., Kirby, M., Robertson, E., Scott, A.P., Stewart, C., and Thain, J.E., Studies of endocrine disruption in marine fish — progress with the EDMAR programme, *Int. Council Explorat. Sea.*, ICES CM 2000/S:06, 2000.

154. Robinson, C.D., Craft, J., Davies, I.M., Moffat, C.F., Pirie, D., Robertson, F., Brown, E., Stagg, R.M., and Struthers, S., Effects of ethynyl oestradiol and sewage effluent upon maturation, molecular markers of oestrogenic exposure, and reproductive success in a marine teleost (*Pomatoschistus minutus,* Pallas), FRS Report Mar. 1, 2001.

155. Scott, A.P., Stewart, C., Allen, Y., and Matthiessen, P., 17β-oestradiol in male flatfish, in *Proceedings of the 6th International Symposium on the Reproductive Physiology of Fish*, Norberg, B., Kjesbu, O.S., Taranger, G.L., Andersson, E. and Stefansson, S.O., Eds., Institute of Marine Research and University of Bergen, 2000.

156. Thomas, K. V., Hurst, M.R., Matthiessen, P., and Waldock, M.J., Identification of oestrogenic compounds in surface and sediment pore water samples collected from industrialised UK estuaries, *Environ. Toxicol. Chem.*, 20, 2165, 2001.

157. Endocrine Disruption in the Marine Environment (EDMAR), Report of the 2nd Annual Seminar, DETR, Great Minster House, London, 2000, pp. 26.

158. van Aerle, R., Nolan, M., Jobling, S., Christiansen, L.B., Sumpter, J.P., and Tyler, C. R., Sexual disruption in a second species of wild cyprinid fish (the gudgeon, *Gobio gobio*) in United Kingdom freshwaters, *Environ. Toxicol. Chem.*, 20, 2841, 2001.

159. Young, W.F., Whitehouse, P., Johnson, I., and Sorokin, N., Proposed Predicted-No-Effect-Concentrations (PNECs) for natural and synthetic steroid oestrogens in surface waters, Environment Agency, R&D Technical Report P2-T04/1, 2002, pp. 170.

160. Huet, M.C., OECD activity on endocrine disrupters test guidelines development, *Ecotoxicology,* 9, 77, 2000.

161. Whittemore, A.S., Prostate cancer, *Cancer Surv.*, 19–20, 1117, 1994.

162. Baskin, L. S., Himes, K., and Colborn, T., Hypospadias and endocrine disruption: Is there a connection?, *Environ. Health Perspect.*, 109, 1175, 2001.

163. Fernandez, M.F., Pedraza, V., and Olea, N., Estrogens in the environment: is there a breast cancer connection?, *Cancer J.*, 11, 11, 1998.

2 Sources of Endocrine Disrupters

J.W. Birkett

CONTENTS

2.1 INTRODUCTION

Endocrine disrupting chemicals (EDCs), as well as other pollutants, have a variety of sources. These sources may have implications for human exposure, effects in wildlife, and effects or accumulation within the environment. Moreover, many EDCs have been found to be ubiquitous in the environment. Generally, sources of pollution fall into two main categories:

1-56670-601-7/03/$0.00+$1.50
© 2003 by CRC Press LLC

1. Point sources
2. Nonpoint sources (or diffuse sources)

Table 2.1 summarizes point and nonpoint sources

2.1.1 POINT SOURCES

A point source is a definitive point of entry of a pollutant into an environmental medium (generally a watercourse). It could be an effluent discharge pipe, a stormwater overflow, or a known point where waste is repeatedly dumped. Point sources tend to be easier to control than nonpoint sources, so, as far as possible, nonpoint sources are converted to point sources.

Point sources vary according to the specific catchment area under study. In a river system, sewage treatment works tend to constitute the main sources of pollution. This fact illustrates the success of the authorities and regulatory bodies in minimizing other sources.

2.1.2 NONPOINT SOURCES

Nonpoint sources (sometimes called diffuse sources) do not have a definitive point of entry. Good examples of nonpoint sources are atmospheric deposition and runoff. The actual point where the pollutants enter depends on the type of source, its location, and on the physical form of the pollutants. If the pollutants are gases or fine airborne particles, they can fall to the ground directly with rain. The rain can also wash particles that have been deposited on surfaces into nearby watercourses. If the pollutants are soluble, they can be transported long distances in the water. During storms, large particles, including soil, can be washed down from the land into water bodies. These may have pollutants such as pesticides attached to them.

TABLE 2.1
Point and Nonpoint Sources

Point Sources	Nonpoint Sources
Discharges from sewage treatment works to rivers	Runoff and underdrainage from agricultural land into rivers
Discharges of industrial wastewaters to rivers	General contamination of recharge rainfall to outcropping aquifers
Discharges of farm effluents to rivers	Septic tank soakaways into permeable strata
Discharges from small domestic sewage treatment plants to rivers	Washoff of litter, dust, and dry fallout from urban roads to rivers
Discharges by means of wells or boreholes into underground strata	General entry of sporadic and widespread losses of contaminants to rivers
Discharges of collected landfill leachate to rivers	Seepage of landfill leachate to underground strata and rivers

From Lester, J.N. and Birkett, J.W., *Microbiology and Chemistry for Environmental Scientists and Engineers*, 2nd ed. E & F.N. Spon, London, 1999. With permission.

Nonpoint source pollution transport processes in urban areas are likely to be different than those in rural areas. There are several reasons for this:[1]

1. Quite a large part of urban areas are covered by impermeable materials. Thus, quite a lot of the rainwater runs off and eventually enters drains and sewers.
2. In urban areas, less soil is exposed so less erosion, and hence transport of soil particles into surface waters, can be expected.
3. In urban areas, pollutant loadings are mainly affected by the accumulation of litter, fallout, and road traffic. In rural areas, most of the pollution is due to the erosion of soils.
4. In the long term, almost all of the pollutants deposited on impermeable surfaces in urban areas (except those removed by external processes such as street cleaning) will eventually end up as surface runoff. In rural areas, deposits can be incorporated into the soil where their removal rate is reduced.

2.1.3 HYDROLOGICAL CONTROL

This refers to man's interference in the movement of water through natural channels. The result is a nonpoint source pollution that can directly or indirectly affect water quality. Examples are the construction of drainage and irrigation systems, dredging in order to create or maintain navigable stretches of water, and the construction of dams and reservoirs.

Dredging can cause water pollution because pollutants that were previously held tightly within the sediment can be redissolved or resuspended in the water during the dredging operation. Dredging spoil is often dumped back into the water at a point remote from where it was taken. This can have the effect of transporting any pollution from one place to another.

2.1.4 GROUNDWATER POLLUTION

Most human activities at the land's surface cause some change in the quality of water in the aquifer beneath them. The importance of the effect of a particular activity is related to the amounts and types of contaminants released. The severity of an occurrence is also related to the ability of the soil and groundwater system to degrade or dilute the contaminants, and the degree to which the contamination will interfere with uses of the water. Contamination is usually more serious in a drinking water supply than in water for other uses.

Groundwater is a good source of drinking water because of the properties of the soils that purify rainwater as it percolates through to the aquifer. This particularly applies to suspended matter, which is effectively filtered out by the strata overlying the aquifer. Except where contaminated water is injected directly into an aquifer, essentially all groundwater pollutants enter the aquifer through recharge water from the land surface.

The main sources of groundwater pollution arise from industrial, domestic, and agricultural sources. Industrial sources are industrial effluents, accidents (e.g., leakage from pipes and tanks), and rainwater that infiltrates and percolates through solid waste deposits. Solid wastes are deposited in landfill sites. Landfills that take hazardous waste are often sited in areas where there are impermeable strata, such as clay, or where clay linings have been deliberately installed to prevent leachate from escaping.

Domestic sources of groundwater pollution fall into more or less the same categories as the industrial sources. Leakage from septic tanks and percolation of rainwater through landfills containing domestic refuse are the main risks.

Agricultural sources are potentially the most dangerous because they are nonpoint sources, and percolation through soils into the groundwater can occur over wide areas. As a result, fertilizers, minerals, herbicides, and pesticides can all enter aquifers. There is also a potential risk from the disposal of sewage sludge on agricultural land, which, in addition to the sea, is another major disposal outlet for this material.

Endocrine disruption manifesting as reproductive and behavioral abnormalities observed in fish has been primarily attributed to EDCs in sewage effluent discharged into watercourses. In addition to the types of discharges already mentioned, contamination of the aquatic environment can be caused by numerous other sources. These sources of EDCs are summarized in Table 2.2.

2.1.5 OTHER SOURCES

For the human population, the most important source of EDCs by far is food. Some of these compounds are utilized in food production and packaging, which has prompted concern that diet may be an exposure route for EDCs.[20] Examples of this include pesticide residues on fruit and vegetables, use of plastic wrapping, and the leaching of compounds from can linings.[21] These, together with other sources for EDCs, are highlighted in the following section.

2.2 SPECIFIC SOURCES RELATING TO ENDOCRINE DISRUPTING CHEMICALS (EDCS)

2.2.1 STEROIDS

The steroids estradiol, progesterone, and testosterone all produce growth effects in humans and animals. Because of this property, exogenous steroids have been used in meat-producing animals in the United States for almost 50 years. Several synthetic chemicals are also used as growth enhancers in cattle. A report by the Food and Agricultural Organization/World Health Organization Joint Expert Committee on Food Additives (JECFA) [22] found that levels of estradiol, estrone, progesterone, and testosterone in animal tissue were all significantly increased (at least twofold) in treated cattle compared with untreated herds. If the use of hormones increases steroid levels in edible tissues, then it is probable that there will be an increase in the steroid intake to humans from the consumption of such products. In Europe, the use of such steroids in meat production was banned in 1989.[20] Table 2.3 highlights the sources of steroid hormones, predominantly the estrogens.

TABLE 2.2
Sources of Endocrine Disrupting Chemicals (EDCs) Entering Watercourses

Source of EDCs	Receiving Waters	Source Method	EDCs Likely to Be Present
Domestic sewage effluent	Surface water[a]	Point	Steroid estrogens,[2] Surfactants,[3] PAEs, BPA[4]
	Groundwater	Nonpoint	
	Groundwater	Point (recharge)	
Industrial sewage effluent	Surface water	Point	Surfactants,[5,6] PAHs,[7] PCBs,[7] PBDEs,[8] pesticides,[7] PAEs,[4] BPA
	Groundwater	Nonpoint	
Industrial discharges	Surface water	Point	Dioxins, PBDEs,[8] TBBA,[9] PAEs, PCBs,[7] PAHs,[7] pesticides,[7] BPA[4]
	Groundwater	Nonpoint	
Paint applied to boats	Surface water	Point	TBT[10]
Agricultural runoff (crops)	Surface water	Nonpoint	Pesticides,[7] APs, APEs,[11] PBDEs,[12] PAHs[13,14]
	Groundwater	Nonpoint	
Agricultural runoff (animals)	Surface water	Nonpoint	Steroid estrogens[15,16]
	Groundwater	Nonpoint	
Recreational/Urban runoff	Surface water	Nonpoint	Pesticides, PAHs[7]
	Groundwater	Nonpoint	
Leachate from waste dumps	Groundwater	Nonpoint	PAHs,[17] PBDEs, TBBA,[9] BPA, PAEs
Deposition from the air	Surface water	Nonpoint	PAHs,[7] PCBs, PCDDs, PCDFs, PBDEs,[18]
	Groundwater	Nonpoint	TBBA, pesticides
Natural	Surface water	Nonpoint	PAHs, steroid estrogens (natural)[19]
	Groundwater	Nonpoint	

Phthalate acid esters (PAEs); bisphenol A (BPA); polyaromatic hydrocarbons (PAHs); polychlorinated biphenyls (PCBs); polybrominated diphenyl ethers (PBDEs); tetrabromobisphenol A (TBBA); tributyltin (TBT); polychlorinated dibenzo-p-dioxins (PCDDs) and polychlorinated dibenzofurans (PCDFs)

[a] Surface waters include streams, rivers, estuaries, and seas

TABLE 2.3
Sources of Steroid Hormones

Source	Steroid Hormone
Food — meat, fish, eggs, pork, dairy products	Estradiol, estrone, progesterone, testosterone
Sewage treatment work effluents	Estradiol, estrone, estriol, ethinyl estradiol
Sewage sludge	Estradiol, estrone, estriol, ethinyl estradiol
Oral contraception	Ethinyl estradiol, mestranol
Hormone replacement therapy (HRT)	Conjugated estrone, 17α and 17β-estradiol
Runoff	Estradiol, estrone
Agricultural waste	Estradiol, estrone

Detectable concentrations of steroid hormones (e.g., estradiol, estrone, progesterone, testosterone) have also been found in fish, poultry, eggs, pork, cheese, milk, and milk products.[23] However, in comparing the production of these hormones in adults and children with their daily intake, it is found that children, who have the lowest hormone production rate, show levels of progesterone to be 20 times greater then their intake, while estradiol and testosterone production levels are 1000 times higher.[23] It is likely that no hormonal effects in humans can be expected from naturally occurring steroid compounds in food. However, use of synthetic EDCs and the presence of xenoestrogens in foodstuffs may be more problematic.

Within the steroid hormone group, it is perhaps the natural (estrone [E1], 17β-estradiol [E2], estriol [E3]) and synthetic (ethinylestradiol [EE2], mestranol) steroid estrogens that have received the most scientific attention. These compounds are the major contributors to the estrogenic activity observed in sewage effluent[24,25] and the receiving water body (see Chapter 6). Their presence in the aquatic environment is attributed to their incomplete removal during the sewage treatment process.[26] Although concentrations of steroid estrogens have been reported in the low ng L^{-1} levels, their estrogenic potency warrants cause for concern, as EE2 has been shown to induce vitellogenin (VTG) production (a female yolk protein) in male fish at 0.2 ng L^{-1}.[26]

The presence of steroid estrogens in STW effluent arises from mammalian excretion, in particular females of reproductive age and those who are pregnant. Women excrete 10 to 100 μg of estrogen per day depending on the phase of their cycle, and pregnant women can excrete up to 30 mg per day.[27] The major excretion product is estriol, with daily excretion rates for women at a maximum of 64 μg per day. The excretion rate of estrone and 17β-estradiol is 3 to 20 μg and 0.5 to 5 μg per day, respectively.[28] The majority of these estrogens are excreted from the human body within urine in a biologically inactive, conjugated form (predominantly as glucuronides and sulfates). However, because free estrogens have been observed in STW effluent, this implies that deconjugation has occurred at some stage during or prior to sewage treatment.[29] Moreover, the amount of natural and synthetic estrogens entering the STW is unlikely to decrease due to the origin and use. Steroid estrogens have also been detected in sewage sludge. Concentrations of estradiol, estrone, and estriol in sewage sludge from 14 different STWs in the United States range from

0.01 to 0.08 ng L^{-1}, while mean concentrations of ethinyl estradiol in raw and treated sewage were 1.21 ng L^{-1} and 0.81 ng L^{-1}, respectively.[19] The application of such sludge to agricultural land is likely to increase the estrogen content from other sources, such as runoff.

The main uses of synthetic estrogens are in oral contraception and hormone replacement therapy (HRT). In 1990, of the 58 million women in the United States who actively practiced contraception, 10 million used oral contraception.[19] These contraceptives contain a combination of estrogen and progestin, with the active ingredient being ethinyl estradiol (EE2), or sometimes mestranol. Typical concentrations of EE2 range from 20 to 50 µg, with 35 µg the most commonly prescribed. It is estimated that of the 40 million menopausal American women, 5 to 13 million are using HRT.[30] The active ingredients in HRT drugs are the conjugated equine estrogens, as well as conjugated estrone and 17 α and 17β-estradiol,[19] with a typical daily dose of conjugated estrogens of 0.625 mg. Arcand-Hoy et al.[19] produced an estimated introduction concentration (EIC) of EE2 into the aquatic environment, based on the amount of pharmaceuticals sold in the United States. The EIC for EE2 was found to be 2.16 ng L^{-1}.

Intensive farming practices also produce appreciable concentrations of estrogens. The concentration of estradiol in runoff and in the soil from land fertilized with broiler litter has been measured and demonstrates sizeable "edge of field" losses.[16] A study of effluent concentrations[31] detected the highest concentrations of estrogens from agricultural settlements.

2.2.2 PHYTOESTROGENS

Phytoestrogens are plant-derived compounds that possess estrogenic activity. The term *phytoestrogen* is used to define a class of compounds that are nonsteroidal and are generally of plant origin, or are produced by the *in vivo* metabolism of precursors that are present in several plants consumed by humans.[32] This metabolism process results in the production of heterocyclic phenols, with a structure similar to the steroid estrogens.[33] Isoflavones and lignans are the two main classes of these compounds. The main source of phytoestrogens is food. Isoflavones are found in a variety of plants, including fruits and vegetables, and are especially abundant in soy products.[34] Lignans are present in whole grains, legumes, vegetables, and seeds, with high concentrations of lignans found in flaxseed. Table 2.4 indicates the main food sources for both these phytoestrogens.

Phytoestrogens have many structural similarities to 17β-estradiol and are more potent *in vitro* than many of the man-made chemicals tested to date.[35] This has raised concern over exposure to phytoestrogens from dietary regimes, particularly those with a high proportion of soy products, typical of diets in Asia and the far East. On consuming products containing phytoestrogens, there are several possible pathways:

1. Excretion
2. Absorption
3. Transformation to other compounds

TABLE 2.4
Sources of Phytoestrogens

Isoflavones	Lignans
Soybeans	Flaxseed
Lentils	Wheat
Beans	Oats
haricot	Bran
broad	Garlic
kidney	Asparagus
lima	Carrot
Chick peas	Broccoli
Wheat	Mushroom
Barley	Pear
Hops	Plum
Rye	Banana
Bran	Orange
Oats	Apple
Rice	Strawberry

It appears that these compounds are easily metabolized, spend little time in the body, and are not stored in tissues, unlike some synthetic EDCs (e.g., PCBs). Moreover, diets rich in phytoestrogens have been linked to beneficial health effects. Evidence shows that phytoestrogens may potentially confer health benefits related to cardiovascular diseases, cancer, osteoporosis, and menopausal symptoms.[34] These potential health benefits are consistent with the epidemiological evidence that rates of heart disease, various cancers, osteoporotic fractures, and menopausal symptoms are lower among populations that consume plant-based diets, particularly among cultures with diets that are traditionally high in soy products.[34] For example, Asian populations suffer significantly lower rates of hormone-dependent cancers compared to Westerners.[36] This suggests that phytoestrogens may have some beneficial effects against such cancers. Certainly as diet patterns change, the risk for certain hormonally related diseases will also change. The difference in diet patterns in the East and West with regard to consumption of phytoestrogens is significant. Asian populations consume between 20 and 80 mg/day of genistein, which is almost entirely derived from soy, whereas intakes have been estimated at 1 to 3 mg/day in the United States.[37]

2.2.3 ORGANO OXYGEN COMPOUNDS

2.2.3.1 Bisphenol A

Bisphenol A (BPA) is manufactured in large quantities, with over 90% being used in the plastics industry for the production of polycarbonate and epoxy resins, unsaturated polyester–styrene resins, and flame retardants.[38] The plastics produced are used in food and drink packaging, such as for lining metal food cans, for bottle tops, and water supply pipes.[4] Other uses include additives in thermal paper, powder paints, in dentistry,[39] and as antioxidants in plastics.[38,40] Almost 30% of the world's production is within the European Union, with reportedly 210,000 tons produced in Germany in 1995.[38]

Studies[21] have shown that bisphenol A, when present in food can linings, can leach into the product and acquire estrogenic activity. Cans containing the highest amount of bisphenol A result in about 80 µg kg^{-1} of the canned food.[21] The experimental values are clearly below the EU limit of migration for bisphenol A in food cans of 3 mg kg^{-1}. Other studies have also shown bisphenol A to leach from cans into vegetables.[41,42]

Because bisphenol A is used widely in households and industry, it is expected to be present in raw sewage, wastewater effluents, and in sewage sludge.[4] Fromme et al. [38] found low concentrations of BPA in surface waters (0.0005 to 0.41 µg L^{-1}), sewage effluents (0.018 to 0.702 µg L^{-1}), sediments (0.01 to 0.19 mg kg^{-1}), and sewage sludge (0.004 to 1.363 mg kg^{-1}). The release of BPA into the environment can occur during manufacturing processes and by leaching from the final product. With mean concentrations in landfill leachates of 269 µg L^{-1}, they may be a significant source of the bisphenol A found in the environment.[43] Industrial sources, however, are thought to be the major source of this chemical environment.[4]

2.2.3.2 Phthalates

Phthalates have been in use for almost 40 years and are used in the manufacture of PVC and other resins, as well as plasticizers and insect repellents.[38] Plasticizers are used in building materials, home furnishings, transportation, clothing, and, to a limited extent, in food packaging and medicinal products.[44] There is also concern regarding the potential health effects of several phthalates because these compounds are used to impart softness and flexibility to normally rigid PVC products in children's toys.[45] Phthalates can enter the environment through losses during manufacture and by leaching from the final product. This is because they are not chemically bonded to the polymeric matrix.[38] These compounds have low water solubilities and tend to adsorb to sediments and suspended solids.

Of the phthalates, di (2-ethylhexyl) phthalate (DEHP) is one of the most significant because it is used as a plasticizer in large quantities. For example, in Germany in 1994/1995, 400,000 tons of phthalates (DEHP 250,000 tons; dibutylphthalate [DBP] 21,000 tons; butylbenzylphthalate [BBP] 9000 tons) were produced.[38] Regarding environmental concentrations, DEHP tends to give the highest phthalate concentrations, ranging from 0.33 to 97.8 µg L^{-1} in surface waters, 1.74 to 182 µg L^{-1} in sewage effluents, 27.9 to 154 mg kg^{-1} in sewage sludge, and 0.21 to 8.44 mg kg^{-1} in sediments.[38] Concentrations of DBP and BBP have been detected in much lower concentrations. Other sources exhibiting low concentrations of these substances, such as runoff from municipal and industrial dump sites, should not be overlooked.

2.2.3.3 Dioxins

The term *dioxins* is used to represent seven of the polychlorinated dibenzodioxins (PCDDs) and 10 of the polychlorinated dibenzofurans (PCDFs). These substances are not produced commercially but are formed as by-products of various industrial and combustion processes.[46] Globally, the most important source is incineration of municipal, hospital, and hazardous wastes.[46] Other sources include fuel combustion, the steel and nonferrous metal industry, and the paper and pulp, cement, and glass industries.[46] According to Stangroom et al.,[7] there are four major sources of PCDDs and PCDFs involving incineration processes: waste incineration, power generation by coal burning, domestic heating with wood or coal, and motor vehicle internal combustion. In 1995, the PCDD/F air emissions in Europe (17 countries) were estimated to have been 6500 g I-TEQ (international toxic equivalents) per year.[46]

Dioxin exposure to humans arises from food, specifically from eating the fats associated with consuming beef, poultry, pork, fish, and dairy products. However, concentrations of these compounds are low and current average daily intakes for Europeans and Americans appear to be within the WHO recommended values of 1 to 4 pg of I-TEQ kg^{-1} of body weight.[47] The main route by which dioxins reach most watercourses is through soil erosion and stormwater runoff from urban areas. The strong binding of dioxins to organic matter[48] also supports the hypothesis that transport to watercourses may occur in association with soil particles. Industrial discharges can also significantly increase water concentrations near the point of discharge to rivers.

2.2.4 ORGANOTIN COMPOUNDS

Organotins are a group of organometallic compounds which were first produced in the 1930s. Since the 1960s, the commercial production and use of these compounds, including tributyltin (TBT) has increased dramatically, particularly for use in antifouling agents applied to boats. Despite legislative restrictions on the use of organotins in numerous countries, these pollutants have been detected in all environmental compartments and still represent a risk to the aquatic and terrestrial ecosystems.

One of the most important applications of organotins is the use of the mono- and di-alkyl tin compounds (MBT and DBT) as heat and light stabilizers in PVC processing, as their addition prevents the discoloration and embrittlement of the PVC polymer.[49] The biocidal properties of mainly TBT species have also led to their widespread use as an antifouling agent and as a general wood preservative. Other uses and potential sources arise from agrochemical use, materials and textiles processing, household products, and foodstuff packaging.[49] Table 2.5 gives some examples of organotin compound sources.

The world's production of organotins has been estimated to be 50,000 tons per year.[50] About 70% of this is utilized in the plastics industry as stabilizers and as catalysts for polyurethane foams and silicones.[49] Several studies have shown that leaching of these organotin components from PVC materials has resulted in the contamination of food, beverages, drinking water, municipal water, and sewage sludge.[51–53] Another study[54] analyzed plastic household products and found butyltin compounds in 50% of these products. Baking parchments, polyurethane gloves, sponges, and cellophane wrap were among those "contaminated." Moreover, these compounds were also seen to be transferred to food, by the analysis of cookies

---

TABLE 2.5
Sources of Organotin Compounds

Sources	Organotin Compounds Used
PVC stabilizers	MBT, DBT
Antifouling formulations	TBT
Wood preservation	TBT
Agrochemical	TBT, TPhT (triphenyltin)
Materials protection (leather, paper)	TBT
Textiles	TPhT
Household products	MBT, DBT, TBT
Poultry farming	DBT

---

baked on the tested parchment. This has obvious implications for human exposure to these compounds.

As a result of the use of antifouling formulations containing TBT, the leaching of this compound from ships has produced severe effects on wildlife in the marine environment (see Chapter 1, Section 1.1). For example, a large commercial ship, leaching TBT at a constant rate, can release more than 200 g TBT per day into the immediate waters, and this can rise to 600 g if the vessel has been freshly painted.[55] This can result in a TBT concentration of 100 to 200 ng Sn l^{-1}. Smaller estuaries and marinas have typical TBT concentrations of 10 to 70 ng Sn l^{-1}.

A significant proportion of these compounds in the environment probably comes from agricultural and biocidal applications. They are generally applied by spraying, which results in contamination of the soil and surface waters due to leaching and runoff. The use of organotins as wood preservatives involves the application of a 1 to 3% solution in an organic solvent.[49] Processes used include dipping, spraying, and brushing, which may leave the wood prone to leaching these compounds. However, the timber industry often utilizes a technique called double vacuum impregnation, which produces negligible leaching.[49]

Concentration of these compounds and their metabolites has been observed in all parts of the aquatic environment, with compounds entering the environment by various routes, as discussed previously. Atmospheric concentrations of organotins are negligible, so this source is unlikely to be significant. Many of these compounds have industrial applications, and therefore influents at STWs may contain concentrations arising from non-antifouling uses of organotins, as well as possible runoff from pesticide application. Thus, inputs from wastewater and sewage sludge as well as the leachate from landfills must be considered as sources of these compounds.

2.2.5 SURFACTANTS

Surfactants represent a group of chemicals whose predominant use is in detergent formulations. Regarding endocrine disruption, the main surfactants of interest are the alkylphenols (APs) and their ethoxylates (APEOs), particularly the nonylphenol (NP) compounds. These compounds are used in many industrial, commercial, and household functions, including detergents, lubrication, defoamers, emulsifiers, cleaners for machinery and materials, paints, pesticides, textiles, and in metal working and personal products. Table 2.6 indicates the sources and examples of concentrations of NP and its ethoxylates.

Table 2.6 shows that, in addition to industrial uses, there are many products containing NP and NPEOs that are used for domestic purposes. Thus, there are many possible entry routes into the environment for these substances during their manufacture, use, and disposal.[57]

The range of uses of these compounds may also represent a direct source of human exposure. Examples include possible residues of NP and NPEOs in food as a result of the use of pesticides, vegetable and fruit waxes, and detergents and disinfectants used in food packaging.[57] A recent study[58] has found 4-nonylphenol in a range of foodstuffs, including breast milk and formula milk. The presence of this, and other xenoestrogens, in food provides further cause for concern because many of these synthetic chemicals have a tendency to bioaccumulate.

Cosmetic products such as makeup, skin creams, hair care products, and bathing products may also be direct sources of exposure to these compounds. An example is Nonoxynol – 9 (NP_9EO), used as a spermicide in contraceptive products.[57]

There are no known natural sources of NP and NPEOs, and thus their ubiquitous presence in the environment is a result of anthropogenic activity. Annual worldwide production of alkylphenol ethoxylates is around 650,000 tons.[58]

Most of the NPEOs used in commercial and industrial formulations will be disposed of via sewers for treatment at sewage treatment works. During treatment, degradation processes lead to the production of metabolites and a shortening of the ethoxylate chain. These ethoxylates can be degraded to produce NP, which is considered to be more persistent and toxic than the ethoxylates. This degradation process is discussed in greater detail in Chapter 4.

TABLE 2.6
Sources and Concentrations of Nonylphenol and Its Ethoxylates

Source	Concentration
Detergents	0–28%
Deodorant	1–3%
Makeup	0.1–10%
Hair products	1–30%
Paints	0.6–3%
Paper processing effluents	NP (0.02–26.2 $\mu g\ L^{-1}$), NPEOs (0.1 to 35.6 $\mu g\ L^{-1}$)
STW effluents	<0.02–330 $\mu g\ L^{-1}$
Pesticides	<1–20%
Food	Nonylphenol (0.1–19.4 $\mu g\ kg^{-1}$)

Data from Whitehouse, P., Wilkinson, M., Fawell, J. K., and Sutton, A., Research and development Report No. Technical report P42, 1998; Environment Canada, Health Canada. Priority Substances List Assessment Report: Nonylphenol and its ethoxylates, 2000, pp. 134; and Guenther, K., Heinke, V., Theile, B., Kleist, E., Prast, H., and Raecker, T., Endocrine disrupting nonylphenols are ubiquitous in food, *Environ. Sci. Technol. 36*, 1676, 2002.

Within the aquatic environment, concentrations of AP and APEOs have been detected. Concentrations of NP in the air of the New York–New Jersey Bight ranged from 2.2 to 70 ng m^{-3}.[57]

No data on the levels on NPEOs in air were identified, although as these compounds are far less volatile than NP, it is expected that they would not partition to the atmosphere. Levels of NPEOs and NP in effluents and waters vary greatly. For example, concentrations of NP in textile mill and paper mill effluents ranged from 2.68 to 13.3 µg L^{-1} and <0.02 to 26.2 µg L^{-1}, respectively.[57] Concentrations of NPEOs from these industries were 2.07 to 8811 µg L^{-1} and 0.1 to 35.6 µg L^{-1}, respectively. Moreover, highest freshwater concentrations of NP were observed in areas near STW discharges, pulp mill discharges, or regions of heavy industry. In the United Kingdom, effluent concentrations from STW range from <0.02 to 330 µg L^{-1}.[56] Surface water concentrations of NP vary dramatically across the U.K. from <0.02 to 53 µg L^{-1}.[56] The highest value is from the River Aire and is thought to be related to the textile industry in the area. Within the sedimentary compartment, alkylphenols and their mono- and di-ethoxylates exhibit a significant association. In sediments below effluent discharges, concentrations of up to 13.1 mg kg^{-1} of NP and up to 25 mg kg^{-1} of NP, nonylphenol ethoxylate (NP$_1$EO), and nonylphenol diethoxylate (NP$_2$EO) were detected.[6,56] In U.S. rivers, sediments monitored for NP from 30 sites, downstream from industrial and municipal wastewater outfalls, ranged from not detectable to 2.96 mg kg^{-1}. Much lower concentrations of NP$_1$EO were found (<0.0023 to 0.175 mg kg^{-1}).[59] Substantial concentrations of NP and other products are found in the sludges from STWs. Disposal of this sludge to sea could reintroduce these compounds into the aquatic environment. Moreover, the leaching of APs and ethoxylates from sludge disposal sites and the application of such sludge to agricultural land may result in potential contamination of the terrestrial environment.[60]

2.2.6 POLYAROMATIC COMPOUNDS

2.2.6.1 Polyaromatic Hydrocarbons (PAHs)

PAHs are formed from both natural and anthropogenic sources, largely by the incomplete combustion of organic materials. Natural sources of PAHs include forest fires,[61] volcanic activity,[61] stubble burning, and the release of petroleum hydrocarbons by marine seeps. According to Brown et al.,[62] anthropogenic sources of PAHs can be classified as stationary or mobile. The stationary category incorporates a wide range of activities, such as residential and commercial heating and industrial processes (e.g., aluminum production and coke manufacture). Within the mobile category, petrol and diesel-engined vehicles are the predominant sources.

Stationary sources accounted for between 80 and 90% of total PAH emission prior to the 1980s, but recently mobile sources seemed to be the major contributors in urban and suburban areas.[63] Anthropogenic sources are considered to be more important than natural sources as a contributor to the environment, and 95% of U.K. PAH emissions are from anthropogenic sources.[64] Estimates in the United States have shown the annual emission of benzo(a)pyrene, a very toxic PAH, to be approximately 1300 tons.[65] In the aquatic environment, PAHs can be formed by microbial synthesis from certain precursors.[66] Like many semivolatile compounds, PAHs are widely distributed in the environment. Possible routes of entry to the aquatic environment include wet and dry deposition, direct and indirect discharges, and surface runoff. Total global PAH inputs to water from all sources has been estimated at > 80,000 tons per year.[67]

The principal sources of anthropogenic PAHs are the combustion of fossil fuels, aluminum smelting, refuse burning, coke ovens, petroleum processing, and vehicle emissions.[61] Concentration of PAHs will also be related to the types of industries and other sources present, high levels being associated with heavy coal burning industries.[65] In general, concentrations in urban and industrial areas are 10 to 100 times higher than those in rural areas.[65].

Other sources of PAHs include sewage, sewage sludge, cigarette smoke, gas leakage, and fuel spills and leakage. Raw sewage or treated effluent may serve as the major point source of PAHs in lowland river systems.[68] In raw sewage, PAHs are derived from three main sources: industrial and domestic effluent, urban runoff, and atmospheric pollution.[69] Concentrations of individual PAHs in domestic raw sewage vary between 1.0 and 3520 ng L^{-1} during dry weather periods, and between 1840 and 16,350 ng L^{-1} during heavy rainfall, when urban runoff may increase PAH concentrations by up to two orders of magnitude.[70] PAH concentrations found in sewage sludge vary between 1.6 and 6 mg kg^{-1}. A survey of sewage sludge from 12 U.K. STWs found concentrations of individual PAHs ranging from 0.08 to 11.4 mg kg^{-1}. The highest PAH concentrations are present in sewage sludge from treatment works serving industrialized areas.[70]

2.2.6.2 Polychlorinated Biphenyls (PCBs)

PCBs have been used in a wide variety of industrial applications because of their high stability and electrical resistance.[7] They are used in dielectric fluids in transformers and capacitors, plasticizers and components of cement, hydraulic lubricants, cutting oils, flame retardants, plastics, paint and adhesives.[46,71] Even though PCB production has been banned in most countries since the 1970s and 1980s, it is estimated that over 1 million ton of PCBs have been generated.[7] About one third of this quantity is thought to be circulating in the environment.[46]

These compounds are known to be ubiquitous in the environment. Concentrations in the atmosphere are usually given in pg m^{-3}, surface waters in ng L^{-1}, birds eggs and fish fat in mg kg^{-1}, and μg kg^{-1} in sediments.[72] For example, concentrations in sediments are variable and range from 10s to 1000s μg kg^{-1}, these values depending on numerous factors, not least the industrial activity within the area. Because PCBs are expected to be sorbed to particulate matter,[71] concentrations in the sediment compartment are more likely to be detectable. Sources to STWs are mainly from the atmosphere via wet deposition processes. Further discussion of the fate of PCBs in STW processes is given in Chapter 4.

Due to their high biostability and lipophilicity, PCBs have accumulated in food chains, especially in aquatic and marine species.[73] Thus, sources of human exposure to PCBs are generally from food, particularly fish, fish products, and animal fats.

2.2.6.3 Brominated Flame Retardants

Flame retardants are used in plastics, textiles, electronic circuitry, and other materials to prevent fires.[74] Brominated compounds are used because of their ability to generate halogen atoms from the thermal degradation of the compound to chemically reduce and "retard" the development of the fire.[75] The types of brominated compounds utilized for this purpose are polybrominated biphenyls (PBB), polybrominated diphenylethers (PBDE), hexabromocyclododecane (HBCD), and tetrabromobisphenol A (TBBPA). These substances are persistent, lipophilic, and have been shown to bioaccumulate.[76] Some of these additives do not chemically bind to the plastics or textiles and, thus, may leach from the products into the environment. TBBPA however, does bind chemically to plastics and textiles.

These compounds are used worldwide and in vast quantities. Annual world production of flame retardants is approximately 600,000 tons, with 60,000 tons being chlorinated and 150,000 tons being brominated compounds.[75] Of the brominated compounds, about one third contain TBBPA, another third contains various bromine compounds including PBBs, and the final third contains the PBDEs, predominantly decaBDE.[75] In recent years, there has been a tendency to use higher order brominated compounds, such as the PBDEs. In these, the commercial formulations consist predominantly of mixtures of penta, octa, and decabromodiphenyl ethers. According to the International Programme on Chemical Safety published by the World Health Organization,[77] there are eight world manufacturers of PBDEs. Regarding the annual consumption of these compounds, the majority is in the form of deca-BDE (30,000 tons), followed by octa-BDE (6000 tons), and penta-BDE (4000 tons).[78] Table 2.7 summarizes the potential sources of brominated flame retardants.

The main source of flame retardants, particularly the PBDEs, is likely to come from waste from the products they are used in. This waste is either incinerated or disposed of at landfills. Despite data gaps in this area of study, incineration and landfills are thought to be potential sources of release of PBDEs into the environment.[75]

TABLE 2.7
Sources of Brominated Flame Retardants

Source	Brominated Flame Retardant
Plastics	PBDE, HBCD, TBBPA
Textiles	PBDE, HBCD, TBBPA
Electronic appliances	PBDE, HBCD, TBBPA
STW discharges and sewage sludge	PBDE, HBCD, TBBPA
Paint	PBDE
Cars	PBDE, TBBPA
Breast milk	PBDE
Meat, eggs, fish	PBDE, TBBPA

Data from Sellstrom U, *Determination of some polybrominated flame retardants in biota, sediment and sewage sludge*, PhD dissertation, Stockholm University, 1999; Oberg, K., Warman, K., and Oberg, T., Distribution and levels of brominated flame retardants in sewage sludge, *Chemosphere,* in press, 2002.

Brominated flame retardants are used in the plastics industry to produce a variety of products, including common plastics, polymers, and resins. Concentrations of these additive flame retardants vary between 5 and 30%, depending on the application.[75] Many of these compounds are also used in electronic components such as circuit boards, computers, televisions, electrical cables, and capacitors, most of which are found in homes. These substances are also present in upholstered furniture, textiles, car cushions, insulation blocks, house wall materials, and packaging materials.[76] Daily contact with these substances raises concern over the level of exposure of humans to brominated flame retardants. Watanabe et al.[81] found mainly deca-BDE in airborne dust from the Osaka region in Japan, and various tri-, tetra-, penta-, and hexa-BDEs were detected in samples from Taiwan and Japan, taken in the vicinity of metal recycling plants.[82] Concentrations ranged from 23 to 53 pg m^{-3} in Taiwan and 7.1 to 21 pg m^{-3} in Japan. A study of indoor air quality[83] revealed that PBDEs were detected in an electronics plant and office environments. Concentrations in the electronics plant were 400 to 4000 times higher than in the office environments. The presence of these substances in air shows that brominated flame retardants are leaking into the indoor environment from electronic appliances and exposing humans.

In the aquatic environment, flame retardants tend to accumulate in the sediments because they are highly lipophilic and have low water solubilities. A study of river, estuarine, and marine sediments[84] found deca-BDE, octa-BDE, hexa-BDE, penta-BDE, and tetra-BDE to be present in all samples. Concentrations of deca-BDE were in the range <25 to 11,600 µg kg^{-1} dry weight. The other compounds present ranged from below the limit of detection to 70 µg kg^{-1} dry weight. Analysis of sediments from European rivers revealed high concentrations of PBDEs in the River Mersey (1700 µg kg^{-1}) and in Belgium on the River Schelde (200 µg kg^{-1}). Most samples had levels below 20 µg kg^{-1}. In the United States, Lake Michigan is considered to be heavily contaminated with PBDEs.[85] Analysis of fish from the lake[86] revealed concentrations of PBDEs from 44 to 149 µg kg^{-1}. In southern Virginia several species of fish from the Roanoke and Dan rivers contained PBDE concentrations greater than 1 part per million (ppm).[87] Analysis of fish from Europe has also shown the presence of PBDEs at concentrations similar to and higher than those found in the U.S.[75] These levels in fish clearly indicate that brominated flame retardants are an environmental concern and raise issues regarding the advisability of consuming fish and other foodstuffs that may be contaminated. Levels of PBDEs have also been detected in dairy products, breast milk, meat, and eggs.[75]

Sewage sludge concentrations of brominated flame retardants reflect the usage and exposure within society, because sources may be households, industries, traffic, and other nonpoint sources.[79] All the brominated flame retardants (i.e., PBDE, HBCD, TBBPA) have been found in sewage sludge samples.[79] Oberg et al.[80] analyzed 116 sewage sludge samples from 22 wastewater treatment plants for brominated flame retardants. Concentrations of PBDEs were in the range of not detectable (n.d.) to 450 µg kg^{-1} and TBBPA between n.d. and 220 µg kg^{-1}. Significant variation between plants was observed, indicating influence from industries and other local sources.

According to Wenning,[88] there may be as many as five different sources of PBDEs:

1. Releases from manufacturing processes using PBDEs currently manufactured by industry
2. Releases from past manufacturing activities using PBDEs that are no longer manufactured, but may be persistent in the environment
3. Natural debromination of the higher brominated congener mixtures
4. Natural sources of brominated compounds[78]
5. Anthropogenic inputs of organobromine compounds from sources other than the flame retardant industry, such as mineral ore mining and deep-well injection in the petroleum industry

These sources, together with the other potential sources discussed, are responsible for the ubiquitous nature of brominated flame retardants in the environment today.

2.2.7 Pesticides

Due to the great number of pesticides that are known to exhibit endocrine activity,[89–93] it is beyond the scope of this text to discuss each individual pesticide source in detail. Therefore, a generic overview of pesticide sources will be presented.

The term *pesticides* in the context of this text also includes insecticides, herbicides, and fungicides. Table 2.8 gives some examples of the various subgroups of pesticides, their uses, and sources to the environment.

An estimated 5 billion ton of pesticides per annum are used worldwide.[67] From an economic viewpoint, pesticides play an important role in a country's economy. For example, in 1994, the cost of pesticide application in the United States was $4 billion. However, this resulted in an estimated savings of $12 billion in crops that would have been destroyed without pesticides.[67]

The majority of pesticides are used in agricultural applications, although they are also used in heavy industry as additives to cooling fluids in metal machining operations.[67] As much as one third of pesticide use may be for nonagricultural purposes. Within agricultural practices, plants and crops can either absorb these chemicals directly through their leaves or indirectly from the soil.[46] The plants may

TABLE 2.8
Uses and Sources of Pesticides

Pesticide	Uses	Source to Environment
Organochlorines	Industrial and domestic use, seed and soil insecticides	Agricultural runoff, industrial effluents, STW effluents
Synthetic pyrethroids	Moth proofing, arable crops	Agricultural runoff, industrial discharges
Organophosphates	Arable crops, sheep dips	Agricultural runoff, sheep dip disposal
Carbamates	Arable crops	Agricultural runoff
Chlorophenoxy herbicides	Arable crops	Agricultural runoff
Triazines	Arable crops	Agricultural runoff

then be eaten by herbivores, and the pesticides will accumulate to produce relatively high levels in meat and dairy products. Pesticide residues in foodstuffs (i.e., meat, fruit, and vegetables) is an important source of exposure to humans. Organochlorine pesticide residues have been found in human fat and breast milk,[94] with their presence thought to be due to factors such as diet, smoking, and place of residence.

Pesticides used in agriculture that do not bind to the soil will enter watercourses through agricultural runoff and have the potential to accumulate in the aquatic food chain, due to their persistence and lipophilicity. Many of these compounds are also semivolatile, and this will allow them to be dispersed into the air. Due to the cycling of these compounds between environmental compartments, it can be difficult to distinguish between sources and sinks.[46]

REFERENCES

1. Lester, J.N. and Birkett, J.W., *Microbiology and Chemistry for Environmental Scientists and Engineers*, 2nd ed., E & F.N. Spon, London, 1999.
2. Solé, M., Lopez de Alda, M.J., Castillo, M., Porte, C., Ladegaard Pedersen, K. and Barceló, D., Estrogenicity determination in sewage treatment plants and surface waters from the Catalonian area (NE Spain), *Environ. Sci. Technol.*, 34, 5076 2000.
3. Endocrine-disrupting substances in the environment: what should be done?, *Environment Agency*, UK, 1998, pp. 13.
4. Furhacker, M., Scharf, S. and Weber, H., Bisphenol A: emissions from point sources, *Chemosphere*, 41, 751, 2000.
5. Ferguson, P.L., Iden, C.R., McElroy, A.E. and Brownawell, B.J., Determination of Steroid Estrogens in Wastewater by Immunoaffinity Extraction Coupled with HPLC-Electrospray-MS, *Anal. Chem.*, 73, 3890, 2001.
6. Ahel, M., Giger, W. and Schaffner, C., Behaviour of alkylphenol polyethoxylate surfactants in the aquatic environment. II. Occurrence and transformation in rivers, *Water Res.*, 28, 1143, 1994.
7. Stangroom, S.J., Collins, C.D. and Lester, J.N., Sources of organic micropollutants to lowland rivers, *Environ. Technol.*, 19, 643, 1998.
8. de Boer, J., van der Horst, A. and Wester, P.G., PBDEs and PBBs in suspended particulate matter, sediments, sewage treatment plant in- and effluents and biota from the Netherlands, *Organohalogen Compd.*, 47, 85, 2000.
9. Sellström, U. and Jansson, B., Analysis of tetrabromobisphenol A in a product and environmental samples, *Chemosphere*, 31, 3085, 1995.
10. Voulvoulis, N., Scrimshaw, M.D. and Lester, J.N., Partitioning of selected antifouling biocides in the aquatic environment, *Marine Environ. Res.*, 53, 1, 2001.
11. Hawrelak, M., Bennett, E. and Metcalfe, C., The environmental fate of the primary degradation products of alkylphenol ethoxylate surfactants in recycled paper sludge, *Chemosphere*, 39, 745, 1999.
12. Hale, R.C., La Guardia, M.J., Harvey, E.P., Gaylor, M.O., Mainor, T.M. and Duff, W.H., Flame retardants — persistent pollutants in land applied sludge, *Nature*, 412, 140, 2001.
13. Harms, H.H., Bioaccumulation and metabolic fate of sewage sludge derived organic xenobiotics in plants, *Sci. Tot. Environ.*, 185, 83, 1986.
14. Smith, K.E.C., Green, M., Thomas, G.O. and Jones, K.C., Behavior of sewage sludge-derived PAHs on pasture, *Environ. Sci. Technol.*, 35, 2141, 2001.

15. Nichols, D.J., Daniel, T.C., Moore, P.A., Edwards, D.R. and Pote, D.H., Runoff of oestrogen hormone 17B-estradiol from poultry litter applied to pasture., *J. Environ. Qual.*, 26, 1002, 1997.

16. Finlay-Moore, O., Hartel, P.G. and Cabrera, M.L., 17β–estradiol and testosterone in soil and runoff from grasslands amended with broiler litter, *J. Environ. Qual.*, 29, 1604, 2000.

17. Pereira Netto, A.D., Sisinno, C.L.S., Moreira, J.C., Arbilla, G. and Dufrayer, M.C., Polycyclic aromatic hydrocarbons in leachate from a municipal solid waste dump of Nierói City, RJ, Brazil, *Bull. Environ. Contam. Toxicol.*, 68, 148, 2002.

18. Peters, A.J., Coleman, P. and Jones, K.C., Organochlorine pesticides in UK air, *Organohalogen Compd.*, 41, 447, 1999.

19. Arcand-Hoy, L.D., Nimrod, A.C. and Bensen, W.H., Endocrine-modulating substances in the environment: estrogenic effects of pharmaceutical products, *Int. J. Toxicol.*, 17, 139, 1998.

20. Andersson, A.M. and Skakkebaek, N.E., Exposure to exogenous estrogens in food: possible impact on human development and health, *Eur. J. Endocrinol.*, 140, 477, 1999.

21. Brotons, J.A., Olea-Serrano, M.F., Villalobos, M., Pedraza, V. and Olea, N., Xenoestrogens released from lacquer coatings in food cans, *Environ. Health Perspect.*, 103, 608, 1995.

22. The Joint Food and Agricultural Organization/World Health Organization Expert Committee on Food Additives, Residues of some veterinary drugs in animals and foods, Food and Agriculture Organization of the United Nations, FAO Food and Nutrition paper 41, 1988.

23. Hartmann, S., Lacorn, M. and Steinhart, H., Natural occurrence of steroid hormones in food, *Food Chem.*, 62, 7, 1998.

24. Desbrow, C., Routledge, E., Brighty, G.C., Sumpter, J.P. and Waldock, M., Identification of estrogenic chemicals in STW effluent. 1. Chemical fractionation and *in vitro* biological screening, *Environ. Sci. Technol.*, 32, 1549, 1998.

25. Rodgers-Gray, T.P., Jobling, S., Morris, S., Kelly, C., Kirby, S., Janbakhsh, A., Harries, J. E., Waldock, M. J., Sumpter, J.P. and Tyler, C.R., Long-term temporal changes in the estrogenic composition of treated sewage effluent and its biological effects on fish, *Environ. Sci. Technol.*, 34, 1521, 2000.

26. Purdom, C.E., Hardiman, P.A., Bye, V.J., Eno, N.C., Tyler, C.R. and Sumpter, J. P., Estrogenic effects of effluents from sewage treatment works, *Chemical Ecology*, 8, 275, 1994.

27. Tyler, C.R., Jobling, S. and Sumpter, J.P., Endocrine disruption in wildlife: a critical review of the evidence, *Crit. Rev. Toxicol.*, 28, 319, 1998.

28. Ternes, T.A., Stumpf, A., Mueller, J., Haberer, K., Wilken, R.D. and Servos, M., Behaviour and occurrence of estrogens in municipal sewage treatment plants. I. Investigations in Germany, Canada and Brazil, *Sci. Total Environ.*, 225, 81, 1999.

29. Sumpter, J.P., Xenoendocrine disrupters-environmental impacts, *Toxicol. Lett.*, 102-103, 337, 1998.

30. Andrews, W.C., Continuous combined estrogen/progestin hormone replacement therapy, *Clin. Ther.*, 17, 812, 1995.

31. Shore, L.S., Gurevitz, M. and Shernesh, M., Estrogen as an environmental pollutant, *Bull. Environ. Contam. Toxicol.*, 51, 361, 1993.

32. Brzezinski, A. and Debi, A., Phytoestrogens: the "natural" selective estrogen receptor modulators?, *Eur. J. Obstet. Gynecol. Reprod. Biol.*, 85, 47, 1999.

33. Murkies, A.L., Wilcox, G. and Davis, S.R., Clinical review 92 — Phytoestrogens, *J. Clin. Endocrinol. Metab.*, 83, 297, 1998.

34. Tham, D.M., Gardner, C.D. and Haskell, W.L., Clinical review 97 - Potential health benefits of dietary phytoestrogens: a review of the clinical, epidemiological, and mechanistic evidence, *J. Clin. Endocrinol. Metab.*, 83, 2223, 1998.

35. Phillips, B. and Harrison, P., Overview of the endocrine disrupters issue, in *Endocrine Disrupting Chemicals*, Hester, R.E. and Harrison, R.M., Eds., The Royal Society of Chemistry, Cambridge, 1999, p. 1.

36. Barrett, J., Phytoestrogens: Friends or foes?, *Environ. Health Perspect.*, 104, 478, 1996.

37. Barnes, S., Peterson, T.G. and Coward, L., Rationale for the use of genistein-containing soy matrices in chemoprevention trials for breast and prostate cancer, *J. Cell Biochem.*, 22, 181, 1995.

38. Fromme, H., Kuchler, T., Otto, T., Pilz, K., Muller, J. and Wenzel, A., Occurrence of phthalates and bisphenol A and F in the environment, *Water Res.*, 36, 1429, 2002.

39. Olea, N., Pulgar, R., Perez, P., OleaSerrano, F., Rivas, A., NovilloFertrell, A., Pedraza, V., Soto, A.M. and Sonnenschein, C., Estrogenicity of resin-based composites and sealants used in dentistry, *Environ. Health Perspect.*, 104, 298, 1996.

40. Staples, C.A., Dorn, P.B., Klecka, G.M., O'Block, S.T. and Harris, L.R., A review of the environmental fate, effects and exposures of bisphenol A, *Chemosphere*, 36, 2149, 1998.

41. Howe, S. R., Borodinsky, L. and Lyon, R.S., Potential exposure to Bisphenol A from food-contact use of epoxy coated cans, *J. Coatings Technol.*, 70, 69, 1998.

42. Biles, J.E., McNeal, T.P., Begley, T.H. and Hollifield, H. C., Determination of Bisphenol A in reusable polycarbonate food-contact plastics and migration to food simulating liquids, *J. Agri. Food. Chem.*, 45, 3541, 1997.

43. Yamamoto, T., Yasuhara, A., Shiraishi, H. and Nakasugi, O., Bisphenol A in hazardous waste landfill leachates, *Chemosphere*, 42, 415, 2001.

44. Staples, C.A., Peterson, D.R., Parkerton, T.F. and Adams, W.J., The environmental fate of phthalate esters: a literature review, *Chemosphere*, 35, 667, 1997.

45. Wilkinson, C.F. and Lamb, J.C., The potential health effects of phthalate esters in children's toys: a review and risk assessment, *Regul. Toxicol. Pharmacol.*, 30, 140, 1999.

46. Vallack, H.W., Bakker, D.J., Brandt, I., Brostrom-Lunden, E., Brouwer, A., Bull, K.R., Gough, C., Guardans, R., Holoubek, I., Jansson, B., Koch, R., Kuylenstierna, J., Lecloux, A., Mackay, D., McCutcheon, P., Mocarelli, P. and Taalman, R.D.F., Controlling persistent organic pollutants — what next?, *Environ. Toxicol. Pharmacol.*, 6, 143, 1998.

47. Huwe, J.K., Dioxins in food: a modern agricultural perspective, *J. Agric. Food Chem.*, 50, 1739, 2002.

48. Arienti, M., Wilk, L., Jasinski, M. and Prominski, N., Dioxin-containing wastes, Treatment Technologies, *Pollut. Technol. Reviews*, 160, 6, 1988.

49. Hoch, M., Organotin compounds in the environment-an overview, *Appl. Geochem.*, 16, 719, 2001.

50. Fent, K., Ecotoxicology of organotin compounds, *Crit. Rev. Toxicol.*, 26, 1, 1996.

51. Forsyth, D.S., Weber, D. and Cleroux, C., Determination of butyltin, cyclohexyltin and phenyltin compounds in beers and wines, *Food Add. Contam.*, 9, 161, 1992.

52. Forsyth, D.S., Sun, W.F. and Dalglish, K., Survey of organotin compounds in blended wines, *Food Add. Contam.*, 11, 343, 1994.

53. Forsyth, D.S. and Jay, B., Organotin leachates in drinking water from chlorinated polyvinyl chloride (CPVC) pipe, *Appl. Organomet. Chem.*, 11, 551, 1997.

54. Takahashi, S., Mukai, H., Tanabe, S., Sakayama, K., Miyazaki, T. and Masuno, H., Butyltin residues in livers of humans and wild terrestrial mammals and in plastic products, *Environ. Pollut.*, 106, 213, 1999.

55. Batley, G., The distribution and fate of tributyltin in the marine environment, in *Tributyltin: Case Study of an Environmental Contaminant*, de Mora, S.J., Ed., Cambridge University Press, London, 1996, pp. 139.

56. Whitehouse, P., Wilkinson, M., Fawell, J.K. and Sutton, A., Research and development Report No. Technical report P42, 1998.

57. Environment Canada, Health Canada. Priority Substances List Assessment Report: Nonylphenol and Its Ethoxylates, 2000, pp. 134.

58. Guenther, K., Heinke, V., Theile, B., Kleist, E., Prast, H. and Raecker, T., Endocrine disrupting nonylphenols are ubiquitous in food, *Environ. Sci. Technol.*, 36, 1676, 2002.

59. Naylor, C.G., Mieure, J.P., Adams, W.J., Weeks, J.A., Castaldi, F.J. and Ogle, L. D., Alkylphenol ethoxylates in the environment, *J. Amer. Oil Chem. Soc.*, 69, 695, 1992.

60. Maguire, R.J., Review of the persistence of nonylphenol and nonylphenol ethoxylates in aquatic environments, *Water Qual. Res. J. Canada*, 34, 37, 1999.

61. Baek, S.O., Field, R.A., Goldstone, M.E., Kirk, P.W., Lester, J.N. and Perry, R., A review of atmospheric polycyclic aromatic hydrocarbons: sources, fate and behaviour, *Water, Air and Soil Pollut.*, 60, 279, 1991.

62. Brown, J.R., Field, R.A., Goldstone, M.E., Lester, J.N. and Perry, R., Polycyclic aromatic hydrocarbons in central London air during 1991 and 1992., *Sci. Total Environ.*, 177, 73, 1996.

63. Harrison, R.M., Smith, D.J.T. and Luhana, L., Source apportionment of atmospheric polycyclic aromatic hydrocarbons collected from and urban location in Birmingham, UK, *Environ. Sci. Technol.*, 30, 825, 1996.

64. Wild, S. R. and Jones, K. C., Polynuclear aromatic carbons in the United Kingdom environment: a preliminary source inventory and budget, *Environ. Pollut.*, 88, 91, 1995.

65. Ko, T.E., *A Source Apportionment Study of Polycyclic Aromatic Hydrocarbons (PAHs) Pollution in the Blackwater Estuary*, MSc., Imperial College of Science, Technology and Medicine, London, UK, 1999.

66. Grevao, B., Jones, K.C. and Hamilton-Taylor, J., Polycyclic aromatic hydrocarbon (PAH) deposition to and processing in a small rural lake, Cumbria, UK, *Sci. Total Environ.*, 215, 231, 1998.

67. Wright, D.A. and Welborn, P., *Environmental Toxicology*, Cambridge University Press, London, 2002, pp. 630.

68. Bedding, N.D., Taylor, P.N. and Lester, J.N., Physicochemical behaviour of polynuclear aromatic hydrocarbons in primary sedimentation. I. Batch studies, *Environ. Technol.*, 16, 801, 1995.

69. Manoli, E. and Samara, C., Occurrence and mass balance of polycyclic aromatic hydrocarbons in the Thessaloniki sewage treatment plant, *J. Environ. Qual.*, 28, 176, 1999.

70. Bedding, N.D., McIntyre, A.E., Perry, R. and Lester, J.N., Organic contaminants in the aquatic environment. I. Sources and occurrence, *Sci. Total Environ.*, 25, 143, 1982.

71. Jones, K.C., Burnett, V., Duarte-Davidson, R. and Waterhouse, K.S., PCBs in the environment, *Chem. in Brit.*, May 1991.

72. Lang, V., Polychlorinated biphenyls in the environment, *J. Chromatogr.*, 595, 1, 1992.

73. Brouwer, A., Ahlborg, U.G., van Leeuwen, F.X.R. and Feeley, M.M., Report of the WHO working group on the assessment of health risks for human infants from exposure to PCDDs, PCDFs and PCBs, *Chemosphere*, 37, 1627, 1998.

74. de Wit, C., Brominated Flame Retardants., Swedish EPA (Report 5065), Stockholm, 2000, pp. 94.

75. Darnerud, P.O., Eriksen, G.S., Johannesson, T., Larsen, P.B. and Viluksela, M., Polybrominated diphenyl ethers: Occurrence, dietary exposure and toxicology, *Environ. Health Perspect.*, 109, 49, 2001.

76. de Wit, C.A., An overview of brominated flame retardants in the environment, *Chemosphere*, 46, 583, 2002.

77. World Health Organization, Brominated diphenyl ethers, *Environ. Health Crit.*, 62, 1994.

78. Rahman, F., Langford, K.H., Scrimshaw, M.D. and Lester, J.N., Polybrominated diphenyl ether (PBDE) flame retardants, *Sci. Total Environ.*, 275, 1, 2001.

79. Sellstrom U, Determination of Some Polybrominated Flame Retardants in Biota, Sediment and Sewage Sludge, Ph.D. thesis, Stockholm University, 1999.

80. Oberg, K., Warman, K. and Oberg, T., Distribution and levels of brominated flame retardants in sewage sludge, *Chemosphere*, in press, 2002.

81. Watanabe, I., Kawano, M. and Tatsukawa, R., Polybrominated and mixed polybromo/chlorinated dibenzo-*p*-dioxins and dibenzofurans in the Japanese environment, *Organohalogen Compd.*, 24, 337, 1995.

82. Watanabe, I., Kawano, M., Wang, Y., Chen, Y. and Tatsukawa, R., Polybrominated dibenzo-*p*-dioxins (PBDDs) and -dibenzofurans (PBDFs) in atmospheric air in Taiwan and Japan, *Organohalogen Compd.*, 9, 309, 1992.

83. Sjodin, A., Carlsson, H., Thuresson, K., Sjodin, S., Bergman, A. and Ostman, C., Flame retardants in indoor air at an electronic plant and at other work environments, *Environ. Sci. Technol.*, 35, 448, 2001.

84. Japan Environment Agency, Chemicals in the environment. Report on environmental survey and wildlife monitoring of chemicals in F.Y., 1988 and 1989, Tokyo: Environ Agency Japan, 1991.

85. Schaefer, A., Lake Michigan heavily contaminated with PBDEs, *Environ. Sci. Technol.*, 35, 139A, 2001.

86. Manchester-Neesvig, J.B., Valters, K. and Sonzogni, W.C., Comparison of polybrominated diphenyl ethers (PBDEs) and polychlorinated biphenyls (PCBs) in Lake Michigan Salmonids, *Environ. Sci. Technol.*, 35, 1072, 2001.

87. Renner, R., Flame retardant levels in Virginia fish are among the highest found, *Environ. Sci. Technol.*, 34, 163A, 2000.

88. Wenning, R.J., Uncertainties and data needs risk assessment of three commercial polybrominated diphenyl ethers: probabilistic exposure analysis and comparison with European Commission results, *Chemosphere*, 46, 779, 2002.

89. Bulger, W.H. and Kupfer, D., Estrogenic action of DDT analogs, *Am. J. Ind. Med.*, 4, 163, 1983.

90. Kupfer, D., Effects of pesticides and related compounds on steroid metabolism and function, *Crit. Rev. Toxicol. 4*, 83, 1975.

91. Nelson, J.A., Struck, R.F. and James, R.E., Estrogenic activities of chlorinated hydrocarbons, *J. Toxicol. Env. Health*, 4, 325, 1978.

92. Ireland, J.S., Mukku, V.R., Robison, A.K. and Stancel, G.M., Stimulation of uterine deoxyribonucleic acid synthesis by 1,1,1,-trichloro-2-(*p*-chlorophenyl)-2-(*o*-chlorophenyl)ethane (*o,p'*-DDT), *Biochem. Pharmacol.*, 24, 1469, 1980.

93. Bulger, W.H., Muccitelli, R.M. and Kupfer, D., Studies ion the *in vivo* and *in vitro* estrogenic activities of methoxychlor and its metabolites. Role of hepatic mono-oxygenase in methoxychlor activation, *Biochem. Pharmacol.*, 27, 2417, 1978.
94. Harris, C.A., Woolridge, M.W. and Hay, A.W.M., Factors affecting the transfer of organochlorine pesticide residues to breastmilk, *Chemosphere*, 43, 243, 2001.

3 Methods for the Determination of Endocrine Disrupters

N. Voulvoulis and M.D. Scrimshaw

CONTENTS

There is a wide range of methods available for the determination of endocrine disrupting compounds, and publications have been dedicated to the subject (e.g., Keith et al., 2000)[1]; however, in selecting methods, it is necessary to consider what is to be determined and for what purpose the analysis is being undertaken. If the objective of the work is to evaluate if a particular compound or a mixture of compounds, possibly in the form of an effluent being discharged to the environment, displays any activity in terms of endocrine disruption, then there is a range of *in vitro* or *in vivo* tests, or assays, that could be utilized to screen for this. However, if the aim is to identify and quantify particular compounds that are present, then a chemical analysis is likely to be more appropriate.

For any chemical or mixture, the initial question must be whether or not it exhibits any estrogenic (or other endocrine disrupting) activity.[2] A number of studies have concluded that for cost effectiveness and to facilitate the relatively rapid screening of a large number of compounds, *in vitro* test methods are most appropriate.[3,4] However, because *in vitro* tests do have limitations that might result in unreliable predictions, a combination of test methods, including *in vivo* assays, which assess both the receptor- and nonreceptor-mediated mechanisms of action has been suggested as most appropriate to determine the endocrine disrupting activities of chemicals.[5] There is no consensus on the use and absolute validity of toxicity tests for the evaluation of the toxic hazard to humans[5,6] and, in general, no single assay is certain to be best suited to determine the estrogenicity of compounds.[3] Modes of action that do not involve the estrogen receptor (ER) may produce positive results in even the most complex of assays. The mouse uterine bioassay, which is considered a hallmark of estrogenic activity,[7] has been reported to be affected by CCl_4; however, the effect was due to damage to the liver by CCl_4, which reduced the rate of metabolism of estradiol.[8] Although it causes disruption to the system, the mode of action of CCl4 is not estrogenic. The Organization for Economic Cooperation and Development (OECD) has reported on approaches to screening techniques. While a range of tests were identified as offering potential, no specific recommendations were made.[9,10]

Although bioassays are used mostly to identify if a compound or mixture of compounds is estrogenic rather than for direct analysis of endocrine disrupters within wastewater and sludge treatment processes, it is important to discuss them for a number of reasons. An appreciation of the basic methods will facilitate an understanding of the term *endocrine disrupters* and allow for a critical appraisal of the significance of compounds called endocrine disrupters. In addition, as more chemicals are classified as endocrine disrupters, the array of chemical analyses required to monitor waste streams for all of those that might be present will become unmanageable. It is therefore probable that screening techniques, based on bioassays, will be used to identify whether influents and effluents cause endocrine disruption, and if tests are positive, steps will be taken to identify the causative agents. Such an approach has been proposed by an expert workshop,[11] which concluded that since it has not yet been established which of the myriad substances in the environment are causing an endocrine problem, biological effects measures are more appropriate because they give information on endocrine effects. If effects are identified, the use of a chemical evaluation to find the cause should be utilized in an analytical chemistry/toxicity identification evaluation (TIE).[11] There is a range of assays available for testing for endocrine activity; many of these have been listed in a compendium of methods[12] or have been the subjects of reviews.[4,5,10,13–15]

3.1 *IN VITRO* ASSAYS

Major advantages of *in vitro* assays are that they follow well-understood pathways and modes of action and that the end points are usually clearly defined; however, they may suffer from a range of limitations and responses measured by *in vitro* assays should be confirmed *in vivo* before compounds are listed as endocrine disrupters.[3]

3.1.1 COMPETITIVE LIGAND BINDING

These assays are based on the primary mode of action of estrogens (and xenoestrogens), which is binding to the ER. The implication is that binding to the receptor results in a subsequent effect on biological activity. These assays have the potential for high-throughput screening and are under consideration for assisting with the prioritization of chemicals by the U.S. EPA.[16]

Measurements utilize the appropriate radiolabeled hormones, predominantly [^{3}H]17β-estradiol in the case of the ER, which is used as a control to confirm the presence of the receptor and to evaluate the effects of competitive binding to that receptor through addition of the compounds being tested. Assays for estrogenic compounds involve the extraction of the ER from a cell line such as MCF-7[17] and subsequent incubation with [^{3}H]17β-estradiol, either with or without the synthetic estrogen diethylstilbestrol (DES) as a control, or the test compounds in increasing concentrations. The bound, labeled [^{3}H]17β-estradiol is isolated with hydroxyapatite, then freed by incubation with charcoal before extraction with ethanol for scintillation counting.[6] If less of the [^{3}H]17β-estradiol is bound to the ER, this is because the test compounds have demonstrated an ability to compete for the binding sites. The greatest limitation of these assays, however, is that although compounds may bind to the receptor, the tests do not distinguish between subsequent agonistic or antagonistic effects.[7,10] Most binding assays rely on the isolation of the receptors from natural cell lines; however, bacterially expressed receptors for high-throughput testing have been developed for the human hERα, a reptilian receptor from the liver of the green anole (*Anolis carolinensis*), aER and from rainbow trout (*Oncorhynchus mykiss*), rtER.[16] The binding of a range of PCB, hydroxylated PCB, and arochlors was also evaluated; however, compounds tested generally exhibited greater affinity for the rtER than either of the other receptors.

The range of available ligand binding assays and the lack of recognized standards for reporting data make intercomparison difficult; however, a number of tests and compounds evaluated are summarized in Table 3.1. In addition to the ER, binding assays for other receptors, such as the androgen receptor (AR), have also been developed,[18] with *p,p'*-DDE being identified as a potent AR antagonist.[19] Such assays usually require overnight incubation at around 4°C with AR isolated from rat reproductive tissues, such as the epididymis and prostate.[14]

Ligand binding assays have been used to identify estrogenic activity in a range of compounds, and some workers have compared the effectiveness of tests, for example the binding of PAH to the α and β isoforms of the human estrogen receptor (hERα and hERβ).[20] This work indicated that, although the tested compounds

TABLE 3.1
Compounds that Have Been Tested Using Competitive Ligand Binding with IC50 Values

Receptor	Compounds Tested
hERα	p-nonylphenol (7.2 µM)[21]
	PAH (nb-28 nM)[20]
	Organochlorines
	o,p'-DDT (485 µM),[21] dieldrin (>50 mM),[22]
	endosulphan (631 µM),[21] (>50 mM),[22]
	toxaphene (470 µM),[21] (>50 mM)[22]
	PCB
	Tetra-ortho PCB104, 184, 188 (>10 µM),[16]
	2′,3′,4′,5′-tetrachloro–4-biphenylol (0.1 µM),[16]
	2,6,2′,6′-tetrachloro–4-biphenylol (0.5 µM)[16]
hERβ	PAH (nb-29 nM)[20]
Fish ER	Alkylphenols
	octylphenol (0.1 µM)[23]
	nonylphenol (0.5 µM)[23]
	NP_1EC (200 µM)[23]
	Phthalates
	benzylbutylphthalate (2–10 nM)[24]
	di-n-butylphthalate (2–10 nM)[24]
	bis(2-ethylhexyl)phthalate (2–10 nM)[24]
	benzophenone (2–10 nM)[24]
	n-butylbenzene (2–10 nM)[24]
	4-nitrotoluene (2–10 nM)[24]
	butylated hydroxy anisole (2–10 nM)[24]
	Chlorophenols
	2,4-dichlorophenol (2–10 nM)[24]
	PCB
	Tetra-ortho PCB54 (>10 µM)[16]
	Tetra-ortho PCB104, 188 (1.3 µM)[16]
	Tetra-ortho PCB184 (0.4 µM)[16]
	2′,3′,4′,5′-tetrachloro–4-biphenylol (0.27 µM)[16]
	2,6,2′,6′-tetrachloro–4-biphenylol (0.3 µM)[16]
Mouse ER (B6C3F₁)	Whiskey, red and white wine*[25]
	Organochlorine mixture (DDT; DDE; methoxychlor; endosulphan; toxaphene)*[25]
Reptilian, aER (green anole)	PCB
	Tetra-ortho PCB104, 184, 188 (>10 µM)[16]
	2′,3′,4′,5′-tetrachloro–4-biphenylol (0.25 µM)[16]
	2,6,2′,6′-tetrachloro–4-biphenylol (0.5 µM)[16]

nb, non-binder; BBP, butylbenzyl phthalate; DBP, di-n-butyl phthalate; DEHP, bis(2-ethylhexyl)phthalate; NP_1EO4-nonylphenoxycarboxylic acid

* Reported as 100% displacement of labeled estradiol for red and white wine; no effect for whiskey and OC mix

exhibited similar affinity for binding to both forms of the receptor, additional use of a recombinant receptor–reporter assay using MCF-7 cells indicated that the capacity for transcriptional activation was isoform specific. Parent PAH did not exhibit any binding to either the α or β forms of the receptor; however, monohydroxylated PAH bound to different degrees.

3.1.2 Cell Proliferation Techniques

These approaches are predominantly based on human-derived cell lines and utilize a number of end points to measure the cell proliferation induced through exposure to estrogenic compounds. Commonly used strains of cells are estrogen-responsive MCF-7 or T47-D human breast cancer cells.[7] The E-screen assay, developed for this purpose, is based on the increased growth of MCF-7 cells in the presence of estrogens.[21] When a range of concentrations are tested, the method can differentiate between agonists, partial agonists, and inactive compounds.[8] The E-screen assay developed by Soto et al.[26] compares cell yields in both positive and negative controls with those from samples exposed to test compounds. A large range of chemicals were evaluated using the E-screen by Soto et al.[21], with relative proliferative effects of up to 100% observed for a number of compounds at concentrations of around 10 µM (Table 3.2). Polychlorinated biphenyl congeners and hydroxylated PCB were also evaluated, with the 2',4',6'-trichloro-4-hydroxybiphenyl demonstrating the greatest effect of 99.8% at a concentration of 100 nM, in comparison to reduced effects for other congeners at concentrations two orders of magnitude greater. The end point of the E-screen has been modified by Körner et al.,[27] who, rather than counting cells or nuclei, utilized a colorimetric end point which was claimed to be faster and easier to perform. A range of other means of quantifying cell growth has also been reported, with the alamar blue (AB) and [3H]thymidine incorporation assays exhibiting greater sensitivity than cell counting, DNA, and MTT assays.[28] The AB assay was also described as quicker and less expensive. This test has been applied to the evaluation of effluents from several STWs in Germany and in evaluating the effect of mixtures of xenoestrogens which included bisphenol A, octyl and nonylphenol, butylphthalate, and 4-hydroxybiphenyl.[2]

However, a range of cell lines for MCF-7 exists, and these display different responses that could cause problems with reproducibility. This was highlighted for response to estradiol, p-nonylphenol, and bisphenol A, using four cell lines, BUS, ATCC, BB, and BB104, with BUS exhibiting the greatest sensitivity.[30] Another cell line, the E3, has also been compared to wild-type (WT) MCF-7 cells and demonstrated a more proliferative (7 compared to 6 reported for the BUS line) and less variable response to 17β-estradiol than the wild-type.[28] Experimental protocols for proliferation of the MCF-7 (and other estrogen responsive lines) require that media used for growth be stripped of steroids with dextran charcoal. Some workers have found that cell lines exhibit proliferation in estrogen-free media.[31,32] Issues such as this present difficulties with regard to evaluation of the technique and in setting up validated and standard methods for evaluating the estrogenicity of compounds; however, they are not unique to the MCF-7 cell line, nor to cell proliferation assays.

TABLE 3.2
Compounds and Materials Tested for Estrogenic Activity Utilizing Cell Proliferation Assays

Cell Line	Compounds Tested
MCF-7 (E-screen)	Alkylphenols and alkylphenol ethoxylates
	nonylphenol (103%),[30] (70%),[23] (100%),*[31] (100%),[21] (105%)[27]
	octylphenol (+ive),[28] (~80% at 1.0 μM[23] (100%),[21] (97%)[27]
	butylphenol (71–76%),[21] (78%)[27]
	NP$_2$EO (> 100% at 10 μM)[23], NP1EC (~ 80% at 1.0 μM)[23]
	Bisphenols[38]
	bisphenol A,[38] (97%),[30] (~ 80%),[34] (82%),[39] (97%)[27]
	bisphenol A dimethacrylate (~ 80%)**[34]
	tetrabromo-bisphenol A (52%)[27]
	Chlorophenols
	Tris-(4-chlorophenyl) methanol,[28] 4-chloro-3-methylphenol (44%),[27] 4-chloro-2-methylphenol (25%)[27]
	Organochlorine compounds
	o,p'-DDT (86%),[21] p,p'-DDT (71%),[21] o,p'-DDE (26%), (n/e),[28] o,p'-DDD (79%),[39] dieldrin (55%),[21] endosulphan (81%)[21]
	HCB (62%),[39] methoxychlor (57%),[21] toxaphene (52%)[21]
	isopropyl benzene[28]
	naphthalene (halowax 1041)
	PCB, OH-PCB
	4-hydroxybiphenyl (87%, 10 μM),[21] (71%)[27]
	4,4'-dihydroxybiphenyl (84%, 10 μM)[21]
	2-monochlorobiphenyl (4%) (87%, 10 μM)[21]
	2,2',4,5-tetrachlorobiphenyl (62%, 10 μM) (87%, 10 μM)[21]
	4-phenylphenol (99%)[39]
	2-phenylphenol (30%)[39]
	Phthalates
	di-n-butylphthalate (62%)[27]
	benzylbutylphthalate (90%),[21,39] (80%)[27]
	sewage effluent Germany (33–90%)[29]
	dental sealant (100% 5 μg ml^{-1})[34]
ZR-75 (human breast cancer)	Phthalates
	butylbenzylphthalate (10^{-5} M),[24] di-n-butylphthalate (>10^{-4} M)[24]
	bis(2-ethylhexyl)phthalate (10^{-5} M)[24]
	butylated hydroxy anisole (10^{-5} M)[24]
	Alkylphenols and alkylphenol ethoxylates
	p-nonylphenol (+ive),[23] octylphenol (10^{-6} M)[23]
	NP$_2$EO (+ive),[23] NP$_1$EC (+ive)[23]

(n/e) no effect

* Control cells in dextran charcoal stripped medium also exhibited growth.

** Response had not plateaued at highest concentration tested (10^{-6} M)

Values are reported as relative proliferative effect (RPE) (%) for simplicity, or the lowest concentration at which effects were observed.

Other proliferative cell lines that have been utilized in testing for estrogenic compounds include the ZR-75 human breast cancer line, with a range of compounds evaluated by Jobling et al.[24] These included a range of phthalates, two antioxidants (butylated hydroxyanisole and butylated hydroxytoluene), caffeine, benzoic acids, methylphenols, and benzophenone. Most of these compounds exhibited only weak activity at concentrations below 10^{-4} M, with the exceptions reported in Table 3.2. Both MCF-7 and T47D cell lines were used to determine the estrogenic effect of wines and spirits in comparison to a mixture of organochlorine compounds.[25] In addition, the estrogenicity of a range of more commonly used products, ranging from sun screens[33] to resin-based composites and sealants used in dentistry, has been evaluated by this technique.[34] Such applications demonstrate that the tests are potentially versatile and may well find future use in evaluation of effluents and in longer-term monitoring schemes.

There is however, a range of disadvantages associated with cell proliferation techniques, in addition to the experimental procedures that are relate to reproducibility. Mammalian cells, which are best for understanding specific mechanism-based details,[35] exhibit tissue-specific expression of receptor subtypes. In addition to their use in cell proliferation techniques, the MCF-7 strain has been developed, and either stable or transient transfections of the cells have been obtained with recombinant estrogen receptor/reporter genes. The use of a recombinant yeast cell bioassay (RCBA) has been compared to the MCF-7 E-screen test and was found to be twice as sensitive to estradiol.[36]

3.1.3 RECOMBINANT RECEPTOR–REPORTER ASSAYS

These techniques have been the method of choice for first-pass screening, but their scale-up to techniques suitable for high throughput has proved difficult.[37] The tests are undertaken with genetically engineered mammalian cells or strains of yeast, with cells transformed (transfected) by introducing vectors containing DNA sequences for the receptor, along with response elements linked to promotor regions for a reporter gene, and the reporter gene itself. A number of recombinant assays are available, with cells either transiently or stably transfected with receptors and expression plasmids. The reporter gene is selected to facilitate measurement of a specific end point if the receptor-mediated mechanism is activated by the test chemicals.[6]

A number of researchers have used yeast cells for this purpose. *Saccharomyces cerevisiae* has been stably transfected with the human estrogen receptor (hER) gene and expression plasmids with an estrogen responsive element (ERE) with the reporter gene *lacZ*, which encodes for the enzyme galactosidase.[40] The end point is the production of the enzyme that metabolizes the chromogenic substrate (chlorophenol red-β-D-galactopyranoside [CPRG]), which is yellow in color, to a red product which is measured by absorbance at 540 nm (Figure 3.1). In addition, it is known that steroid receptor antagonists, such as ICI 164,384 exhibit positive activity in the yeast system (possibly because yeast does not contain the repressor proteins necessary for antagonism), and it is therefore a useful screen for detecting all chemicals, whether they are agonists or antagonists.[41]

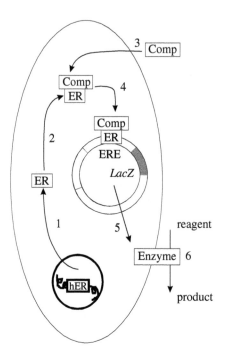

1. The hER is incorporated within the genome

2. The expressed receptor [ER] is activated by estrogenic compounds [Comp] which enter the cell (3) and bind with it

3.

4. The activated estrogen receptor binds to the estrogen responsive element (ERE)

5. Binding to the ERE initiates transcription which results in expression of the proteins (in this case the enzyme β-galactosidase)

6. The enzyme metabolises the reagent and the change is measured by absorbance

FIGURE 3.1 Outline of the estrogen expression system in yeast. (From Routledge, E.J. and Sumpter, J.P., Estrogenic activity of surfactants and some of their degradation products assessed using a recombinant yeast screen, *Environ. Toxicol. Chem.*, 15, 241, 1996. With permission.)

The breast cancer cell line MCF-7 expresses the estrogen receptor (hERα), the glucocorticoid receptor (GR), and progesterone receptor (PR). Similar constructs have been introduced into this cell line,[46] and this has become a routine method for measuring ER transcriptional activation.[14] The cells may be transiently transfected, although permanent cell lines, such as the MVLN/MCF-7 have been developed.[47] Similarly, other stable transfected cell lines have been developed from the MCF-7, the MELN, and MMLN.[48] In addition to identifying estrogenic activity,[49] these lines have also been used to investigate activity of PCB77, arochlor 1254, DDE, and atrazine relative to the progestin receptor agonist R5020 and androgen receptor agonist R1881. All these compounds were observed to act as antagonists to both the AR and PR; however, no antiglucocorticiod activity was observed.[50]

It is sometimes difficult to compare directly the results from tests undertaken by different workers because of the way they report their results. When estrogenic activity is considered, 17β-estradiol is most frequently used as a (positive) control, and activity is reported relative to this. This may be related to a concentration of the test compound at which the observed response is equivalent to the maximum observed for 17β-estradiol and is usually reported in terms of the test compound being "x" times less active in terms of concentration. Where induced response did not reach the maximum observed for 17β-estradiol at any test concentration, data have been reported as submaximal.[3] A different approach, again with a recombinant yeast assay, was to report EC50 values, where relative potency of test compounds was reported as the concentrations of E2 and the test compound that provided 50%

of maximum induction observed for E2. Such results are reported as potency relative to E2 (which has a value of 100%).[36] The results from a range of recombinant assays are summarized in Table 3.3, with an indication of relative activities where possible. In cases where data were produced graphically, and the values were not written by the authors, values have been omitted from Table 3.3; however, inclusion of a compound in Table 3.3 indicates that activity was detected.

Although a considerable number of compounds have been evaluated utilizing recombinant receptor–reporter assays within the environment and in discharges from STW, compounds are likely to occur as complex mixtures. Testing effluents and extracts from a range of complex solid matrices gives some information on the activity of such mixtures, but it is not possible to test all combinations over an infinite range of concentrations. There are some indications that synergistic effects between EDCs do occur. A range of organochlorine compounds were tested using the yeast screen by Arnold et al. (> 33 µM)[22] (Table 3.2). In this work, chlordane did not exhibit any estrogenic effect, but it did demonstrate synergistic activity in combination with endosulphan by decreasing the EC50 value from > 33 µM to 0.2 µM. Further work on synergistic effects has demonstrated that they can be predicted. However, the assay used needs to be fully characterized, and the endpoint selected (molecular or biological at cellular/organism level) may be important, the maximal effects may need to be similar to allow for accurate predictions, and the (synergistic) impact of relatively weak xenoestrogens in systems where endogenous steroid estrogens are naturally present also needs to be addressed.[51] Understanding and interpreting the results of *in vitro* assays have been considered by Beresford et al.[3] The study was aimed at experience gained over a number of years from the use of a yeast-based assay,[40] and issues raised were considered to be applicable to other *in vitro* assays and may account for some of the inconsistent results reported in the literature. Overall, screens developed from human cell lines, or a battery of such screens, may appear to resolve many issues relating to testing for chemical effects relevant to humans. However, some problems persist, such as differences in metabolism between *in vitro* and *in vivo* cells, cyclical or lifecycle changes in sensitivity of organs to exposure to estrogens (e.g. fetal exposure), and the applicability of end-points to the whole organism.[10]

3.2 *IN VIVO* ASSAYS

These assays are necessary for the evaluation of impacts on the endocrine system as a whole, and it has been proposed that for full understanding, multigeneration reproduction studies are the ultimate test for identifying adverse effects and should be undertaken with toxicity studies.[35] The most widely used *in vivo* assay is the rodent uterotrophic assay,[3] which is based on the ability of chemicals to stimulate uterine growth.[15,52] An increase in the uterine weight measured in ovariectomized or immature female rats or mice is considered to be the "gold standard" of estrogenicity.[8,14] In comparing the yeast assay with a uterotrophic assay, prepubertal (18-day-old) mice were exposed to test compounds dissolved in corn oil through subcutaneous injection over 3 consecutive days.[36] Although at the highest doses, bisphenol-A and nonylphenol demonstrated acute toxicity, nonylphenol did, at a lower dose (5 mg), result in a significant increase in uterine weight. In a similar evaluation of the MCVF-7 growth screen, immature female rats were used.[28]

TABLE 3.3
Recombinant Assays Used to Determine Activity for a Range of Receptors

Reporter Cell Line	Receptor Gene	Compounds Tested
MCF-7 luc (transiently transfected) (MVLN) (MELN)	hERα	Alkylphenols and alkylphenol ethoxylates nonylphenol mixture (50 nM)[49] nonylphenol (1–10 μM),[23] (1–10 μM)[49] octylphenol (0.1 μM),[23] (1000 ×<)[24] NP$_1$EC (1–10 μM),[23] NP$_2$EO (1–10 mM)[23] Organochlorine compounds Arochlor 1254 (inactive)[50] o,p'-DDE (1–10 μm),[49] p,p'-DDE (1–10 μm)[49] PCB77 (inactive),[50] mixture (12 μg ml^{-1}, inactive)[25] PAH (inactive–160 nM)[20]
		Atrazine (inactive)[50]
		Phthalates butylbenzylphthalate (10^{-6}–10^{-4} M),[24] di-n-butylphthalate (10^{-5}–10^{-4} M)[24] bis(2-ethylhexyl)phthalate (>10^{-4} M)[24] butylated hydroxytoluene (>10^{-4} M)[24]
		Environmental samples Effluents ($\equiv$5 – >30 pM)[50], Sediment ($\equiv$1–5 pM),[50] Water ($\equiv$1–5 pM)[50] Foods whiskey, red and white wine [25]
	hERβ	PAH (inactive–40 nM)[20]
MCF-7-luc (MMLN)	PR; GR	Organochlorine compounds Arochlor 1254 (PR50%; AR40%),*[50] DDE (PR60%; AR40%)*[50] PCB77 (PR40%; AR50%)*[50] Pesticides/herbicides atrazine (PR55%)*[50]
T47D	hER	Alkyphenols nonylphenol (260 nM)[44] Organochlorines o,p'-DDT (660 nM),[44] chlordane (6240 nM),[44] dieldrin (24,490 nM),[44] endosulphan (5920 nM),[44] methoxychlor (5720 nM)[44] bisphenol A (770 nM)[44]
		Foods whiskey, red and white wine [25]
PC3 (human prostatic) (PALM)	AR	arochlor, DDE, PCB77, and atrazine (no antiandrogenic effect)[50]

(continued)

TABLE 3.3 (continued)
Recombinant Assays Used to Determine Activity for a Range of Receptors

Reporter Cell Line	Receptor Gene	Compounds Tested
Yeast	hERα	Alkylphenols
		nonylphenol (10 μM),[43] (4000 ×<),[3] (15,000 ×<),[40] (0.005%),[36] (5000 ×<)[41]
		octylphenol (10 μM),[43] (0.003%),[36] (7000 ×<),[40] (0.003%)[36]
		NP$_2$EO (25,000 ×<),[40] NP$_2$EC (5 × 105 ×<)[40]
		butylated hydroxytoluene (0%)[36]
		bisphenol-A (1 mM),[43] (10,000 ×<),[3] (0.005%),[36] (15,000 ×<)[41]
		Organochlorine compounds
		methoxychlor (0.0033%),[36] (80,000 ×<),[3] (5 × 10^6 ×<)[41]
		o,p'-DDT (0.00011%),[36] (8 × 10^6 ×<),[41] o,p'-DDD (15 × 10^6 ×<)[41]
		o,p'-DDE (0.00004%),[36] (24 × 10^6 ×<),[41] dieldrin (>33 μM)[22]
		endosulphan (>33 μM),[22] toxaphene (>33 μM)[22]
		Phthalates
		benzylbutylphthalate (20% sub. max)[3] (0.0004%)[36]
		dibutylphthalate (0%)[36]
		PCB/Dioxins
		4'-chloro-4-biphenylol (0.06%),[36] 2',5'-dichloro-4-biphenol (0.62%),[36] 2,3,7,8 TCDD (0.26%)[36]
	hPR	bisphenol A, nonylphenol, methoxychlor, DDT, DDD, DDE (all inactive)[41]
	hAR	p,p'-DDE (2.5 x 10^6 ×<)[41]

* Represents % inhibition of activity relative to presence of the progestin R5020 (PR) or the androgen receptor agonist R1881 (AR).

Unless otherwise indicated, results are reported as EC$_{50}$ values in terms of concentration (e.g., nM), as a percentage relative to E2 (%), as the number of times less potent than E2 (×<), or as E2 equivalents (≡).

Overall, it is unlikely that *in vivo* tests will be utilized for monitoring the environment or discharges to or from STW, nor are they primarily designed as screening tests for compounds exhibiting estrogenic activity because they are expensive and time consuming[35] and their use for such purposes might raise ethical questions.[53] Their application is in validating other techniques, either for screening or monitoring, and they will therefore not be given further consideration in this work. As assays become more commonly utilized, and as methods for testing strategies come under scrutiny for their suitability, a range of issues regarding the

TABLE 3.4
Performance of Assays with Respect to Their Response to 17β-estradiol

Assay	Detection Limit	EC50	Induction
Recombinant			
T47-D luciferase[44]	0.5 pM	6.0 pM	100
Yeast cell (hER)[36]	100 fM		
Yeast cell (hER)[40]	7.3 pM		
Yeast cell (hER)[41]		225 pM	
Yeast cell (AR)[41]		3.5 nM*	
Yeast cell (PR)[41]	100 pM**		
Cell proliferation			
MCF-7[36]	1 pM		
MCF-7[29]	1 pM	5–10 pM	0.7–3
in-vivo			
Mouse uterotrophic[36]	18 pM		
Roach vitellogenin induction[54]	365 pM (100 ng l⁻¹)*		400
Trout vitellogenin induction[54]	365 pM (100 ng l⁻¹)***		5.2×10^4

2ng l⁻¹ estradiol = 7.3 pM

* Response to dihydroxytestosterone

** Response to progesterone

*** Response at 10 ng l⁻¹ was not statistically significant

performance of different methods in terms of sensitivity and reproducibility is likely to arise. The performance of a range of different assay types with respect to sensitivity to estradiol is presented in Table 3.4.

3.3 SCREENING SYSTEMS

Evaluation of all the available options for *in vitro* and *in vivo* evaluation of endocrine disruption leads to the conclusion that no one test is suitable to determine if any compound (or mixture) is or is not (likely to be) an endocrine disrupter. Therefore, proposed methods for testing compounds have focused on identifying a group of techniques and a range of end points that will be robust enough to evaluate compounds for different modes of action and to minimize the possibility of false positive or negative results through built-in redundancy. While many of these tests are being utilized for evaluation of individual compounds, it is possible that when they are further characterized, they will be adapted for screening of waste waters and other discharges to the environment.

The main driver for the development of screening systems in the United States has been the Food Quality Protection Act (FQPA) of 3rd August 1996 and amendments to the Safe Drinking Water Act, which required the EPA[55] to:

Develop a screening program, using appropriate validated test systems and other scientifically relevant information, to determine whether certain substances may have an effect in humans that is similar to an effect produced by a naturally occurring estrogen, or other such endocrine effect as the Administrator may designate.

In response to this legislation, the EPA has proposed a two-tier screening system that consists of a Tier 1 screening battery of eight tests and a Tier 2 test battery of five multigeneration tests to confirm and characterize effects.[37] However, the number of individual chemical substances that should be considered for endocrine disrupter screening and testing exceeds 87,000, and, in a report to Congress in August 2000, the EPA stated that[56]

Following validation of the assays, EPA plans to publish final test guidelines and a *Federal Register* notice setting forth the final policy and procedures for implementation of the EDSP. EPA will publish a proposed list of chemicals for Phase I screening approximately one year in advance of the date screening would be required to begin. EPA anticipates requiring screening of pesticide active ingredients and other pesticide formulation ingredients with high production volume beginning in 2003.

Such a time scale, since the original legislation of 1996, is a reflection of the complexities relating to validation and suitability of assays for endocrine disrupters.

In addition to the approach made by the U.S. EPA, others have proposed testing strategies or screening systems that incorporate both *in vivo* and *in vitro* assays. Workers at DuPont reported the development of a Tier I screening battery that consisted of two *in vivo* tests, one on female and one on male rats, which monitor a range of end points, and a third, *in vitro* test, that utilized the yeast transactivation assay.[57] This test battery does not just involve identification of estrogenic activity. The female test battery, undertaken on ovariectomized females, is designed to identify estrogenicity, increases in serum prolactin, and neuroendocrine end points; the male battery identifies androgen receptor (AR) antagonists, dopamine agonists and antagonists, thyroid effects, impact on steroid biosynthesis, and 5 reductase inhibitors, while the yeast assay gives information on agonist/antagonist activity on the receptors (ER, AR, and PR). Further development of this procedure has focused on its sensitivity in relation to *in utero* exposure undertaken by other workers.[58] Data generated through using 17β-estradiol as a control indicated that the female test battery, using uterine stomal cell proliferation as an end point, was as sensitive (0.001 mg kg^{-1} day^{-1}) as 90-day/one-generation reproduction studies using accelerated time to vaginal opening as an end point.[59,60] The end points for males were also more sensitive, again measured at 0.001 mg kg^{-1} day^{-1} in comparison to *in utero* exposure.[59,61] Further development of the tiered testing scheme to include developmental reproductive screens (Tier II), investigations into the pharmacokinetics (Tiers III and IV) linked to risk assessment has also been described by Cook et al.[57]

In Europe, attention has focused more on evaluating the environmental risk through activities of the European Centre for Ecotoxicology and Toxicology of Chemicals (ECETOC), which established the Environmental Estrogens Task Force.[62] This group has reviewed the EDSAC proposals,[55] in particular with respect to the scientific rationale and ethical use of animals for ecotoxicity assessment.[53] On an international level, the Organization for Economic Cooperation and Development (OECD) is also involved in the development of tests for EDCs and guidelines for the validation of such tests.[9] This work focuses on an initial framework of tests for endocrine substances, but also for evaluating effects on wildlife.

In summary, bioassays are predominantly being used as tools to identify and confirm whether or not individual chemical compounds are EDCs. However, as the techniques mature, they are likely to find use in monitoring discharges to the environment for any biological effect and will complement the range of available chemical techniques. Such an approach has already been utilized by a number of workers to fractionate mixtures (e.g., final effluents from STW), to isolate the estrogenically active components, and subsequently identify the active compounds.

3.4 CHEMICAL TECHNIQUES

The use of chemical techniques allows for monitoring the occurrence of compounds identified as estrogenic within the environment (water, air, soils, and sediments), within organisms, in foodstuffs and associated packaging, and during experimental work where this is deemed necessary. However, this section will focus on techniques that have been used for the determination of endocrine disrupters in wastewater treatment processes. Determination of concentrations of target compounds at around 1 ng l^{-1} or below in a complex matrix such as wastewaters, which may contain many compounds that can interfere with analysis, is an important analytical challenge.[63] This challenge begins with the sampling of materials to be analyzed and may be further subdivided into an extraction and sample preparation stage which includes sample cleanup, and for some analyses, derivatization, concentration, and finally a quantification step. The predominant method of quantification for estrogenic compounds has been either gas chromatography (GC) or liquid chromatography (LC). The role of these systems in analysis has been discussed by Grob,[64] and capillary GC has, in many cases, been the method of choice for over 2 decades.[65] However, GC is limited to use with compounds that are volatile and thermally stable, although it is possible to overcome some limitations of volatility and stability through the use of derivatization prior to analysis.[66]

The type of detector used with either GC or LC systems depends on the compounds being determined. However, because of the low concentrations involved, specific, sensitive detectors, such as the electron capture (ECD), flame photometric (FPD), and nitrogen–phosphorus (NPD) detectors, have all been utilized with GC. The flame ionization detector (FID) is still frequently utilized as well. Detectors available for LC usually depend on the presence of a chromophore in the analyte, or its inclusion through derivatization,[66] with fluorescence and ultraviolet (UV; although less specific) most frequently used. However, mass spectrometric (MS) or MS–MS detection is now usually the method of choice with either GC or LC techniques as these instruments, in particular MS–MS, frequently give the highest sensitivity and specificity.

For the quantification of compounds by LC–MS, however, ionization of the analyte must occur before it enters the mass spectrometer, unlike in gas chromatography, where ionization is achieved within the instrument. The interface between the LC and the MS is therefore significant in determining the degree of ionization, and because the interfaces for the instruments are specific to manufacturers, some differences in efficiency are to be expected. There are two interfaces in common use: the electrospray interface (ESI) and atmospheric pressure chemical ionization (APCI). Biological techniques have also been used for the quantification (rather than for identifying compounds as estrogenic) of both natural and xenoestrogens with both enzyme-linked immunosorbent assay (ELISA) and radioimmunoassay (RIA) techniques reported in the literature.

There are effectively two sample matrices of concern in an STW, the liquid phase and solid phase. The objectives of any analysis may determine the treatment of samples and which type of samples are taken. The analysis of samples with low suspended solids (such as final effluents) where only concentrations in the dissolved phase are to be determined is less problematic than that of a sludge where concentrations in the solids may be of interest.

3.4.1 SAMPLING AND HANDLING

The determination of the compounds begins with sample collection and some type of storage to allow for transport to the laboratory. The collection of samples is usually made in amber glass, but some workers have filtered samples and utilized solid phase extraction (SPE) on site.[67] Formaldehyde (1% v/v) has been used as a preservative in conjunction with refrigeration[68] and was also used for influent and effluent samples by Lee and Peart,[69] who also analyzed sludge samples which were unpreserved but air dried on arrival at the laboratory. Other agents used to preserve samples (through preventing microbial activity) include methanol,[70] sulfuric acid,[71] and mercuric chloride.[72] The option of collecting the analytes on the solid phase extraction material and storing the cartridge (or disk) has been evaluated by Baronti et al.,[68] who found that preservation was enhanced by washing the cartridge with 5 ml of methanol and storing at −18°C. These workers used Carbograph material to trap the steroid estrogens. This and other sorbents, such as C18, need evaluation because washing with methanol may elute analytes. Degradation of lower-molecular-weight PAH has been observed in samples of raw sewage stored for up to 56 days at 15°C, with a decline noticeable after 7 days; however, preservation with formaldehyde effectively arrested changes.[73]

Obtaining representative samples is an important requirement and is made more problematic when working with raw sewage or other samples that are not homogeneous. Some workers have homogenized samples in blenders such as an Ultra-Turrax,[74] whereas others working with dried solids have utilized coffee grinders and subsequently sieved materials to ensure homogeneity.[75] Sludge samples are usually taken when determining the fate of hydrophobic contaminants which are likely to be associated with the solids, and because contaminant concentrations are likely to be enhanced, smaller sample volumes are usually taken. Sampling may involve taking an initial bulk sample of kilogram proportions, followed by further sub-sampling of 100 to 250 g after mixing. These may then be preserved by freezing.[76] The size of the final sample taken for analysis is usually between 1 and 50 g.

3.4.2 CHROMATOGRAPHIC ANALYSIS

Both LC and GC are well suited to the simultaneous determination of a number of compounds. Methods that analyze for a number of components simultaneously are termed *multi-residue techniques*. However, the chromatographic analysis is usually the final step in a complex chemical analysis which involves extraction of the compounds of interest from the solid or aqueous phase into organic solvent, followed by concentration and cleanup steps (to remove compounds that may interfere with quantification). Although single methods have been developed for the analysis of around 100 determinants in sludges, such as PCB, PAH, phthalates, chlorinated aromatics, and phenolics, using gel permeation chromatography (GPC) as a cleanup step,[77] most techniques focus on specific groups of compounds. The same approach, focusing on particular classes of compounds, is also usually taken in analysis of aqueous samples. Such samples are often analyzed using solid phase extraction, which is also amenable to the development of complex multi-residue methods.[78]

3.4.2.1 Steroid Estrogens

Some workers extract aqueous samples on-site to enhance sample preservation and to minimize problems associated with handling and storing relatively large volumes (1 to 5 l or more) of liquid samples, but more frequently samples are returned to the laboratory, where extraction has predominantly, for liquids, utilized SPE (Table 3.5). In terms of effectiveness in retaining estrogenic compounds from effluents, a non-endcapped C18 and a polystyrene copolymer resin (ENV+), tested using the E-screen assay, demonstrated no significant differences.[29] Prior to any SPE step, the liquid samples are usually filtered through a range of filter sizes. This in effect means that only compounds in the dissolved phase are analyzed. In some instances, the authors have described a wash of collected particulate matter with solvent, although determination of extraction efficiency for this is not clearly identified. The determination of the steroid estrogens in the solid phase (sludge) has been approached through freeze drying the material before subsequent extraction with methanol and acetone, subsequent cleanup by size-exclusion chromatography and silica gel, and quantification with GC/MS-MS.[79] A similar approach has been used for the determination of steroid estrogens in sediments, and the final quantification was by LC with diode-array and MS detection.[80]

A range of derivatives have been used for making the steroid estrogens more amenable to analysis by GC–MS (Table 3.5). These are predominantly silylanized derivatives used to facilitate determination by GC. Advantages may be offset by additional sample handling, and compromise is required between time required for formation of derivatives, the extent of the reaction, and final stability of the compounds formed. A range of derivatives used for the determination of the steroid estrogens has been included in a review by López de Alda and Barceló.[63] However, the use of other derivatizing reagents offers the opportunity to utilize different detection techniques. Pentafluorophenyl derivatives of sterols have been determined by geochemists to facilitate detection by ECD,[81] and this approach, using pentafluorobenzoyl derivatives, has been used to determine estrogens by negative chemical ionization MS.[82] This type of derivative is also used in pharmacological analysis, as they undergo dissociative electron capture in the gas phase within an APCI interface, with a detection limit of 0.2 pg on-column for estrone using an LC–MS/MS system.[83]

Liquid chromatographic techniques have also been used extensively for analysis of the underivatized steroid estrogens (Table 3.5). Both ESI and APCI have been utilized in LC–MS analyses, which, as they do not require derivatization, have become a method of choice when the equipment is available. The technique has also been used to study the autooxidation and photodegradation of ethinylestradiol.[84] Liquid chromatography with MS has also been used in work that has quantified both conjugated and unconjugated steroid estrogens. One approach first determined the presence of unconjugated compounds, and subsamples were then treated with glucuronidase and arylsulfatase enzymes to give a result for total conjugated and unconjugated compounds, with the concentration of conjugates being evaluated by difference.[85] However, an LC–MS method for direct determination of both conjugated and free steroid estrogens has been reported.[86]

3.4.2.2 Alkylphenols

The alkylphenols (alkylphenol ethoxylates or ethoxylated nonionic surfactants) present a significant analytical challenge. The parent compounds are not classified as highly toxic, but their metabolic products, the mono- and di-ethoxylates (AP_1EO and AP_2EO), parent alkylphenols (nonyl, NP; octyl, OP), and carboxylic derivatives (APECs) are of concern because they persist in the environment and are implicated as EDCs. In addition, the formation of ring halogenated derivatives during chlorination has also been verified.[98] It has been noted that the oligomer distribution of the APEOs may be skewed toward the low end during sample storage.[99] The extraction of the alkylphenols has been undertaken using a steam distillation/solvent extraction procedure, which yields relatively clean extracts, although an alumina column cleanup step was utilized prior to final determination by GC–MS.[100] In the same study, it was also demonstrated that the approach is only suitable for NP, nonylphenol ethoxylate (NP_1EO), and nonylphenol diethoxylate (NP_2EO), with oligomers of nonylphenol triethoxylate NP_3EO and above showing <15% recovery, and the extraction of the higher oligomers requiring liquid/liquid extraction (Table 3.6) or the use of gaseous stripping into ethyl acetate.[101] More recently, aqueous samples have been extracted using SPE, which has largely replaced liquid/liquid extraction (Table 3.6)

The use of mass spectrometers linked to LC systems for the quantification of alkylphenols has become the method of choice, because it provides improved sensitivity and selectivity at low concentrations in difficult matrices. Fluorescence detection is still used by some workers (Table 3.6), and UV detection has also been utilized at 275 nm[102] and 277 nm.[100] The use of both electrospray ionization (ESI) and atmospheric pressure chemical ionization (APCI) techniques for the determination of the alkylphenols has been discussed by Petrovic and Barcelo.[98] With the use of the ESI interface and an aprotic solvent, the NPEOs and OPEOs demonstrate affinity for the Na^+ ion, forming $[M+Na]^+$ ions. The use of protic solvents generates a more extensive range of adducts, with H^+, K^+, NH_4^+ and H_2O. With the APCI source, regardless of the solvent, a range of adduct ions are formed, with some variability in abundance.

TABLE 3.5
Methods Used for the Determination of Steroid Hormones in Wastewater Samples

Sample	Determinants	Volume	Extraction	Separation	Detector	LOD (ng l^{-1})	Recovery (%)	Reference
Inf. and Eff.	E2, EE2	300 ml–2 l	SPE (C18)	GC	MS and MS–MS ELISA	0.1	75–79	Huang and Sedlak[87]
Eff.	E1, E2, E3, EE2	2.5 l	SPE (C18) then PFBCl derivatives	GC	CI–MS	0.2	84–116	Xiao et al.[82]
Eff.	E1, E2, E3, EE2 DES, LEV, NOR, PROG	200 ml	SPE (PLRPS/C18)	RP–HPLC	DAD, MS (–ESI)	2–500	57–112	Lopez de Alda and Barcelo[88] Lopez de Alda and Barcelo[71]
AS liquor	[³H]-EE2		Centrifuge to remove particles	LC	Scintillation			Vader et al.[89]
Inf. and Eff.	E1, E2, E3, EE2	150 ml Inf. 400 ml Eff.	SPE (Carbograph-4)	LC	MS–MS (–ESI)	0.08–0.9	84–91	Baronti et al.[68]
Eff.	E1, E2, E3, EE2		SPE (Envi-Carb)	LC GC	MS–MS	5–10		Lagana et al.[90]
Eff.	E1, E2, E3, EE2	1 l	ENVI-18 (Supelco) pentafluoropropionic acid (for GC)	GC LC	MS and MS/MS (ESI)	unclear	unclear	Croley et al.[91]
Eff.	EE	7 l	SPE (C18)	GC	MS		74	Siegener and Chen[72]
Inf. and Eff.	E1, E2, E3, EE2	1 l	SPE (Envi–Carb)	RP–HPLC	MS–MS (+APCI)	0.5–1.0	87–94	Lagana et al.[90]

Matrix	Analytes	Volume	Extraction	Separation	Detection	LOQ	Recovery (%)	Reference
Eff.	E1, E2, E3, EE2, MES, LEV, NOR-a	20 l	SPE (LiChrolut-EN/C18) trimethylsilyl-ethers	GC	MS	1.0 (LOQ)		Kuch and Ballschmiter[92]
Eff.	E1, E2, EE2	2.5 l	SPE (C18) tert-butyl-dimethylsilyl	GC	MS-MS	1.0	92–100	Kelly[93]
Inf. and Eff.	E1, E2, E3, EE2	0.5–1 l	SPE (Carbograph-4)	HPLC	MS-MS (-ESI)	0.2–0.5	88–97	Johnson et al.[94]
Inf. and Eff.	E1, E2, E3, EE2, MES	1 l	SPE (LiChrolut-EN/C18) trimethylsilyl-ethers	GC	MS-MS		77–90	Ternes et al.[95]
Eff.	E1, E2, EE2	15 l	SPE (Env +) acetylation	GC	MS	0.5–2	n/a	Larsson et al.[96]
Eff.	E2, EE"	5 l	SPE (SDB–XC)	RIA		E2 0.1 EE2 0.05	72–78	Snyder et al.[67]
Eff.	E1, E2, EE2	1 l	SPE (SDB–XC) silylation	GC	MS-MS	0.1–2.4	88–98	Belfroid et al.[97]
Eff.	E1, E2, EE2	20 l	SPE (C18)	GC	MS	0.2		Desbrow et al.[70]

PFBCl = pentafluorobenzoyl chloride

TABLE 3.6
Methods Used for the Determination of Alkylphenols in Wastewater Samples

Sample	Determinants	Volume	Extraction	Separation	Detector	LOD (ng l⁻¹)	Recovery (%)	Reference
Effluent	nonylphenol bisphenol A	15 l	SPE (Env +)	GC	MS	50 / 10		Larsson et al.[96]
Effluent	octylphenol nonylphenol	5 l	SPE (SDB-XC)	HPLC	fluorescence 229/310 nm			Snyder et al.[67]
Effluent	nonylphenol NPEO		SPE liquid/liquid (hexane)	HPLC	fluorescence 228/305 nm	0.1		Ahel et al.[106]
Influents	nonylphenol	200 ml	SPE (C18; LiChrolut EN)	HPLC	APCI-MS	500 50^s	95 $(92)^s$	Castillo et al.[107]
Effluents Sludge	NPEOn NPEC	2 g (sludge)	freeze drying/extraction/ SPE			200 20^s / 400 40^s	94 $(96)^s$ / 87 $(84)^s$	
Sludge	octylphenol nonylphenol NPEO; NPEC bisphenol A	2 g	freeze drying/extraction/ SPE (C18)	HPLC	APCI-MS	140^s / 150^s / 65^s; 75^s / 130^s	88 / 92 / 96; 84 / 94	Petrovic and Barceló[103]
Influent Effluent	dicarboxylated APEOs	25 ml 250 ml	Filtration SPE (Carbograph 4)	LC	ESI-MS (+ fluorescence)		87–93	Di Corcia et al.[108]
Influent	halogenated APEOs	50 ml	SPE (LiChrolut C18)	LC	APCI/ESI-MS	20–100	72–98	Petrovic et al.[105]
Effluent Sludge		200 ml 2 g (sludge)	freeze dry/extraction		(+ diode array)	$5–15^s$	$59–81^s$	
Sludge	nonylphenol	0.2–0.5 g (dw)	Supercritical CO_2/Soxhlet	GC	MS		77	Lin et al.[109]

Sludge	nonylphenol	10 ml	Steam distillation/Chex	HPLC	UV @ 227 nm		82	Sweetman [110]
Effluent	NP_nEO (n = 4 and 6)	200 ml	0.45 µm filter, SPE (C18)	HPLC	APCI-MS	200	92–94	Castillo et al. [78]
Influent Effluent	NP, NP_1EO, NP_2EO	500 ml	simultaneous distillation extraction	GC	MS	4–2122	55–93	Planas et al. [111]
Effluent	NP, NP_1EO, NP_2EO	2 l	simultaneous distillation extraction	HPLC	UV @ 277 nm	500	82–105	Ahel et al. [100]
Sludge		10–50 g		GC	MS	< 1[s]		

[s] for sludge samples, detection limit in µg kg^{-1}

As a result of these differences, the ESI interface offers improved sensitivity for a wider range of NPEO oligomers than APCI.[103] However, the strength of APCI is that the technique will ionize a wider range of compounds, which facilitates the development of multiresidue methods. It has been used to this effect in monitoring for other surfactants along with the APEOs,[104] although the halogenated derivatives only ionized with ESI.[105]

3.4.2.3   Organochlorine Pesticides, Polychlorinated Biphenyls, and Dibenzodioxins/Furans

This large and complex group of chlorinated compounds are considered together because they are co-extracted by the solvent systems used, and it is only in sample cleanup stages or during chromatographic determination that they can be separated. If the pesticides, or more common PCB congeners, which frequently occur at concentrations that are orders of magnitude higher than the PCDD/F are to be determined, sample cleanup may be less robust. However, for analysis of planar PCB congeners and the PCDD/F, more extensive cleanup is usually used.

For the determination of these compounds, sewage sludges have been extracted by high speed mixing with hexane, followed by cleanup with alumina/alumina silver nitrate ($AgNO_3$) to remove lipids and sulfides,[112] using microwave assisted extraction, which was compared to the more traditional Soxhlet technique,[113] both with a silica cleanup and by supercritical fluid extraction (Table 3.7). Clean up techniques are often complex, the addition of silver nitrate to material used in adsorption columns,[114] or of copper to remove sulfur[113,] are frequently used and the presence of relatively large amounts of fat may also necessitate hydrolysis with sulfuric acid.[114] A range of options for removal of sulfur using copper was investigated by Folch et al.,[115] who recommended the use of copper powder with Soxhlet extraction followed by Florisil cleanup when analyzing sludge amended soils.

The initial cleanup step is frequently followed by fractionation of pesticides and PCB on silica, which is first eluted with hexane and, subsequently, with more polar solvent. This firstly elutes the less polar PCBs (although often some of the less polar pesticides, such as p,p'-DDE are also eluted) and then the more polar pesticides.[112] The activity of the silica is controlled by heating at 300 to 500°C and then re-hydrating to a fixed percentage moisture, the control of which, along with solvent selection, is important in controlling fractionation of the compounds.[116]

Early work using GC-ECD for quantification utilized packed glass columns, for example 2 m x 3 mm packed with 1.5% OV-17 and 1.95% QF-1 on 100/120 mesh Suplecoport, which was used for quantifying Arochlor 1260, p,p'-DDE, dieldrin, and γHCH.[112] The PCB were quantified by using the last 4 major peaks in the chromatogram; however, subsequent work has used capillary columns to determine these compounds in wastewaters.[117,118] During the last 2 decades, individual PCB congeners have become available as standards, allowing for more robust quantification methods and the more common use of capillary GC has enabled separation of most congeners.[119–121]

3.4.2.4   Chlorophenoxy Herbicides and Halogenated Phenols

These compounds are acidic due to the presence of the phenolic group. In order to extract them efficiently into organic solvents, it is necessary to suppress their ionization by reducing the pH of the sample. In extracting sludges, sulfuric acid has been used to reduce the pH to 2 and ethyl acetate used to extract the chlorophenoxy herbicides (CPH).[122] Subsequent cleanup of the extract was by back-extraction into a weak aqueous alkali, which was then again acidified and the CPH extracted into dichloromethane. Similar acid-base back-extractions have also been used to determine chlorophenols; however, in this case after back extraction into alkali, the pH was adjusted to 9.9 and samples were derivatized using pentafluorobenzoyl chloride before extracting into hexane.[123] Modifications to this method, which included increasing the ionic strength of the sample to facilitate extraction and incorporation of brominated phenols, achieved detection limits of 0.05 ng l^{-1} in final effluent samples.[124]

Phenolic compounds are usually derivatized prior to determination by GC because of their polar nature. The selection of a halogenated derivatizing reagent will also allow for use of an ECD in quantification, which enhances sensitivity and selectivity. Three derivatization techniques, 2-chloroethylation, 2,2,2-trichloroethylation and pentafluorobenzylation were compared by Hill et al.,[122] who concluded that 2-chloroethylation provided the most reproducible data, however, analysis of extracted sludges by GC-ECD was not possible due to interferences and GC-MS was utilized. In all three derivatization methods tested, a cleanup step was undertaken using 5% deactivated silica.

The determination of CPH is particularly time consuming due to the complex extraction and derivatization procedure. The use of supercritical fluids, although likely to involve high capital costs for the appropriate equipment, has been shown to allow for extraction and *in-situ* derivatization of CPH from sediments and could be applied to sewage sludges.[125]

3.4.2.5 PAH

Analytical techniques for these compounds are well-documented in the literature and consist of solvent extraction; sample cleanup on adsorption columns, frequently with sulphur and fat removal steps; and finally quantification with GC or LC techniques. As recently as 1981, methods based on thin layer chromatography and quantification through visual estimation of fluorescent intensity of spots on the plates have been used.[126] However, along with the presence of compounds to be determined, high concentrations of other material are present and are likely to be extracted. This is especially the case when extracting sludges, where the removal of such co-extracted material is necessary. A number of sulphur species may be found in sludges and these are frequently co-extracted in organic solvents. Activated (through washing with nitric acid) copper powder is frequently used to clean up extracts.[76] The presence of relatively large amounts of fat and fatty acids in sludge extracts has been addressed by the use of a saponification step, utilizing potassium hydroxide and methanol.[76] It is possible to determine the PAH with other approaches, although some cleanup or fractionation is still required. They have been extracted with steam distillation and then separated from co-extracted compounds by normal-phase chromatography prior to quantification by GC-FID.[110]

TABLE 3.7
Examples of Methods Used for the Determination of Hydrophobic Xenoestrogens in Sludges

Determinants	Mass	Extraction	Separation	Detector	LOD (mg kg⁻¹)	Recovery (%)	References
PAH	1 g (wet)	Soxhlet (DCM/methanol); Saponification; activated copper; Silica /alumina	GC	FID/MS	0.03		Moreda et al.[76]
	1 g (dm) wet	Soxhlet (toluene, cyclohexane or DCM)	LC	DAD/ fluorescence		63–98	Miege et al.[127]
	1 g (dry)	Supercritical CO_2 + modifiers (5%) solvent				57–112[a]	
	0.5 g (dry)	Subcritical water (150°C) and acetonitrile cleanup	GC	MS		55–106	McGowin et al.[75]
	1 g (dry)	Supercritical CO_2 + modifiers (1–4%) solvent; Silica/alumina cleanup	GC	MS		104–125	Berset et al.[141]
	2 g (dry)	Sonication (DCM/MeOH); alumina cleanup	GC	MS	0.001–0.01	78–113	Perez et al.[142]
Chlorophenols			LC				Cass et al.[143]
	50 ml	Acidify, DCM extraction, PFBCl derivatization	GC	ECD		41–86	Buisson et al.[123]
CPH	50 ml	Acidify, ethyl acetate extraction, acid-base back extraction, 2-chloroethylation	GC	ECD/MS		34–232	Hill et al.[122]
PCB	1 g (dry)	Microwave assisted and Soxhlet (hexane/acetone) + activated Cu. Silica cleanup.	GC	MS			Dupont et al.[113]

Analyte	Sample	Extraction/cleanup		Detection	Recovery	Reference
PCB	1 g (dry)	Supercritical CO_2. Silica/alumina cleanup	GC	MS	75–106	Berset et al.[141]
PCB (planar)	Freeze dried	Soxhlet (toluene); H_2SO_4; silica/$AgNO_3$; florisil; Envi-Carb	GC	HRMS	70–113	Molina et al.[114]
Phthalates	5 ml	Blended with 15% DCM in hexane	GC	ECD (250°C)	63–90	Ziogou et al.[134]
Organophosphorus pesticides	50 ml	Alumina/$AgNO_3$ cleanup DCM/hexane (1:1) in separating funnel or blender Alumina cleanup	GC	FPD	71–81	McIntyre et al.[131]
Organochlorine pesticides		Extraction into hexane, alumina/silica cleanup	GC	ECD	85	Garcia-Gutierrez et al.[116]
γHCH; aldrin; dieldrin; endrin	20 g (wet)	Extraction into acetone/hexane, silica cleanup; GPC for enantiomer specific analysis	GC	MS ECNI		Buser et al.[144]
p,p'-DDE toxaphene	3 g (dry)	Supercritical CO_2 diatomaceous earth/florisil cleanup	GC	MS	56–121	Berset et al.[141]
PBDE	15–20 g (wet)	Acetone extraction; $NaCl/H_3PO_4$ cleanup; sulfur removal.	GC	MS ECNI	71–86	Nylund et al.[145]

dm = dry matter; [a]SFE with 5% toluene; ECNI = electron-capture negative ionization

The use of LC with fluorescence detection for determination of PAH does lead to a degree of selectivity that facilitates analysis with minimal sample cleanup. In combination with diode-array detection, which generates a characteristic UV spectra that can be used to confirm identity of the compounds, this approach was used to compare Soxhlet with the use of supercritical fluid extraction[127] and has been used by other workers.[128] It has not been possible to utilize LC-MS to determine PAH because the techniques available for ionization, ESI, and APCI are not effective with these nonpolar compounds, although the introduction of photoionization techniques may facilitate LC-MS analysis. The PAH are also amenable to determination by GC and two methods of detection, FID and MS, have been shown to give similar accuracy in determining their concentrations in raw sewage samples.[129] Flame ionization detectors are well suited to determination of PAH as specified by the U.S. EPA Method 8100. Potter et al. used this method to study their degradation during composting.[130]

3.4.2.6   Other Pesticides and Herbicides

The effectiveness of two different extraction methods for determination of the organophosphorus pesticides (OP) diazinon, malathion and parathion demonstrated that mixing with a laboratory homogenizer was more effective at extracting compounds into the solvent than simple shaking.[131] The solvents used for extraction were DCM/hexane (1:1), and alumina (5% deactivated) was used to clean up sample extracts.[131]

Subcritical water has been used to extract a range of compounds from solid samples, such as PAH and the herbicides ametryne, atrazine, carbaryl, chlorpyrifos and trifluralin from compost.[75] The use of molecularly imprinted polymers (MIP) has been utilized in conjunction with a restricted access material (RAM) loaded with C18 on the inner pore surface to determine triazine herbicides atrazine, simazine, and propazine from river water samples.[132] While these approaches have not been used for determination of these compounds in wastewater samples, they may be useful for the cleanup of more complex matrices. Recovery for atrazine was 102%, with a detection limit of 0.03 µg l^{-1} using LC-MS.

3.4.2.7   Other Compounds

Sludges, influents, and effluents have been determined for a large range of compounds, and many are classed as potential EDCs. General issues related to their analysis have been addressed by some workers,[70,133] and some more specific examples are given for sludges (Table 3.7) and influents and effluents (Table 3.8). Phthalates have been determined in sludges by solvent extraction and alumina cleanup followed by GC with ECD.[134] The use of ECD improves sensitivity by over two orders of magnitude. However, at temperatures above 300°C the response of phthalate esters falls off rapidly.[135] Aqueous (effluent) samples were extracted using C18 SPE cartridges which were subsequently eluted with a sequential extraction with 2 × 5 ml hexane (A), 2 × 5 ml DCM/hexane (4:1) (B), and 2 × 5 ml MeOH/DCM (9:1) (C). The phthalates were eluted in fraction (C) and subsequently determined by LC-MS (Table 3.8).

One major group of compounds with demonstrated effects as EDCs within the environment are the organotins. Their analysis by GC involves the requirement for derivatization to volatile hydrides, methyl, or ethyl derivatives that may be combined with an extraction step. As tin species may be methylated in some environmental situations, the use of ethylation for analysis is preferred since it facilitates the analysis of any methylated compounds present.[136] The use of sodium tetraethylborate generates ethyl derivatives. However, the sample requires buffering at ~pH 4.6 with an acetate buffer for optimum reaction.[137] Quantification of the derivatives has often used flame photometric detection (FPD) with a 610 nm cut off-filter, which monitors emission from the excited Sn-H bonds formed in hydrogen rich flames.[136] The use of a pulsed FPD has improved sensitivity for the organotins between 25 to 50 times.[138]

Sewage sludges have been extracted by SPE using C18 cartridges modified with tropolone as a complexing agent to increase recovery of the more polar, water soluble, organotin compounds with fewer and shorter alkyl groups attached to the tin atom.[136] Extracted samples were then ethylated with the Grignard reagent, ethyl magnesium bromide and determined by both GC-FPD and MS in the positive chemical ionization mode, which was shown to be more sensitive than electron impact (EI) mode.

Electron-capture negative ionization MS has been utilized for the determination of PBDE in a survey of over 100 sewage sludge samples in Sweden.[139] This technique results in the formation of bromide ions, which are monitored at m/z -79 and -81, with ammonia as the reaction gas, and has previously been used to monitor for the compounds in sediment and fish tissue.[140]

3.5 OTHER TECHNIQUES FOR QUANTIFICATION

There are two other analytical approaches used to quantify EDCs within wastewater matrices. These are enzyme immunoassay (EIA) and radioimmunoassay (RIA) techniques. Immunoassays are characterized by the following reaction:[153]

$$Ab + Ag_{free} + Ag^* \leftrightarrow AbAg_{bound} + AbAg^*_{bound}$$

where Ab = the antibody, Ag = the antigen, and Ag* = the labeled antigen.

The antigen is the analyte to be determined that will bind to the antibody (protein). The labeled antigen is linked to a marker, or in the case of RIA labeled with a radioisotope.[154] The antibodies are produced by immune systems after immunization, and may be of polyclonal origin or from a cell culture medium. The enzyme-linked immunosorbent assay (ELISA), a solid-phase enzyme immunoassay, is the most commonly used assay for pesticides and other environmental contaminants.[153] The utility and cost-effectiveness of this technique for determining a range of organic compounds is well-established in research and clinical diagnostic applications. It has also been used to detect pesticides and other toxic substances in biological and environmental matrices.[155,156] In particular, the technique has found a niche for the measurement of herbicides, with a range of commercially available test kits.[157]

TABLE 3.8
Examples of Methods Used for the Determination of Hydrophobic Xenoestrogens in Influent and Effluent Samples

Determinants	Volume[a]	Extraction	Separation	Detector	LOD (µg l⁻¹)	Recovery (%)	Reference
PAH	I 1 l	LLE (cyclohexane) alumina/silica gel clean up	GC	MS	0.1–0.5	69–127%	Bedding et al.[73,129]
	I 3 l	LLE (hexane/DCM)	LC		0.5–2.0		Blanchard et al.[146]
(also PCB, benzenes, atraxine)	20 d in-situ	Triolein semi-permeable membrane device (SMPD) cyclohexane extraction, silica cleanup	GC	MS			Wang et al.[147]
Chlorophenols	E 500 ml	acidify, DCM extraction, PFBCl derivatization	GC	ECD		60–107	Buisson et al.[123]
	E 500 ml	acidify (pH2) +12g NaCl, DCM extraction, PFBCl	GC	ECD		73–83	Booth and Lester[124]
	E 200 ml	0.45 µm filter, SPE (C18), sequential extraction	HPLC	APCI MS	0.37	90	Castillo et al.[78]
	E 20 ml	solid phase microextraction (SPME)	GC	MS			dos Santos et al.[148]
PCB	I 3 l	LLE (hexane/DCM); sulfuric acid cleanup. Cu and Hg to remove sulfur	GC	ECD			Blanchard et al.[146]
PCDD/F	E 500 ml	addition of NaOH. liquid liquid (hexane/DCM); alumina cleanup	GC	MS/MS (ion trap)	0.01–0.05		Kuchler and Brzezinski[149]
Phthalates	E 200 ml	0.45 µm filter, SPE (C18), sequential extraction	HPLC	APCI MS	0.06–0.1	69–71	Castillo et al.[78]

Phenyl ureas	E 1 l	SPE (Carbopack B), elution with methanol/DCM, derivatization with iodoethane	GC	MS			Gerecke et al.[150]
Bisphenol A	IE 1 l	Sample pH adjusted to 2.5, SPE (ENV+)	GC	MS			Körner et al.[29]
	IE 250 ml	1.2 μm filter, SPE (C18) derivatized (PFPA)	GC	MS	103–105	0.4 pg (injected)	Lee and Peart[69]
	E 1–10 l	47 mm; C18 Speeddisks ethanol toluene eluent	GC	HRMS			Pujadas et al.[151]
musk xylene	IE 0.5–1 l	C18 Speeddisks, DCM eluent	GC	MS	77–127		Simonich et al.[152]

aSample type, I = influent; E = final effluent; PFBCl = pentafluorobenzoyl chloride; PFPA = Pentafluoropropionic acid

A range of assays has been reviewed; however, their application has predominantly been in the analysis of relatively clean matrices, such as river, ground, and drinking waters. Test kits covered in this work were available for some compounds implicated as EDCs: aldrin, dieldrin, endosulphan, trifluralin, and triazines including atrazine.[153] Application of the technique for the analysis of contaminants in wastewater has included determination of pentachlorophenol and PAH[158]; benzene; toluene; ethylbenzene; and o-, m-, and p-xylene.[159] More recently, an assay has been developed for 4-nitrophenol.[160] The ELISA for 4-nitrophenol exhibited cross-reactivity with nonylphenol at <0.01%, and it would appear likely that development of specific assays for other xenoestrogens is possible. Test kits for the determination of 17β-estradiol and 17α-ethinyl estradiol are also available and have been utilized to analyze wastewaters and compared to GC-MS/MS analysis.[87] Sample preparation prior to the ELISA included preconcentration (SPE) by 225 to 1,500 times, followed by HPLC fractionation. In comparison to the GC-MS/MS technique, it was concluded that the ELISA techniques had lower detection limits, required smaller concentration factors, and were less susceptible to matrix interference. Confirmation with GC-MS/MS was still recommended.

Disadvantages of the enzyme immunoassay techniques include cross-reacting compounds and matrix effects. Another drawback is the fact that test kits are, in general, designed for single compounds that limit multi-residue analyses. This last issue was demonstrated in Huang and Sedlak's work,[87] where two assays from different manufacturers were used for the determination of 17β-estradiol and 17α-ethinyl estradiol. False positives are caused by interactions of other compounds with the antibody; these may be structurally similar compounds and metabolites, or co-extracted compounds. Although such effects can be minimized by sample preparation (such as SPE or liquid–liquid extraction and further cleanup),[87,157] such effort reduces the ease of use and cost effectiveness of such determinations. However, it has been concluded that the use of an ELISA assay prescreen for determining atrazine at concentrations above 1 μg l^{-1}, reduced the sample load by 71% with 2.4% false negative and <1% false positive results.[161]

3.6 APPLICATION OF METHODS

The use of a combination of assays and chemical techniques has been utilized in studies on the effluents from STW to both confirm that they exhibit estrogenic activity and to identify the compounds that were responsible. Such toxicity identification and evaluation (TIE) was undertaken in the United Kingdom, following the observation that sewage effluent was estrogenic to fish, in an attempt to identify the causative agents.[70] Such approaches have been used by a number of workers to evaluate industrial[162] and sewage effluents.[163] The general scheme utilized consists of fractionation of the sample and the use of a test (in this case a yeast screen assay) to identify the fractions of the sample that contain the toxicity; the active fractions were then analyzed using either GC-MS or HPLC-MS to enable elucidation of the compounds present (Figure 3.2).

The work undertaken by Desbrow et al.[70] identified that over 80% of estrogenic activity was associated with midpolar organic material retained on a C18 SPE

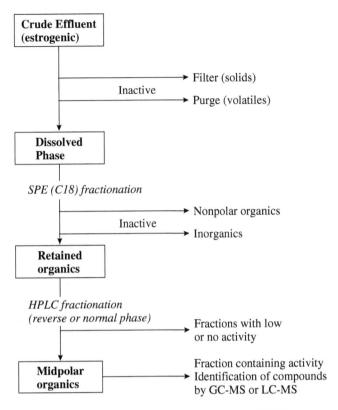

FIGURE 3.2   Schematic representation of the approach used for TIE for compounds responsible for estrogenic effects.

cartridge. Further fractionation by HPLC and subsequent analysis by GC-MS identified steroid estrogens as the active components. The work undertaken on analyzing the contaminants present in U.K. effluents is significant because it was backed up with *in vivo* data. These data were obtained from exposing male rainbow trout (*Oncorhynchus mykiss*) and adult roach (*Rutilus rutilus*) to the contaminants, estradiol and octylphenol, and confirmed that these compounds induce vitellogenin synthesis.[54] A similar chemical approach has been used to identify components in effluents from STW in Michigan.[164] Again, samples were collected onto SPE disks and the organic compounds eluted, with subsequent normal phase HPLC fractionation. The least polar fraction did not exhibit estrogenic activity, while a midpolar fraction containing nonylphenol and octylphenol did demonstrate some activity. The most polar fraction, which contained steroid estrogens, was the most active. The actual conditions used by different workers to elute SPE cartridges and their descriptions of fractions as polar, midpolar, etc. are not standardized and may at times be misleading if the full methodology is not available. Other TIE tests, with endpoints for toxicity (the echinoderm fertilization assay) rather than for estrogenic effects, have found other contaminants, such as copper,[163] in effluents and identified them as the toxic component. An adaptation of the yeast assay has also been used to demonstrate that extracts from sewage sludges are estrogenic. However, the compounds responsible were not identified.[43]

Human exposure to xenobiotic estrogens has been evaluated by extraction of fat tissue from patients surgically treated for breast cancer and a matching group of controls.[165] The method involved extraction, cleanup on alumina, followed by normal-phase liquid chromatography and subsequent GC analysis of three fractions. Estrogenic activity in the fractions was measured using the E-screen, and the greatest activity was observed in the fraction containing lipophilic organochlorine pesticides.

3.7 CONCLUSION

Despite the activities at national and international levels, including the OECD, all issues related to testing for EDCs are not being investigated. In particular, it was noted that certain areas required development in addition to that being undertaken by the Endocrine Disrupter Screening Program in the United States.[11] These included development of tests for invertebrates, which should be based on greater understanding of their endocrinology; the use of alternative techniques, such as SAR and new *in vitro* tests that focus on models other than receptor binding; and the use of genomics, transcription profiling, proteomics, and metabonomics.[166] These are rapidly developing technologies that enable researchers to understand biological events at the genetic level (genomics) and their subsequent expression in organisms. Expression can be studied at the stage of transfer of genetic information (transcriptomics), the stage of formation of proteins (proteomics), or the determination of metabolites resulting from the activities of those proteins (metabonomics). It is not that these techniques are new, but that the ability to apply them in such a way that massive amounts of data can be generated, characterizing changes in the presence and amounts of potentially thousands of biomolecules simultaneously, is now possible.[167]

There are few methods to determine the occurrence and significance of exposure pathways, for a multitude of compounds in more complex matrices such as sediments.[11] Furthermore, there would be little doubt that sewage influents, effluents, and sludges would fall within the description of complex matrices. The development of analytical methods for difficult matrices, which will then allow the understanding of exposure pathways to allow for risk assessment, has been identified as a research priority by a number of workers.[168] As analytical work continues, it appears likely that some form of assay-based screening, such as ELISA[87] or other techniques, may be useful tools in screening large numbers of samples prior to more complex and expensive analysis by chromatographic methods with mass spectrometric detectors for confirmation of compound identity.

REFERENCES

1. Keith, L.H., Jones, T.L., and Needham, L.L., Eds., *Analysis of Environmental Endocrine Disruptors*, American Chemical Society, 2000.
2. Sonnenschein, C. and Soto, A.M., An updated review of environmental estrogen and androgen mimics and antagonists, *J. Steroid Biochem. Mol. Biol.*, 65, 143, 1998.

3. Beresford, N., Routledge, E.J., Harris, C.A., and Sumpter, J.P., Issues arising when interpreting results from an in vitro assay for estrogenic activity, *Toxicol. Appl. Pharmacol.*, 162, 22, 2000.

4. Andersen, H.R., Andersson, A.M., Arnold, S.F., Autrup, H., Barfoed, M., Beresford, N.A., Bjerregaard, P., Christiansen, L.B., Gissel, B., Hummel, R., Jorgensen, E.B., Korsgaard, B., Le Guevel, R., Leffers, H., McLachlan, J., Moller, A., Nielsen, J.B., Olea, N., Oles-Karasko, A., Pakdel, F., Pedersen, K.L., Perez, P., Skakkeboek, N.E., Sonnenschein, C., Soto, A.M., Sumpter, J.P., Thorpe, S.M., and Grandjean, P., Comparison of short-term estrogenicity tests for identification of hormone-disrupting chemicals, *Environ. Health Perspect.*, 107, 89, 1999.

5. Baker, V.A., Endocrine disrupters: testing strategies to assess human hazard, *Toxicol. Vitro*, 15, 413, 2001.

6. Combes, R.D., Endocrine disruptors: a critical review of *in vitro* and *in vivo* testing strategies for assessing their toxic hazard to humans, *Altern. Lab. Anim.*, 28, 81, 2000.

7. Zacharewski, T., *In vitro* bioassays for assessing estrogenic substances, *Environ. Sci. Technol.*, 31, 613, 1997.

8. Korach, K.S. and McLachlan, J.A., Techniques for detection of estrogenicity, *Environ. Health Perspect.*, 103, 5, 1995.

9. Huet, M.C., OECD activity on endocrine disrupters test guidelines development, *Ecotoxicology*, 9, 77, 2000.

10. Holmes, P., Humphrey, C., and Scullion, M., Appraisal of test methods for sex-hormone disrupting chemicals capable of affecting the reproductive process, http://www.oecd.org/ehs/test/monos.htm, 1998.

11. European Commission, European Workshop on Endocrine Disruptors, workshop report, *European Workshop on Endocrine Disruptors Aronsborg, Sweden*, European Commission, 2001.

12. European Centre for Ecotoxicology and Toxicology of Chemicals, Environmental Oestrogens: A Compendium of Test Methods, Document No. 33, 1996.

13. Ashby, J., Houthoff, E., Kennedy, S.J., Stevens, J., Bars, R., Jekat, F.W., Campbell, P., VanMiller, J., Carpanini, F. M., and Randall, G.L.P., The challenge posed by endocrine-disrupting chemicals, *Environ. Health Perspect.*, 105, 164, 1997.

14. Gray, L.E., Kelce, W.R., Wiese, T., Tyl, R., Gaido, K., Cook, J., Klinefelter, G., Desaulniers, D., Wilson, E., Zacharewski, T., Waller, C., Foster, P., Laskey, J., Reel, J., Giesy, J., Laws, S., McLachlan, J., Breslin, W., Cooper, R., DiGiulio, R., Johnson, R., Purdy, R., Mihaich, E., Safe, S., Sonnenschein, C., Welshons, W., Miller, R., McMaster, S., and Colborn, T., Endocrine screening methods workshop report: detection of estrogenic and androgenic hormonal and antihormonal activity for chemicals that act via receptor or steroidogenic enzyme mechanisms, *Reprod. Toxicol.*, 11, 719, 1997.

15. Shelby, M.D., Newbold, R.R., Tully, D.B., Chae, K., and Davis, V.L., Assessing environmental chemicals for estrogenicity using a combination of *in vitro* and *in vivo* assays, *Environ. Health Perspect.*, 104, 1296, 1996.

16. Matthews, J. and Zacharewski, T., Differential binding affinities of PCBs, HO-PCBs, and aroclors with recombinant human, rainbow trout (*Oncorhynchus mykiss*), and green anole (Anolis carolinensis) estrogen receptors, using a semi-high throughput competitive binding assay, *Toxicol. Sci.*, 53, 326, 2000.

17. Taylor, C.M., Blanchard, B., and Zava, D.T., A simple method to determine whole cell uptake of rediolabelled oestrogen and progesterone and their subcellular localisation in breast cancer cell lines in monolayer culture, *J. Steroid Biochem.*, 20, 1083, 1984.

18. Kelce, W.R., Monosson, E., Gamesik, M.P., Laws, S.C., and Gray, J.L.E., Environmental hormone disrupters: evidence that vinclozolin developmental toxicity is mediated by antiandrogenic metabolites, *Toxicol. Appl. Pharmacol.*, 126, 275, 1994.

19. Kelce, W.R., Stone, C., Laws, S., Gray, J.L.E., Kemppainen, J., and Wilson, E., Persistent DDT metabolite p,p'-DDE is a potent androgen receptor agonist, *Nature*, 375, 581, 1995.

20. Fertuck, K.C., Kumar, S., Sikka, H.C., Matthews, J.B., and Zacharewski, T.R., Interaction of PAH-related compounds with the alpha and beta isoforms of the estrogen receptor, *Toxicol. Lett.*, 121, 167, 2001.

21. Soto, A.M., Sonnenschein, C., Chung, K.L., Fernandez, M.F., Olea, N., and Serrano, F.O., The E-Screen assay as a tool to identify estrogens: an update on estrogenic environmental-pollutants, *Environ. Health Perspect.*, 103 (Suppl. 7), 113, 1995.

22. Arnold, S.F., Klotz, D.M., Collins, B.M., Vonier, P.M., Guillette L.J., Jr., and McLachlan, J.A., Synergistic activation of estrogen receptor with combinations of environmental chemicals, *Science*, 272, 1489, 1996.

23. White, R., Jobling, S., Hoare, S.A., Sumpter, J.P., and Parker, M.G., Environmentally persistent alkylphenolic compounds are estrogenic, *Endocrinology*, 135, 175, 1994.

24. Jobling, S., Reynolds, T., White, R., Parker, M.G., and Sumpter, J.P., A variety of environmentally persistent chemicals, including some phthalate plasticizers, are weakly estrogenic, *Environ. Health Perspect.*, 103, 582, 1995.

25. Gaido, K., Dohme, L., Wang, F., Chen, I., Blankvoort, B., Ramamoorthy, K., and Safe, S., Comparative estrogenic activity of wine extracts and organochlorine pesticide residues in food, *Environ. Health Perspect.*, 106, 1347, 1998.

26. Soto, A.M., Lin, T.M., Justicia, H., Silvia, R.M., and Sonnenschein, C., An "in culture" bioassay to assess the estrogenicity of xenobiotics (E-screen), in *Advances in Modern Environmental Toxicology: Chemically-induced Alterations in Sexual and Functional Development: The Wildlife/Human Connection*, Colborn, T., and Clement, C., Eds., Princeton Scientific, Princeton, NJ, 1992.

27. Körner, W., Hanf, V., Schuller, W., Bartsch, H., Zwirner, M., and Hagenmaier, H., Validation and application of a rapid *in vitro* assay for assessing the estrogenic potency of halogenated phenolic chemicals, *Chemosphere*, 37, 2395, 1998.

28. Desaulniers, D., Leingartner, K., Zacharewski, T., and Foster, W.G., Optimization of an MCF7-E3 cell proliferation assay and effects of environmental pollutants and industrial chemicals, *Toxicol. Vitro*, 12, 409, 1998.

29. Körner, W., Hanf, V., Schuller, W., Kempter, C., Metzger, J., and Hagenmaier, H., Development of a sensitive E-screen assay for quantitative analysis of estrogenic activity in municipal sewage plant effluents, *Sci. Total Environ.*, 225, 33, 1999.

30. Villalobos, M., Olea, N., Brotons, J.A., Oleaserrano, M.F., Dealmodovar, J. M. R., and Pedraza, V., The E-Screen Assay: a Comparison of Different MCF-7 Cell Stocks, *Environ. Health Perspect.*, 103, 844, 1995.

31. Odum, J., Tittensor, S., and Ashby, J., Limitations of the MCF-7 cell proliferation assay for detecting xenobiotic oestrogens, *Toxicol. Vitro*, 12, 273, 1998.

32. Katzenellenbogen, B.S., Kendra, K.L., Norman, M.J., and Berthois, Y., Proliferation, hormonal responsiveness and estrogen receptor content of MCF-7 human breast cancer cells grown in the short term and long term absence of estrogens, *Cancer Res.*, 47, 4355, 1987.

33. Schlumpf, M., Cotton, B., Conscience, M., Haller, V., Steinmann, B., and Lichtensteiger, W., *In vitro* and *in vivo* estrogenicity of UV screens, *Environ. Health Perspect.*, 109, 239, 2001.

34. Olea, N., Pulgar, R., Perez, P., Olea Serrano, F., Rivas, A., Novillo Fertrell, A., Pedraza, V., Soto, A.M., and Sonnenschein, C., Estrogenicity of resin-based composites and sealants used in dentistry, *Environ. Health Perspect.*, 104, 298, 1996.

35. O'Connor, J.C., Frame, S.R., Biegel, L.B., Cook, J.C., and Davis, L.G., Sensitivity of a tier I screening battery compared to an in utero exposure for detecting the estrogen receptor agonist 17 beta-estradiol, *Toxicol. Sci.*, 44, 169, 1998.

36. Coldham, N.G., Dave, M., Sivapathasundaram, S., McDonnell, D.P., Connor, C., and Sauer, M.J., Evaluation of a recombinant yeast cell estrogen screening assay, *Environ. Health Perspect.*, 105, 734, 1997.

37. Fenner-Crisp, P.A., Maciorowski, A.F., and Timm, G.E., The endocrine disruptor screening program developed by the US Environmental Protection Agency, *Ecotoxicology*, 9, 85, 2000.

38. Perez, P., Pulgar, R., Olea-Serrano, F., Villalobos, M., Rivas, A., Metzler, M., Pedraza, V., and Olea, N., The estrogenicity of bisphenol A-related diphenylalkanes with various substituents at the central carbon and the hydroxy groups, *Environ. Health Perspect.*, 106, 167, 1998.

39. Soto, A.M., Fernandez, M.F., Luizzi, M.F., Karasko, A.S.O., and Sonnenschein, C., Developing a marker of exposure to xenoestrogen mixtures in human serum, *Environ. Health Perspect.*, 105 (Suppl. 3), 647, 1997.

40. Routledge, E.J. and Sumpter, J.P., Estrogenic activity of surfactants and some of their degradation products assessed using a recombinant yeast screen, *Environ. Toxicol. Chem.*, 15, 241, 1996.

41. Gaido, K.W., Leonard, L.S., Lovell, S., Gould, J.C., Babai, D., Portier, C.J., and McDonnell, D.P., Evaluation of chemicals with endocrine modulating activity in a yeast-based steroid hormone receptor gene transcription assay, *Toxicol. Appl. Pharmacol.*, 143, 205, 1997.

42. Klein, K.O., Baron, J., Colli, M.J., McDonnell, D.P., and Cutler G.B., Jr., Estrogen levels in childhood determined by an ultrasensitive recombinant cell bioassay, *J. Clin. Inv.*, 94, 2475, 1994.

43. Rehmann, K., Schramm, K.W., and Kettrup, A.A., Applicability of a yeast oestrogen screen for the detection of oestrogen-like activities in environmental samples, *Chemosphere*, 38, 3303, 1999.

44. Legler, J., van den Brink, C.E., Brouwer, A., Murk, A.J., van der Saag, P.T., Vethaak, A.D., and van der Burg, P., Development of a stably transfected estrogen receptor-mediated luciferase reporter gene assay in the human T47D breast cancer cell line, *Toxicol. Sci.*, 48, 55, 1999.

45. Hoogenboom, L.A.P., Hamers, A.R.M., and Bovee, T.F.H., Bioassays for the detection of growth-promoting agents, veterinary drugs and environmental contaminants in food, *Analyst*, 124, 79, 1999.

46. Pons, M., Gagne, D., Nicolas, J.C., and Mehtali, M., A new cellular-model of response to estrogens: a bioluminescent test to characterize (anti)estrogen molecules, *Biotechniques*, 9, 450, 1990.

47. Gagne, D., Balaguer, P., Demirpence, E., Chabret, C., Trousse, F., Nicolas, J.C., and Pons, M., Stable luciferase transfected cells for studying steroid-receptor biological-activity, *J. Biolumin. Chemilumin.*, 9, 201, 1994.

48. Joyeux, A., Balaguer, P., Germain, P., Boussioux, A.M., Pons, M., and Nicolas, J.C., Engineered cell lines as a tool for monitoring biological activity of hormone analogs, *Anal. Biochem.*, 249, 119, 1997.

49. Balaguer, P., Francois, F., Comunale, F., Fenet, H., Boussioux, A.M., Pons, M., Nicolas, J.C., and Casellas, C., Reporter cell lines to study the estrogenic effects of xenoestrogens, *Sci. Total Environ.*, 233, 47, 1999.

50. Balaguer, P., Fenet, H., Georget, V., Comunale, F., Terouanne, B., Gilbin, R., Gomez, E., Boussioux, A.M., Sultan, C., Pons, M., Nicolas, J.C., and Casellas, C., Reporter cell lines to monitor steroid and antisteroid potential of environmental samples, *Ecotoxicology*, 9, 105, 2000.

51. Silva, E., Rajapakse, N., and Skortenkamp, A., Something from "nothing": eight weak estrogenic chemicals combined at concentrations below NOECs produce significant mixture effects, *Environ. Sci. Technol.*, 36, 1751, 2002.

52. Odum, J., Lefevre, P.A., Tittensor, S., Paton, D., Routledge, E.J., Beresford, N.A., Sumpter, J.P., and Ashby, J., The rodent uterotrophic assay: critical protocol features, studies with nonyl phenols, and comparison with a yeast estrogenicity assay, *Regul. Toxicol. Pharmacol.*, 25, 176, 1997.

53. European Centre for Ecotoxicology and Toxicology of Chemicals, Screening and Testing Methods for Ecotoxicological Effects of Potential Endocrine Disrupters: Response to the EDSTAC. Recommendations and a Proposed Alternative Approach, Document No. 39, 1999.

54. Routledge, E. J., Sheahan, D., Desbrow, C., Brighty, G.C., Waldock, M., and Sumpter, J. P., Identification of estrogenic chemicals in STW effluent. 2. In vivo responses in trout and roach, *Environ. Sci. Technol. 32*, 1559, 1998.

55. EDSTAC, Report to the Endocrine Disrupter Screening and Testing Assessment Committee, Final Report, U.S. Environmental Protection Agency, EPA/743/R-98/003, Washington, DC, August 1998.

56. U.S. Environmental Protection Agency (USEPA), Endocrine Disrupter Screening Program, Report to Congress, August 2000.

57. Cook, J.C., Kaplan, A.M., Davis, L.G., and O'Connor, J.C., Development of a tier I screening battery for detecting endocrine-active compounds (EACs), *Regul. Toxicol. Pharmacol.*, 26, 60, 1997.

58. O'Connor, J.C., Cook, J.C., Slone, T.W., Makovec, G.T., Frame, S.R., and Davis, L.G., An ongoing validation of a Tier I screening battery for detecting endocrine-active compounds (EACs), *Toxicol. Sci.*, 46, 45, 1998.

59. Biegel, L.B., Flaws, J.A., Hirshfield, A.N., O'Connor, J.C., Elliott, G.S., Ladics, G.S., Silbergeld, E.K., Van Pelt, C.S., Hurtt, M.E., Cook, J.C., and Frame, S.R., 90-day feeding and one-generation reproduction study in Crl:CD BR rats with 17 beta-estradiol, *Toxicol. Sci.*, 44, 116, 1998.

60. Biegel, L.B., Cook, J.C., Hurtt, M.E., and O'Connor, J.C., Effects of 17 beta-estradiol on serum hormone concentrations and estrous cycle in female Crl:CD BR rats: effects on parental and first generation rats, *Toxicol. Sci.*, 44, 143, 1998.

61. Cook, J.C., Johnson, L., O'Connor, J.C., Biegel, L.B., Krams, C.H., Frame, S.R., and Hurtt, M. E., Effects of dietary 17 beta-estradiol exposure on serum hormone concentrations and testicular parameters in male Crl:CD BR rats, *Toxicol. Sci.*, 44, 155, 1998.

62. Hutchinson, T.H., Brown, R., Brugger, K.E., Campbell, P.M., Holt, M., Lange, R., McCahon, P., Tattersfield, L.J., and van Egmond, R., Ecological risk assessment of endocrine disruptors, *Environ. Health Perspect.*, 108, 1007, 2000.

63. Lopez de Alda, M.J. and Barcelo, D., Review of analytical methods for the determination of estrogens and progestogens in waste waters, *Fresenius J. Anal. Chem.*, 371, 437, 2001.

64. Grob, R.L., Chromatographic techniques for pollution analysis, *Environ. Monit. Assess.*, 19, 1, 1991.
65. McIntyre, A.E. and Lester, J.N., Organic contaminants in the aquatic environment. IV. analytical techniques, *Sci. Total Environ.*, 27, 201, 1983.
66. Blau, K. and Halket, J.M., *Handbook of Derivatives for Chromatography*, 2nd ed., John Wiley & Sons, Chichester, London, 1993.
67. Snyder, S.A., Keith, T.L., Verbrugge, D.A., Snyder, E.M., Gross, T.S., Kannan, K., and Giesy, J.P., Analytical methods for detection of selected estrogenic compounds in aqueous mixtures, *Environ. Sci. Technol.*, 33, 2814, 1999.
68. Baronti, C., Curini, R., D'Ascenzo, G., Di Corcia, A., Gentili, A., and Samperi, R., Monitoring natural and synthetic estrogens at activated sludge sewage treatment plants and in a receiving river water, *Environ. Sci. Technol.*, 34, 5059, 2000.
69. Lee, H.B. and Peart, T.E., Bisphenol A contamination in Canadian municipal and industrial wastewater and sludge samples, *Water Qual. Res. J. Canada*, 35, 283, 2000.
70. Desbrow, C., Routledge, E.J., Brighty, G.C., Sumpter, J.P., and Waldock, M., Identification of estrogenic chemicals in STW effluent. I. Chemical fractionation and *in vitro* biological screening, *Environ. Sci. Technol.*, 32, 1549, 1998.
71. Lopez de Alda, M.J. and Barcelo, D., Determination of steroid sex hormones and related synthetic compounds considered as endocrine disrupters in water by liquid chromatography-diode array detection-mass spectrometry, *J. Chromatogr.*, A 892, 391, 2000.
72. Siegener, R. and Chen, R.F., Detection of pharmaceuticals entering Boston Harbor, in: *Analysis of Environmental Endocrine Disruptors*, Keith, L.K., Jones, T.L., and Needham, L.L., Eds., American Chemical Society, Washington, DC, 2000.
73. Bedding, N.D., McIntyre, A.E., Lester, J.N., and Perry, R., Analysis of waste-waters for polynuclear aromatic-hydrocarbons. II. Errors, sampling, and storage, *J. Chromatogr. Sci.*, 26, 606, 1988.
74. McIntyre, A.E., Lester, J.N., and Perry, R., Persistence of organo-phosphorus insecticides in sewage sludges, *Environ. Technol. Lett.*, 2, 111, 1981.
75. McGowin, A.E., Adom, K.K., and Obubuafo, A.K., Screening of compost for PAHs and pesticides using static subcritical water extraction, *Chemosphere*, 45, 857, 2001.
76. Moreda, J.M., Arranz, A., De Betono, S.F., Cid, A., and Arranz, J.F., Chromatographic determination of aliphatic hydrocarbons and polyaromatic hydrocarbons (PAHs) in a sewage sludge, *Sci. Total Environ.*, 220, 33, 1998.
77. Lega, R., Ladwig, G., Meresz, O., Clement, R.E., Crawford, G., Salemi, R., and Jones, Y., Quantitative determination of organic priority pollutants in sewage sludge by GC/MS, *Chemosphere*, 34, 1705, 1997.
78. Castillo, M., Alonso, M.C., Riu, J., and Barcelo, D., Identification of polar, ionic, and highly water soluble organic pollutants in untreated industrial wastewaters, *Environ. Sci. Technol.*, 33, 1300, 1999.
79. Ternes, T., Andersen, H., Gilberg, D., and Bonerz, M., Determination of Estrogens in Sludge and Sediments by Liquid Extraction and GC/MS/MS, *Anal. Chem.*, 74, 3498, 2002.
80. Lopez de Alda, M.J. and Barcelo, D., Use of solid-phase extraction in various of its modalities for sample preparation in the determination of estrogens and progestogens in sediment and water, *J. Chromatogr.*, A 938, 145, 2001.
81. Jayasinghe, L.Y., Marriott, P.J., Carpenter, P.D., and Nichols, P.D., Application of pentafluorophenyldimethylsilyl derivatization for gas chromatography-electron-capture detection of supercritically extracted sterols, *J. Chromatogr.*, A 809, 109, 1998.

82. Xiao, X.Y., McCalley, D.V., and McEvoy, J., Analysis of estrogens in river water and effluents using solid-phase extraction and gas chromatography-negative chemical ionisation mass spectrometry of the pentafluorobenzoyl derivatives, *J. Chromatogr.*, A 923, 195, 2001.

83. Singh, G., Gutierrez, A., Xu, K., and Blair, I.A., Liquid chromatographic/electron capture atmospheric pressure chemical ionisation/mass spectrometry: analysis of pentafluorobenzoyl derivatives of biomolecules and drugs in the attomole range, *Anal. Chem.*, 72, 3007, 2000.

84. Segmuller, B.E., Armstrong, B.L., Dunphy, R., and Oyler, A.R., Identification of autooxidation and photodegradation products of ethinylestradiol by on-line HPLC-NMR and HPLC-MS, *J. Pharm. Biomed. Anal.*, 23, 927, 2000.

85. Adler, P., Steger-Hartmann, T., and Kalbfus, W., Distribution of natural and synthetic estrogenic steroid hormones in water samples from Southern and Middle Germany, *Acta Hydrochim. Hydrobiol.*, 29, 227, 2001.

86. Gomes, R.L., Brookes, J., Birkett, J.W., and Lester, J.N., Solid phase extraction and direct determination of free and conjugated steroid estrogens by liquid chromatography/mass spectrometry in aqueous matrices, *Int. J. Environ. Anal. Chem.*, in press.

87. Huang, C.H. and Sedlak, D.L., Analysis of estrogenic hormones in municipal wastewater effluent and surface water using enzyme-linked immunosorbent assay and gas chromatography/tandem mass spectrometry, *Environ. Toxicol. Chem.*, 20, 133, 2001.

88. Lopez de Alda, M.J. and Barcelo, D., Determination of steroid sex hormones and related synthetic compounds considered as endocrine disrupters in water by fully automated on-line solid-phase extraction-liquid chromatography-diode array detection, *J. Chromatogr.*, A 911, 203, 2001.

89. Vader, J.S., van Ginkel, C.G., Sperling, F.M.G.M., de Jong, G., de Boer, W., de Graaf, J.S., van der Most, M., and Stokman, P.G.W., Degradation of ethinyl estradiol by nitrifying activated sludge, *Chemosphere*, 41, 1239, 2000.

90. Lagana, A., Bacaloni, A., Fago, G., and Marino, A., Trace analysis of estrogenic chemicals in sewage effluent using liquid chromatography combined with tandem mass spectrometry, *Rapid Commun. Mass Spectrom.*, 14, 401, 2000.

91. Croley, T.R., Hughes, R.J., Koenig, B.G., Metcalfe, C.D., and March, R.E., Mass spectrometry applied to the analysis of estrogens in the environment, *Rapid Commun. Mass Spectrom. 14*, 1087, 2000.

92. Kuch, H.M. and Ballschmiter, K., Determination of endogenous and exogenous estrogens in effluents from sewage treatment plants at the ng/L-level, *Fresenius J. Anal. Chem.*, 366, 392, 2000.

93. Kelly, C., Analysis of steroids in environmental water samples using solid-phase extraction and ion-trap gas chromatography-mass spectrometry and gas chromatography-tandem mass spectrometry, *J. Chromatogr.*, A 872, 309, 2000.

94. Johnson, A.C., Belfroid, A., and Di Corcia, A., Estimating steroid oestrogen inputs into activated sludge treatment works and observations on their removal from the effluent, *Sci. Total Environ.*, 256, 163, 2000.

95. Ternes, T.A., Stumpf, M., Mueller, J., Haberer, K., Wilken, R.D., and Servos, M., Behavior and occurrence of estrogens in municipal sewage treatment plants. I. Investigations in Germany, Canada and Brazil, *Sci. Total Environ.*, 225, 81, 1999.

96. Larsson, D.G.J., Adolfsson-Erici, M., Parkkonen, J., Pettersson, M., Berg, A.H., Olsson, P.E., and Forlin, L., Ethinyloestradiol — an undesired fish contraceptive?, *Aquat. Toxicol.*, 45, 91, 1999.

97. Belfroid, A.C., Van der Horst, A., Vethaak, A.D., Schafer, A.J., Rijs, G.B.J., Wegener, J., and Cofino, W.P., Analysis and occurrence of estrogenic hormones and their glucuronides in surface water and waste water in The Netherlands, *Sci. Total Environ.*, 225, 101, 1999.
98. Petrovic, M. and Barcelo, D., Analysis of ethoxylated nonionic surfactants and their metabolites by liquid chromatography/atmospheric pressure ionization mass spectrometry, *J. Mass Spectrom.*, 36, 1173, 2001.
99. Kubeck, E. and Naylor, C.G., Trace analysis of alkylphenol ethoxylates, *J. Am. Oil Chem. Soc.*, 67, 400, 1990.
100. Ahel, M. and Giger, W., Determination of alkylphenols and alkylphenol mon- and diethoxylates in environmental samples by high-performance liquid-chromatography, *Anal. Chem.*, 57, 1577, 1985.
101. Ahel, M. and Giger, W., Determination of nonionic surfactants of the alkylphenol polyethoxylate type by high-performance liquid chromatography, *Anal. Chem.*, 57, 2584, 1985.
102. Zhou, C., Bahr, A., and Schwedt, G., Separation and determination of non-ionic surfactants of the nonylphenol polyglycol ether type by liquid chromatography, *Anal. Chim. Acta.*, 236, 273, 1990.
103. Petrovic, M. and Barceló, D., Determination of anionic and nonionic surfactants, their degradation products, and endocrine-disrupting compounds in sewage sludge by liquid chromatography/mass spectrometry, *Anal. Chem.*, 72, 4560, 2000.
104. Castillo, M., Ventura, F., and Barcelo, D., Sequential solid phase extraction protocol followed by liquid chromatography-atmospheric pressure chemical ionization-mass spectrometry for the trace determination of non ionic polyethoxylated surfactants in tannery wastewaters, *Waste Manage.*, 19, 101, 1999.
105. Petrovic, M., Diaz, A., Ventura, F., and Barcelo, D., Simultaneous determination of halogenated derivatives of alkylphenol ethoxylates and their metabolites in sludges, river sediments, and surface, drinking, and wastewaters by liquid chromatography-mass spectrometry, *Anal. Chem.*, 73, 5886, 2001.
106. Ahel, M., Giger, W., Molnar, E., and Ibric, S., Determination of nonylphenol polyethoxylates and their lipophilic metabolites in sewage effluents by normal-phase high-performance liquid chromatography and fluorescence detection, *Croat. Chem. Acta.*, 73, 209, 2000.
107. Castillo, M., Martínez, E., Ginebreda, A., Tirapu, A., and Barceló, D., Determination of non-ionic surfactants and polar degradation products in influent and effluent water samples and sludges of sewage treatment plants by a generic solid-phase extraction protocol, *Analyst*, 125, 1733, 2000.
108. Di Corcia, A., Cavallo, R., Crescenzi, C., and Nazzari, M., Occurrence and abundance of dicarboxylated metabolites of nonylphenol polyethoxylate surfactants in treated sewages, *Environ. Sci. Technol.*, 34, 3914, 2000.
109. Lin, J.G., Arunkumar, R., and Liu, C.H., Efficiency of supercritical fluid extraction for determining 4-nonylphenol in municipal sewage sludge, *J. Chromatogr.*, A 840, 71, 1999.
110. Sweetman, A.J., Development and application of a multi-residue analytical method for the determination of N-alkanes, linear alkylbenzenes, polynuclear aromatic-hydrocarbons and 4-nonylphenol in digested sewage sludges, *Water Res.*, 28, 343, 1994.
111. Planas, C., Guadayol, J.M., Droguet, M., Escalas, A., Rivera, J., and Caixach, J., Degradation of polyethoxylated nonylphenols in a sewage treatment plant: quantitative analysis by isotopic dilution-HRGC/MS, *Water Res.*, 36, 982, 2002.

112. McIntyre, A.E., Perry, R., and Lester, J.N., Development of a method for the analysis of polychlorinated biphenyls and organochlorine insecticides in sewage sludges, *Environ. Technol. Lett.*, 1, 157, 1980.

113. Dupont, G., Delteil, C., Camel, V., and Bermond, A., The determination of polychlorinated biphenyls in municipal sewage sludges using microwave-assisted extraction and gas chromatography mass spectrometry, *Analyst*, 124, 453, 1999.

114. Molina, L., Cabes, M., Diaz-Ferrero, J., Coll, M., Marti, R., Broto-Puig, F., Comellas, L., and Rodriguez-Larena, M.C., Separation of non-ortho polychlorinated biphenyl congeners on per-packed carbon tubes: application to analysis in sewage sludge and soil samples, *Chemosphere*, 40, 921, 2000.

115. Folch, I., Vaquero, M.T., Comellas, L., and Broto-Puig, F., Extraction and clean-up methods for improvement of the chromatographic determination of polychlorinated biphenyls in sewage sludge-amended soils: elimination of lipids and sulphur, *J. Chromatogr.*, A 719, 121, 1996.

116. Garcia-Gutierrez, A., McIntyre, A.E., and Lester, J.N., Analysis of aldrin and endrin in sewage sludges, *Environ. Technol. Lett.*, 3, 541, 1982.

117. Garcia-Gutierrez, A., McIntyre, A.E., Lester, J.N., and Perry, R., Determination of organochlorine compounds in wastewater samples by gas-chromatography and gas chromatography-mass spectrometry using capillary columns, *Environ. Technol. Lett.*, 4, 129, 1983.

118. Garcia-Gutierrez, A., McIntyre, A.E., Lester, J.N., and Perry, R., Comparison of gas-chromatography and gas-chromatography mass-spectrometry for the analysis of poly-chlorinated-biphenyls and organochlorine insecticides in sewage sludges, *Environ. Technol. Lett.*, 4, 521, 1983.

119. Mullin, M.D., Pochini, C.M., McCrindle, S., Romhes, M., Safe, S.H., and Safe, L.M., High resolution PCB analysis: synthesis and chromatographic properties of all 209 PCB congeners, *Environ. Sci. Technol.*, 18, 468, 1984.

120. Schantz, M.M., Parris, R.M., Kurz, J., Ballschmiter, K., and Wise, S.A., Comparison of methods for the gas-chromatographic determination of PCB congeners and chlorinated pesticides in marine reference materials, *Fresenius J. Anal. Chem.*, 346, 766, 1993.

121. Schulz, D.E., Petrick, G., and Duinker, J.C., Complete characterisation of polychlorinated biphenyl congeners in commercial aroclor and clophen mixtures by multidimensional gas chromatography-electron capture detection, *Environ. Sci. Technol.*, 23, 852, 1989.

122. Hill, N.P., McIntyre, A.E., Perry, R., and Lester, J.N., Development of a method for the analysis of chlorophenoxy herbicides in waste-waters and wastewater-sludges, *Int. J. Environ. Anal. Chem.*, 15, 107, 1983.

123. Buisson, R.S.K., Kirk, P.W.W., and Lester, J.N., Determination of chlorinated phenols in water, wastewater, and wastewater-sludge by capillary GC/ECD, *J. Chromatogr. Sci.*, 22, 339, 1984.

124. Booth, R.A. and Lester, J.N., A method for the analysis of phenol and monochlorinated and brominated phenols from complex aqueous samples, *J. Chromatogr. Sci.* 32, 259, 1994.

125. Taylor, P.N., Scrimshaw, M.D., and Lester, J.N., Supercritical fluid extraction of acidic herbicides from sediment, *Int. J. Environ. Anal. Chem.*, 69, 141, 1998.

126. McIntyre, A.E., Perry, R., and Lester, J.N., Analysis of polynuclear aromatic hydrocarbons in sewage sludges, *Anal. Lett.*, 14, 291, 1981.

127. Miege, C., Dugay, J., and Hennion, M.C., Optimization and validation of solvent and supercritical-fluid extractions for the trace-determination of polycyclic aromatic hydrocarbons in sewage sludges by liquid chromatography coupled to diode-array and fluorescence detection, *J. Chromatogr.*, A 823, 219, 1998.

128. Schnaak, W., Kuchler, T., Kujawa, M., Henschel, K.P., Sussenbach, D., and Donau, R., Organic contaminants in sewage sludge and their ecotoxicological significance in the agricultural utilization of sewage sludge, *Chemosphere*, 35, 5, 1997.

129. Bedding, N.D., McIntyre, A.E., Lester, J.N., and Perry, R., Analysis of waste-waters for polynuclear aromatic-hydrocarbons. I. Method development and validation, *J. Chromatogr. Sci.*, 26, 597, 1988.

130. Potter, C.L., Glaser, J.A., Chang, L.W., Meier, J.R., Dosani, M.A., and Herrmann, R.F., Degradation of polynuclear aromatic hydrocarbons under bench-scale compost conditions, *Environ. Sci. Technol.*, 33, 1717, 1999.

131. McIntyre, A.E., Perry, R., and Lester, J.N., Analysis and incidence of organo-phosphorus compounds in sewage sludges, *Bull. Environ. Contam. Toxicol.*, 26, 116, 1981.

132. Koeber, R., Fleischer, C., Lanza, F., Boos, K.S., Sellergren, B., and Barcelo, D., Evaluation of a multidimensional solid-phase extraction platform for highly selective on-line cleanup and high-throughput LC-MS analysis of triazines in river water samples using molecularly imprinted polymers, *Anal. Chem.*, 73, 2437, 2001.

133. Castillo, M. and Barcelo, D., Analysis of industrial effluents to determine endocrine-disrupting chemicals, *Trac-Trends Anal. Chem.*, 16, 574, 1997.

134. Ziogou, K., Kirk, P.W.W., and Lester, J.N., Evaluation of a cleanup procedure for the determination of Phthalic-acid esters in sewage-sludge, *Environ. Technol. Lett.*, 10, 77, 1989.

135. Russell, D.J. and McDuffie, B., *Int. J. Environ. Anal. Chem.*, 15, 165, 1983.

136. Muller, M.D., Comprehensive trace level determination of organotin compounds in environmental-samples using high-resolution gas: chromatography with flame photometric detection, *Anal. Chem.*, 59, 617, 1987.

137. Carlier-Pinasseau, C., Lespes, G., and Astruc, M., Determination of butyltin and phenyltin by GC-FPD following ethylation by NaBEt$_4$, *Appl. Organomet. Chem.*, 10, 505, 1996.

138. Bancon-Montigny, C., Lespes, G., and Potin-Gautier, M., Improved routine speciation of organotin compounds in environmental samples by pulsed flame photometric detection, *J. Chromatogr.*, A 896, 149, 2000.

139. Oberg, K., Warman, K., and Oberg, T., Distribution and levels of brominated flame retardants in sewage sludge, *Chemosphere*, 48, 805, 2002.

140. Sellstrom, U., Kierkegaard, A., de Wit, C., and Jansson, B., Polybrominated diphenyl ethers and hexabromocyclododecane in sediment and fish from a Swedish river, *Environ. Toxicol. Chem.*, 17, 1065, 1998.

141. Berset, J.D. and Holzer, R., Quantitative determination of polycyclic aromatic hydrocarbons, polychlorinated biphenyls and organochlorine pesticides in sewage sludges using supercritical fluid extraction and mass spectrometric detection, *J. Chromatogr.*, A 852, 545, 1999.

142. Perez, S., Guillamon, M., and Barcelo, D., Quantitative analysis of polycyclic aromatic hydrocarbons in sewage sludge from wastewater treatment plants, *J. Chromatogr.*, A 938, 57, 2001.

143. Cass, Q.B., Freitas, L.G., Foresti, E., and Damianovic, M., Development of HPLC method for the analysis of chlorophenols in samples from anaerobic reactors for wastewater treatment, *J. Liq. Chromatogr. Relat. Technol.*, 23, 1089, 2000.

144. Buser, H.R., Haglund, P., Muller, M.D., Poiger, T., and Rappe, C., Rapid anaerobic degradation of toxaphene in sewage sludge, *Chemosphere*, 40, 1213, 2000.

145. Nylund, K., Asplund, L., Jansson, B., Jonsson, P., Litzen, K., and Sellstrom, U., Analysis of some polyhalogenated organic pollutants in sediment and sewage-sludge, *Chemosphere*, 24, 1721, 1992.

146. Blanchard, M., Teil, M.J., Ollivon, D., Garban, B., Chesterikoff, C., and Chevreuil, M., Origin and distribution of polyaromatic hydrocarbons and polychlorobiphenyls in urban effluents to wastewater treatment plants of the Paris area (France), *Water Res.*, 35, 3679, 2001.

147. Wang, Y., Wang, Z.J., Ma, M., Wang, C.X., and Mo, Z., Monitoring priority pollutants in a sewage treatment process by dichloromethane extraction and triolein-semiper-meable membrane device (SPMD), *Chemosphere*, 43, 339, 2001.

148. dos Santos, L.S., Vale, M.G.R., de Araujo, M.B.C., Caramao, E.B., and Oliveira, E. C., Application of SPME in pre-concentration of chlorinated phenolic compounds from cellulose bleaching effluents, *J. Sep. Sci.*, 24, 309, 2001.

149. Kuchler, T. and Brzezinski, H., Application of GC-MS/MS for the analysis of PCDD/Fs in sewage effluents, *Chemosphere*, 40, 213, 2000.

150. Gerecke, A. C., Tixier, C., Bartels, T., Schwarzenbach, R. P., and Muller, S. R., Determination of phenylurea herbicides in natural waters at concentrations below 1 ng 1(−1) using solid-phase extraction, derivatization, and solid-phase microextraction gas chromatography mass spectrometry, *J. Chromatogr.*, A 930, 9, 2001.

151. Pujadas, E., Diaz- Ferrero, J., Marti, R., Broto-Puig, F., Comellas, L., and Rodriguez-Larena, M.C., Application of the new C18 Speedisks™ to the analysis of polychlo-rinated dibenzo-p-dioxins and dibenzofurans in water and effluent samples, *Chemo-sphere*, 43, 449, 2001.

152. Simonich, S.L., Begley, W.M., Debaere, G., and Eckhoff, W.S., Trace analysis of fragrance materials in wastewater and treated wastewater, *Environ. Sci. Technol.*, 34, 959, 2000.

153. Meulenberg, E.P., Mulder, W.H., and Stoks, P.G., Immunoassays for pesticides, *Envi-ron. Sci. Technol.*, 29, 553, 1995.

154. Diaz-Ferrero, J., Rodriguez-Larena, M.C., Comellas, L., and Jimenez, B., Bioanalyt-ical methods applied to endocrine disrupting polychlorinated biphenyls, polychlori-nated dibenzo-p-dioxins and polychlorinated dibenzofurans: a review, *Trac-Trends Anal. Chem.*, 16, 563, 1997.

155. Van Emon, J.M. and Lopezavila, V., Immunochemical Methods for Environmental-Analysis, *Anal. Chem.*, 64, A79, 1992.

156. Vandelaan, M., Stacker, L., Watkins, B., and Roberts, R., *Immunoassays for Moni-toring Human Exposure to Toxic Chemicals in Food and the Environment*, American Chemical Society, Washington, DC, 1990.

157. Gascon, J., Oubina, A., and Barcelo, D., Detection of endocrine-disrupting pesticides by enzyme-linked immunosorbent assay (ELISA): application to atrazine, *Trac-Trends Anal. Chem.*, 16, 554, 1997.

158. Oubina, A., Castillo, M., Gascon, J., and Barcelo, D., Evaluation of polynuclear aromatic hydrocarbons and pentachlorophenol ELISA in contaminated industrial effluents, *Abstr. Pap. Am. Chem. Soc.*, 214, 25, 1997.

159. Castillo, M., Oubina, A., and Barcelo, D., Evaluation of ELISA kits followed by liquid chromatography atmospheric pressure chemical ionization mass spectrometry for the determination of organic pollutants in industrial effluents, *Environ. Sci. Tech-nol.*, 32, 2180, 1998.

160. La Farre, M., Oubina, A., Marco, M.P., Ginebreda, A., Tirapu, L., and Barcelo, D., Evaluation of 4-nitrophenol ELISA kit for assessing the origin of organic pollution in wastewater treatment works, *Environ. Sci. Technol.*, 33, 3898, 1999.

161. Sherry, J.P. and Borgmann, A., Enzyme-immunoassay techniques for the detection of atrazine in water samples — evaluation of a commercial tube based assay, *Chemosphere*, 26, 2173, 1993.

162. Di Giano, F.A., Clarkin, C., Charles, M.J., Maerker, M.J., Francisco, D.E., and Larocca, C., Testing of the EPA toxicity identification evaluation protocol in the textile dye manufacturing-industry, *Water Sci. Technol.*, 25, 55, 1992.

163. Bailey, H.C., Miller, J.L., Miller, M.J., and Dhaliwal, B.S., Application of toxicity identification procedures to the echinoderm fertilization assay to identify toxicity in a municipal effluent, *Environ. Toxicol. Chem.*, 14, 2181, 1995.

164. Snyder, S.A., Villeneuve, D.L., Snyder, E.M., and Giesy, J.P., Identification and quantification of estrogen receptor agonists in wastewater effluents, *Environ. Sci. Technol.*, 35, 3620, 2001.

165. Rivas, A., Olea, N., and Olea-Serrano, F., Human exposure to endocrine-disrupting chemicals: assessing the total estrogenic xenobiotic burden, *Trac-Trends Anal. Chem.*, 16, 613, 1997.

166. DeFur, P.L., Crane, M., Ingersoll, C., and Tattersfield, L.J., Proceedings of the SETAC workshop on endocrine disruption in invertebrates: endocrinology, testing and assessment (EDIETA), in: *SETAC Workshop on Endocrine Disruption in Invertebrates: Endocrinology, Testing and Assessment (EDIETA)*, Noordwijkerhout, The Netherlands, 1999.

167. European Centre for Ecotoxicology and Toxicology of Chemicals, Genomics, Transcript Profiling, Proteomics and Metabonomics (GTPM): An Introduction, Document No. 42, 2000.

168. Lai, K.M., Scrimshaw, M.D., and Lester, J.N., The effects of natural and synthetic steroid estrogens in relation to their environmental occurrence, *Crit. Rev. Toxicol.*, 32, 113, 2002.

4 Fate and Behavior of Endocrine Disrupters in Wastewater Treatment Processes

K.H. Langford and J.N. Lester

CONTENTS

1-56670-601-7/03/$0.00+$1.50

4.1   INTRODUCTION

The nonpolar and hydrophobic nature of many endocrine disrupting chemicals (EDCs) causes them to sorb onto particulates. This suggests that the general effect of wastewater treatment processes would be to concentrate organic pollutants, including EDCs, in the sewage sludge. Mechanical separation techniques, such as sedimentation, would result in significant removal from the aqueous phase to primary and secondary sludges. The removal of endocrine disrupters in wastewater treatment processes is dependent on the inherent physicochemical properties of the pollutants and on the nature of the treatment process involved. The result of this is a treated wastewater discharged relatively free of EDCs and a sewage sludge that contains most of the contamination that entered via the influent.

It is generally recognized that there are four main removal pathways for organic compounds during conventional wastewater treatment:

1. Adsorption onto suspended solids or association with fats and oils
2. Aerobic and anaerobic biodegradation
3. Chemical (abiotic) degradation by processes such as hydrolysis
4. Volatilization

A compound's physicochemical data can be used to predict physical processes, such as sorption, volatilization, and dissolution. The important properties to be considered are octanol/water partition coefficient (K_{ow}), aqueous solubility, acid dissociation constant, and Henry's Law constant (H_c). A knowledge of chemical partitioning between the aqueous and solid phase is needed to assess pathways of EDC transport and transformation. For example, partitioning to the solid phase may increase the likelihood of reductive dehalogenation by anaerobic bacteria.[1-3] In addition, surface-catalyzed hydrolysis, or sorption to settled sludge may reduce the possibility of volatilization or photochemical degradation.[4]

4.1.1   SEWAGE TREATMENT PROCESSES

Conventional sewage treatment is typically a three-stage process consisting of preliminary treatment, primary sedimentation and secondary treatment,[5,6] which is schematically outlined in Figure 4.1. Some form of sludge treatment normally follows as outlined in Chapter 5.

Wastewater treatment commences at the head of the works with preliminary treatment, typically inlet screens. However, some conditioning of the wastewater will have been initiated in the sewer. For modeling purposes, the sewer pipe has been considered to consist of a sediment above which there is a bio-film.[7] Anaerobic

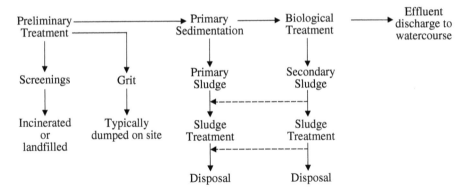

FIGURE 4.1 Schematic diagram of STW. (From Meakins, N.C., Bubb, J.M., and Lester, J.N., The fate and behaviour of organic micro pollutants during waste water treatment, *Intern. J. Environ. Poll.*, 4, 27, 1994. With permission.)

biodegradation may occur in the sewers where bacterial slime accumulates on the walls. In large catchment areas, wastewater can have a significant retention time in a sewage system and allows significant degrees of removal to begin there.

4.1.1.1 Preliminary Treatment

Preliminary treatment involves the initial screening of the raw sewage through parallel bars to remove large floating objects.[5] A small amount of putrescible organic material is removed from the screens and generally landfilled or incinerated. Grit and dense inorganic solids are removed by means of settlement in tanks or constant velocity channels where the lighter organic material remains in suspension. Very little removal of organic micropollutants is observed at this stage.[5]

4.1.1.2 Primary Sedimentation

The raw sewage then passes into the primary sedimentation tanks where the most significant mechanism is adsorption onto solids, which under the influence of gravity settle to form primary sludge. The degree of pollutant removal is largely dependant on suspended solids removal, which is controlled by settling characteristics of the particles (their density, size, and ability to flocculate); the retention time in the tank; and the surface loading. The retention time in the tank is referred to as the sludge retention time (SRT) or sludge age and is a measure of the time bacterial cells remain in the system. It is the total amount of sludge in the system divided by the total rate of sludge leaving the system and is typically 4 to 9 days.[5] Removal of organic compounds can be affected by temperature and the solids content of the effluent is higher during winter months when the temperature is low. Fats, oils, and greases adsorb significant amount of hydrophobic compounds, including many EDCs, and are removed from the surface of the tank and added to the sludge prior to sludge treatment. Figure 4.2 indicates some of the possible removal mechanisms for EDCs during primary sedimentation. Primary sedimentation is used in the majority of sewage treatment works (STW); in some cases

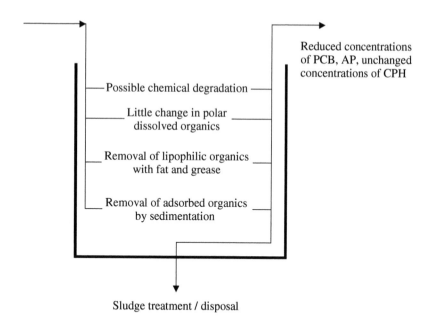

Sludge treatment / disposal

FIGURE 4.2 Removal mechanisms during primary sedimentation for EDC removal. (From Meakins, N.C., Bubb, J.M., and Lester, J.N., The fate and behaviour of organic micro pollutants during waste water treatment, *Intern. J. Environ. Poll.*, 4, 27, 1994. With permission.)

flocculants are added to aid flocculation.[8] Coagulants, such as aluminum and ferric salts have been used to remove organic matter, although their use is often deemed impractical due to the high costs.[9] Where oxidation ditches are employed for wastewater treatment there is no primary sedimentation, and removal of EDCs probably conforms to the mechanisms identified or postulated for activated sludge treatment.[10]

4.1.1.3 Secondary Treatment

Secondary treatment can involve anaerobic biodegradative processes, although almost invariably aerobic processes are responsible.[11] The principle of aerobic secondary treatment is to allow the aerobic bacteria and other microorganisms in mixed liquor contact with oxygen to convert the organic compounds into carbon dioxide and water. Activated sludge or trickling filters are the principal secondary treatment processes employed. Both use two vessels: a reactor containing large populations of microorganisms that oxidize the biochemical oxygen demand (BOD), and a secondary sedimentation tank or clarifier where the microorganisms are removed from the final effluent. In activated sludge, the majority of these microorganisms are recycled to the aerator. In trickling filters, all of the film (humus) in the secondary sedimentation tank is disposed of. Removal pathways for organic pollutants during secondary biological treatment (Figure 4.3) include adsorption onto the microbial flocs and removal in the waste sludge, biological or chemical degradation, and transformation and volatilization during aeration.

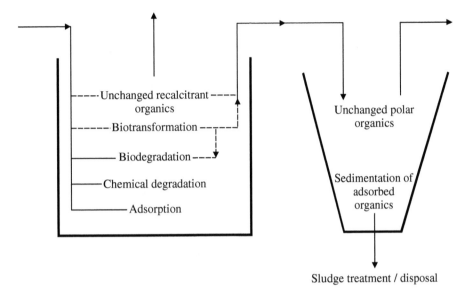

FIGURE 4.3   Removal mechanisms during secondary biological treatment. (From Meakins, N.C., Bubb, J.M., and Lester, J.N., The fate and behaviour of organic micro pollutants during waste water treatment, *Intern. J. Environ. Poll.*, 4, 27, 1994. With permission.)

Nitrification can occur during secondary treatment and can have benefits to the removal of organic contaminants. Nitrification is the conversion of ammonia to nitrate and is a two-stage process involving autotrophic bacteria, generally *Nitrosomonas* spp. and *Nitrobacter* spp. A considerable amount of oxygen is required for nitrification to take place, and a longer SRT is required because of the slow growing nature of these bacteria. This coupled with the high concentration of dissolved oxygen required increases the cost of waste treatment.[5]

Tertiary treatment processes are also increasingly used and include sand filtration and microstrainers and occasionally may include methods such as activated carbon and membrane filtration. This is particularly the case in small domestic plants to polish drinking water, but these latter treatments are costly and pose maintenance problems.[9]

Table 4.1 includes some of the removal process and physical conditions important for high removal efficiencies for some of the EDCs.

4.2 BEHAVIOR OF ENDOCRINE DISRUPTERS IN SEWAGE TREATMENT WORKS

The behavior of EDCs in STWs is dependent on their physicochemical properties (Table 4.2). Aqueous solubility (mg l^{-1}), organic carbon/water partition coefficients (K_{oc}), and K_{ow} influence the partitioning and sorption of a compound during treatment. The H_c is their volatilization potential and is also an indicator of the likelihood of evaporation during treatment.

TABLE 4.1

Treatment Types and Removal Efficiencies from Influent for Selected EDCs

Compound	Process Type	Removal Efficiency
PCB[112]	Biofiltration	90%
	Activated sludge	96%
	Biofiltration/activated sludge	99%
NP[58]	High loading/non-nitrifying	37%
	Low loading/nitrifying	77%
NP$_1$EO	High loading/non-nitrifying	-3% produced as degradation product
	Low loading/nitrifying	31%
NP$_2$EO	High loading/non-nitrifying	-5% produced as degradation product
	Low loading/nitrifying	91%
NP$_6$EO	High loading/non-nitrifying	78%
	Low loading/nitrifying	98%
17β-estradiol/17α-ethinylestradiol[163]	Filtration — sand/microfiltration	70%
		95%
	Advanced treatment — reverse osmosis	
Organotins[101]	Primary effluent	73%
	Secondary effluent	90%
	Tertiary effluent	98%
Triazines[82]	Conventional 2 stage	<40%

4.2.1 ADSORPTION

Sewage sludge is a complex mixture of fats, proteins, amino acids, sugars, carbohydrates, lignin, celluloses, humic material, and fatty acids.[10] In secondary sludge, the large amounts of live and dead microorganisms provide a large surface area (0.8 to 1.7 m^2 g^{-1}).[11] EDCs preferentially adsorb onto these suspended particulates because of their hydrophobic properties. The K_{ow} values often correlate with the degree of association between an organic compound and the solid phase.[12,13] K_{ow} is the concentration ratio at equilibrium of an organic compound partitioned between an organic liquid and water and is one of the quantitative physical properties that correlates best with biological activity. It can be used as a measure of lipophilicity[14] and is therefore used to predict sorption onto solids.[15]

Log K_{ow} values increase with increasing lipophilicity and correlate inversely with solubility. Large log K_{ow} values are characteristic of large hydrophobic molecules that tend to associate with solid organic matter, while smaller hydrophilic molecules have low log K_{ow}. It is a better indication of the extent of adsorption by microorganisms than solubility.[16] Below a log K_{ow} of approximately 4, the removal of EDCs during primary treatment is dominated by advection of the dissolved compounds. A log K_{ow} of less than 2.5 demonstrates a low sorption potential, and a log K_{ow} greater than 4 shows a high sorption potential.[11] At a log K_{ow} greater than 4, the major removal process is sorption to the settled sludge and with that associated with suspended

TABLE 4.2
EDC Chemical and Physical Properties

Compound	Structure	Aqueous Solubility (mg l⁻¹)	Log K_{ow}	H_c (atm m³ mol⁻¹)
17β-Estradiol		12.96^{34}	3.94^{34}	6.3×10^{-7} [164]
Estrone		12.42^{34}	3.43^{34}	6.2×10^{-7} [164]
Estriol		13^{53}	2.81^{53}	2×10^{-11} [164]
17α-ethinylestradiol		4.83^{34}	4.15^{34}	3.8×10^{-7} [164]
4-nonylphenol		5.4^{165}	4.48^{166}	11.02^{167}
Nonylphenol polyethoxylate	C_9H_{19}-⬡-$(OCH_2CH_2)_nOH$			
NP₁EO	(n = 1–40)	3.02^{166}	4.17^{167}	
NP₂EO		3.38^{166}	4.21^{167}	
NP₃EO		5.88^{166}	4.20^{167}	

(continued)

TABLE 4.2 (continued)
EDC Chemical and Physical Properties

Compound	Structure	Aqueous Solubility (mg l⁻¹)	Log K_{ow}	H_c (atm m³ mol⁻¹)
Polybrominated diphenyl ethers		9×10^{-7} [168]		
TeBDE			5.9–6.2[169]	
PeBDE			5.6–7.0[169]	
OcBDE			8.4–8.9[169]	
DeBDE			$10^{[169]}$	
Tetrabromobisphenol A		4.2^{168}	4.5^{170}	
Lindane (1,2,3,4,5,6-hexachlorcyclohexane)		10^{16}	3.72^{16}	4.93×10^{-7} [4]
Atrazine		33^{82}	2.75	6.63×10^{-9}

Polychlorinated biphenyls 9				
2,3,4'-trichlorobiphenyl		0.085	5.74/5.69	0.0049
2,5,2'5'-tetrachlorobiphenyl		0.046	6.26/6.09	0.00031
3,4,3'4'-tetrachlorobipheny		0.18	6.52/5.62	0.00031
2,4,5,2'5'- pentachlorobiphenyl		0.031	6.85/7.07	0.0012
2,4,5,2'4'5'- hexachlorobiphenyl		0.0088	7.44/7.75	0.00069
PAH				
Benzo(k)fluoranthene		0.8×10^{-3} [121]	6.04[121]	2.0×10^{-6} [9]
Benz(a)anthracene		14×10^{-3} [121]	5.91[121]	1.2×10^{-7} [121]
Benzo(a)pyrene		3.8×10^{-3} [121]	6.5[121]	7.5×10^{-6} [121]
2,4-D 89		540–890	2.81	

(continued)

TABLE 4.2 (continued)
EDC Chemical and Physical Properties

Compound	Structure	Aqueous Solubility (mg l^{-1})	Log K_{ow}	H_c (atm m^3 mol^{-1})
2,4-dichlorophenol 89		4500	2.71–3.08	4.2×10^{-5}
2,4,6-trichlorophenol 89		800–900	2.9–3.9	
Bisphenol A 110		120–300	3.4	1×10^{-10}
Di-(2-ethyhexyl)phthalate (DEHP) 104		0.003	7.5	1.71×10^{-5}

matter.[17] The organic carbon content of the solid phase and the significance of the polarity and composition of organic matter have been highlighted as major surface variables influencing sorption for the majority of organic compounds.[18,19]

The K_{oc} is also an important parameter when considering adsorption. K_{oc} is the ratio between the concentration of the organic compound on organic carbon (mg g^{-1}) and its concentration in water (mg l^{-1}) at equilibrium; it can be estimated from log K_{ow} values or solubility. The likelihood that a compound will sorb to organic matter, such as sewage sludge, can be assessed using log K_{oc}. Generally compounds with a high log K_{oc} will tend to adsorb onto sewage sludge, while those with lower values will tend to remain in the aqueous phase.

In addition to sorption onto suspended solids as a removal mechanism, it is possible that compounds may partition onto the nonpolar fat and lipid material in raw sewage. Domestic sewage generally contains fats, mineral oils, greases, and surfactants. Some compounds are resistant to degradation and can reach watercourses via final effluent discharge.[20,21]

Karickhoff et al.[22] looked at the sorption of hydrophobic compounds onto pond and river sediments and demonstrated that partition coefficients are directly related to the organic carbon content for a given particle size. Small particles with a high organic carbon content will be the most effective at adsorbing as demonstrated by the increasing partition coefficient onto the organic phase.[23] An excellent linear relationship was observed between log K_{ow} and log K_{oc}. Dissolved organic matter (DOM) plays an important role in the fate of hydrophobic compounds. A reduction of 50 to 90% in the amount of n-alkanes solubilized was observed when the DOM was removed from water.[22] DOM enhances the apparent solubility of pollutants; they are stabilized in the water phase because of their reduced ability to be removed from the water column by partitioning onto solid phases.

Increasing the SRT reduces the amount of sludge wastage from the activated sludge system. This results in a decrease in compound concentrations in the final effluent because of increased sorption and biodegradation. At low SRTs, the high mass flows of wasted sludge and the wash-out of slow growing specific degraders result in the majority of EDCs being removed in the sorbed phase.[24] Removal of EDCs in settled sludge from primary sedimentation varies as a function of the partitioning behavior to sludge through log K_{ow}, a compound's removal can be influenced by changes in sedimentation efficiency. A reduction in sedimentation efficiency would result for example from an increase in influent flow rate.

Sorption onto inorganic and biological solids is an important removal mechanism because adsorption onto cellular material is often the first stage in biological degradation of these compounds. However, compounds that are strongly bonded onto inorganic particles are less available for degradation and volatilization. Organic compounds are adsorbed onto raw wastewater solids during primary treatment and onto biological sludges during secondary treatment. There are several different mechanisms responsible for the accumulation of compounds in biological sludges. They include

1. Absorption into bacterial lipid structures
2. Sorption onto polysaccharide structures located on the outside of bacterial cells
3. Chemical binding to bacterial proteins and nucleic acids[10]

4.2.2 BIOLOGICAL DEGRADATION AND TRANSFORMATIONS

Biological degradation and transformations occur aerobically by biological oxidation in activated sludge, trickling filters, or anaerobically in the sewage system or anaerobic sludge digesters.

Numerous chemical factors, such as structural properties and environmental factors, influence biodegradation.[25,26] Generally, molecules with highly branched hydrocarbon chains are less amenable to biodegradation than unbranched chains, and shorter chains are not as quickly degraded as longer chains. There are certain substituents on a compound that will make it resistant to degradation. These include:

1. Halogen groups or substitutions in the *meta*-position on a benzene ring
2. Sulfonates
3. Methoxy groups
4. Nitro groups

Molecular mass or size can limit active transport, solubility can result in competitive partitioning, and toxicity can result in cell damage or enzyme inhibition. Oxygen-requiring enzymes may be needed for aromatic compounds or extracellular metabolisms may be required for polymeric compounds. Environmental factors include dissolved oxygen for oxygen-sensitive or oxygen-requiring enzymes. Temperature can also have an effect, since microorganisms are often more active at higher temperatures. Narrow pH ranges are often required for microorganism growth. Light can be an important factor for photochemical enhancement, and nutrient and trace element concentrations are important for microorganism growth.[27]

Biological degradation may occur as a result of intra- and extracellular enzymes. The hydrolysis process, for example, involves regulation of extracellular enzyme synthesis in the cells. It takes place by enzymes secreted by the cells before the substrate can be taken up by the microorganisms and be metabolized.[28] Chemical properties of a compound influence their entry into microbial cells, which is a prerequisite for the induction of intracellular enzymes. Reactions that involve extracellular enzymes are important for large volume molecules and possibly molecules with large molecular masses. These enzymes are excreted from cells into solution or are released when aged cells undergo lysis in low growth conditions.

A high hydraulic retention time (HRT) in STWs allows more time for degradation. A greater SRT could influence biota and the physical nature of floc particles which would have an important effect on their affinity as sorbents.[29] At low SRT, slow growing specific degraders can be washed out before degradation has occurred if their growth rate is less than the SRT. Biodegradation is therefore reduced and sorption becomes the most significant removal mechanism.[24,30] As SRT increases, biodegradation occurs over a longer period of time and has more influence on mass removal. The SRT at which maximum biodegradation occurs is a function of log K_{ow} and the biotransformation rate of the compound. Therefore, long SRTs are required for very hydrophobic compounds.[13]

Unlike naturally occurring compounds, anthropogenic compounds tend to be relatively resistant to biodegradation. This is partly due to the fact that microorganisms lack the necessary enzymes required for transformation, so a longer acclimation period may be required. However, the biodegradation of anthropogenic compounds can be facilitated by co-metabolism.[31–33] Co-metabolic transformations are the transformations of a compound by metabolic reactions that do not contribute carbon or energy to the biological growth of the organism. The organism uses co-substrates to support its growth.

The importance of biotransformation increases with SRT and increasing log K_{ow} is their influence on biosorption. Biotransformation rates increase to a maximum at log K_{ow} 3 to 3.5 (e.g., estrone has a log K_{ow} of 3.43[34]) and then decline rapidly as sorption to sludge dominates the removal mechanisms for more hydrophobic compounds.[14] Compounds subject to such a sorption mechanism include nonylphenol (log K_{ow} 4.48[35]) and polychlorinated biphenyls (PCBs) (log K_{ow} 5.62 to 7.75[9]). Biotransformation has more influence on compounds with log K_{ow} in the range 1.5 to 4, such as estradiol with log K_{ow} 3.94.[34] K_{ow} values also play a role in determining bioconcentration in the food chain. Compounds that are easily metabolized or are polar tend to have lower K_{ow} values and do not easily enter the food chain. Lipophilic compounds with high K_{ow} values however do accumulate.[36]

4.2.3 CHEMICAL DEGRADATION

In addition to biological degradation, chemical reactions can be responsible for the breakdown of compounds. Hydrolysis is usually the most important chemical transformation. Hydrolysis is a nucleophilic displacement reaction that can occur when molecules have linkages separating highly polar groups.[11] Reductive dehalogenation has been identified as an important mechanism for chlorinated and brominated compounds such as PCBs and possibly polybrominated diphenyl ethers (PBDEs) However, it is predominantly a biological process that may occur abiotically via nucleophilic substitutions. For example, reductive dehalogenation of PCBs reduces the levels of chlorination, making them more available for aerobic or anaerobic degradation.[37] Factors such as pH, temperature, moisture, and inorganic matter can also have an effect on chemical degradation rates.[25,26]

4.2.4 VOLATILIZATION

Volatilization is the transfer of a compound from the aqueous phase to the atmosphere from the surface of open tanks such as clarifiers. However, the majority of losses occur through air stripping in aeration vessels. A proportion may be lost during sludge treatment at the dewatering or thickening stage, particularly if the sludge is aerated or agitated. The activated sludge stage of treatment aeration allows air stripping to occur. Low molecular mass, nonpolar compounds with low aqueous solubilities, and low vapor pressures are known to be transferred to the atmosphere during aeration in wastewater treatment.[13,38] However, due to the static nature of sedimentation process, losses by volatilization are small. A peak removal rate is seen for compounds with a log K_{ow} of around 2. At less than 2, increased water solubility inhibits volatilization. H_c can be used to predict losses by volatilization. Generally, compounds with an H_c greater than 10^{-3} mol^{-1} m^{-3} can be removed by volatilization.[38]

Lower H_c and increased partitioning to the organic carbon content reduces volatilization for compounds with log K_{ow} greater than 2.[13] Compounds with H_c greater than 1×10^{-4} and H_c/K_{ow} greater than 1×10^{-9} have a high volatilization potential. Compounds with H_c less than 1×10^{-4} and H_c/K_{ow} less than 1×10^{-9} have a low volatilization potential.[9]

Recent studies have suggested that air stripping of volatile compounds is less significant than their biodegradation during secondary treatment.[39]

4.3. FATE OF ENDOCRINE DISRUPTERS IN SEWAGE TREATMENT WORKS

Removal mechanisms and biotransformation pathways are different for each group of compounds and are determined by their physical and chemical properties (Table 4.2). Table 4.3 contains the primary removal mechanisms as determined by the physical and chemical properties for some EDCs and the effects of parameter changes.

4.3.1 STEROID ESTROGENS

Steroid removal can be influenced by HRTs and SRTs in STWs as demonstrated in Table 4.4. Other factors include the time it takes to reach STW, the nutrient status, type of treatment and the activity and stability of resident biota, and the use of secondary treatment processes (e.g., activated sludge). Johnson et al.[29] found no indication as to whether biodegradation or sorption is the most important removal mechanisms.

Research has shown that the majority of steroid estrogens enter the STW in their conjugated form. In addition, STWs are the primary source of free steroids to the aquatic environment, demonstrating that deconjugation occurs during treatment.[25,40–43] No significant levels of conjugated estrogens have been found in effluent.[44] *Escherichia coli* synthesize large amounts of the enzyme β-glucuronidase, which deconjugates the steroids in the gut.[40,41,45] These bacteria are also present in the STW, so it is expected that they will deconjugate steroids during treatment.

Ternes et al.[46] looked at the behavior and occurrence of estrogens in aerobic batch experiments with activated sludge. 17β-estradiol-17-glucuronide and 17β-estradiol-3-glucuronide were cleaved in contact with the sludge to form 17β-estradiol (E2). E2 was shown to oxidize to estrone (E1) on contact with the activated sludge, but no further degradation occurred and estrone was the final product. The concentration of E2 was immediately reduced. After 1 to 3 hours, 95% had been removed and the level of E1 had risen by 95%. The synthetic hormone mestranol was quickly degraded and 17α-ethinylestradiol (EE2) was produced in small amounts; EE2 was then only slightly reduced demonstrating its persistence. Ternes et al.[47] also found that the removal rates of steroids were less efficient after trickling filter treatment than activated sludge. This research is supported by prior work that found that estrogens, particularly synthetic compounds, are stable enough to withstand the sewage treatment process.[48]

Studies in Japan have shown that E2 accounted for only 34% of the estrogenicity in raw sewage, but accounted for 100% in the final effluent after activated sludge treatment.[49] They concluded that human estrogens are the major estrogenic compounds in sewage and effluent.

TABLE 4.3
Primary Removal Mechanisms of EDCs during Wastewater Treatment

Compound	Primary Removal Mechanism	Effect of Parameter Change
Nonylphenol polyethoxylates	Adsorption	Efficiency increases from high loading/non-nitrifying to low loading/nitrifying.[58]
		Removal increases with increasing temperature and SRT.[58,72]
		At lower temperatures, higher SRT and longer lag phase are required for removal.[72]
Steroids	Degradation	E1 — increasing SRT from 6–11 days at 13–15°C increases removal from 64–94%. Increasing temperature increases removal from 66–98%.
		E2 — increasing SRT from 6–11 days at 13–15°C increases removal from 92–98%. Increasing temperature increases removal from >75->94%.[23]
Organochlorines	Adsorption	Increasing dry weather flow increases removal efficiency.[118]
Lindane	Degradation[89]	Best removal at high sludge loading, worse at high sludge age[89]
Triazines	Adsorption[82]	Changes in suspended solids and retention time have no effect on removal.[82]
	Primary degradation[83]	
Polychlorinated biphenyls	Adsorption[114,117,118]	Increases in sludge age increase removal.[114]
		Removal decreases with higher flow conditions[21]
Organotins	Adsorption[101–103,171]	
Phthalates	Biodegradation[104,106,107,172]	Increasing temperature increased removal efficiency.[107]
Polyaromatic hydrocarbons	Adsorption during primary treatment[38,121–123,125–128]	Low molecular weight PAH show increased removal with increasing SS loading.
	Volatilization/biodegradation during secondary treatment[123,124]	High molecular weight PAH removal was independent of hydraulic loading.[121,122]

TABLE 4.4
Removal Rates of Steroid Estrogens at Varying Retention Times

	13–15 °C		18–19 °C	
	E1% Loss	E2% Loss	E1% Loss	E2% Loss
18-hour HRT 6-day SRT	64%	92%	75%	—
18-hour HRT 11-day SRT	94%	98%	>98%	—
26-hour HRT 20-day SRT	66%	>75%	98%	>94%

Source: From Johnson, A.C., Belfroid, A., and Di Corcia, A., Estimating steroid oestrogen inputs into activated sludge treatment works and observations on their removal from the effluent, *Sci. Total Environ.*, 256, 163, 2000. With permission.

Activated sludge treatment plants in Rome showed removal rates for estriol, E2, E1, and EE2 of 95, 87, 61, and 85%, respectively.[41] The efficiency for estrone is lowest, and at 4 out of 30 sites it exhibited an increase in concentration from influent to effluent. This can partly be explained by the biological oxidation of E2 to E1. Turan[48] reported no change in EE2 concentration in activated sludge over 120 hours.

Removal rates of EE2 in nitrifying activated sludge were at a maximum for the first 2 days with a degradation rate of 1 µg g^{-1} Dry Weather h^{-1}.[50] After this period, the degradation rate slowed. This was most likely due to the affinity of microorganisms for EE2 at lower concentrations or a decrease in the activity of the nongrowing cells. The low concentration may result in the bacteria being in a starved condition (senescent cells) in the phase between death and the breakdown of the osmotic regulatory system (in moribund state).[5] The oxidation of EE2 was confirmed by the formation of hydrophilic organic compounds that were not identified. When hydrazine was added as an external electron donor to provide unlimited reducing energy, degradation of EE2 was slightly higher than without hydrazine addition. This demonstrated that EE2 degradation is mediated by monooxygenase activity. The capacity to transform EE2 by the nitrifying sludge may become a function of the reducing energy.

A study of mass balances of estrogens in STW in Germany[51] demonstrated that most of the estrogenic activity in the wastewater was biodegraded during treatment rather than adsorbed onto suspended solids. There was a 90% reduction in estrogenic load, and less than 3% of the estrogenic activity was found in the sludge. Radiolabelled E2 was used in a study of estrogen fate in STW.[15] At low concentrations the majority of the radiolabelled E2 remained in the liquid phase.

In another study,[52] suspended solids content was an important factor. A higher suspended solids content resulted in a higher removal of estrogens, while an increase in influent estrogen concentration caused a decrease in removal. Salinity and pH also affect the removal. pH changes influence the amount and the type of bonding involved in sorption; a higher sorption rate was observed at neutral pH and at high salinity. Competition for binding sites effects estrogen removal, the addition of estrogen-valerate with a very high K_{ow} (6.41) reduced the amount of estrogen removed, as was also shown by Lai et al.[53]

During a study of Japanese night soil treatment processes, the highest amount of estrogenic activity was found to be in the sludge from the sludge tank, while the effluent contained very much lower concentrations.[54] E2 accounted for 16% of the estrogenic activity, and the estrogenic activity in the aqueous phase decreased by 1000 times during biological treatment. The estrogens were considered to be accumulating in the sludge and passed through the biological treatment.

In municipal biosolids, 84% of estradiol and 85% of estrone were mineralized in 24 hours, compared to less than 4% in industrial biosolids.[55] No correlation between BOD and suspended solids removal with mineralization was observed, although temperature was seen to have an effect. At different temperatures no significant differences in first order rate constants were seen for EE2. However, E2 was significantly different, and even at cold temperatures, it was rapidly removed by biosolids.

4.3.2 SURFACTANTS

The group of surfactants of concern is alkylphenol polyethoxylates (APEOs) and their breakdown products, alkylphenols (APs) and alkylphenol carboxylates (APECs); all have been shown to be estrogenic.[43,49,56,57] In aerobic conditions, the oxidative shortening of the polyethoxylate chain occurs easily and rapidly. However, complete mineralization is poor due to the presence of the highly branched alkyl group on the phenolic ring. The hydrophilic group in ethoxylated compounds contains more abundant carbon than the hydrophobic alkyl group. These moieties are therefore potential sources of bacterial nutrients that become available by the successive removal of ethoxy groups. This chain shortening results in the formation of recalcitrant intermediates such as AP_1EO, AP_2EO, AP, AP_1EC, and AP_2EC. Ultimate biodegradation of these metabolites occurs more slowly, if ever, because of the presence of the benzene ring and their limited water solubility. The frequent occurrence of oxidized intermediates, such as APEC, may indicate oxidative mechanisms.[58,59] It is possible that nonoxidative ether scission dominates APEO biodegradation to shorter homologs. Longer lived intermediates undergo hydroxyl group oxidation as a side chain reaction and are possibly catalyzed by alcohol hydrogenases known to occur ubiquitously in bacteria.

It has been concluded that degradation occurred in two steps. The first step occurred within 10 hours of aeration and was attributed to the cleavage or oxidation of the ethoxylate chain[60,61]; this stage occurs rapidly. The second step required more time because of the need to develop a specific population of enzymes or bacteria that was not initially available. Figure 4.4 shows a theorized degradation pathway for APEOs.

APEO lipophilicity decreases with increasing chain length. As a result, shorter chain compounds have the highest tendency to sorb onto the solid phase and compounds with longer chains appear in final effluent.[62–64] AP's are more lipophilic compounds than APEOs with higher log K_{ow} values (4.48 for 4-nonyl phenol [4NP]).

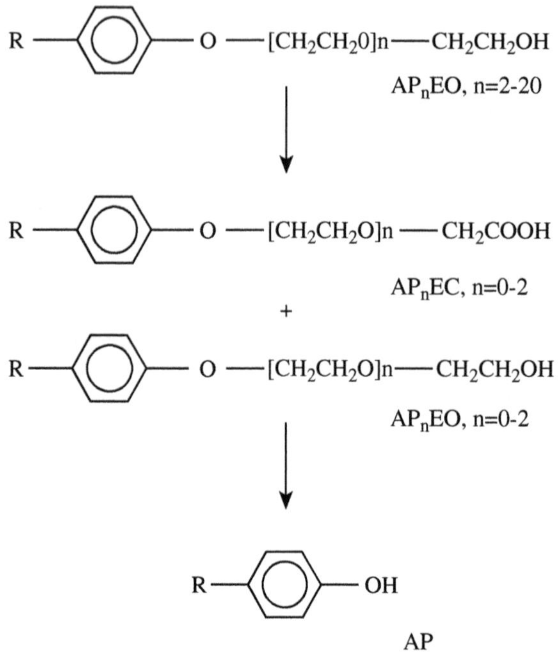

FIGURE 4.4   Theorized degradation pathway for APEO in STW.

Because of this, APs, 4NP in particular, sorb onto the solid phase making them more resistant to biodegradation.[64-67] APECs are more water soluble and have a very limited tendency to be found in the solid phase; they are found in high concentrations in final effluent.[58,68-70]

The most dramatic change in APEO distribution appears during the activated sludge phase of treatment.[58] After treatment, no APEO with greater than 8 ethoxy units were detected. In the primary effluent $NP_{3-20}EO$ were dominant and in secondary effluent the metabolite (NP and $NP_{1-2}EO$) concentration had increased to greater than 70%. Table 4.5 indicates the NP and NPEO concentrations at each stage of treatment in a mechanical-biological process. Table 4.6 contains some influent compared to effluent concentrations at different STW.

In several studies, NPECs were found to be the dominant species in final effluent.[68,69] In one study[68] the concentration of NP_1EC and NP_2EC in secondary effluent (5 to 20 µg l^{-1}) was 5 times more than any other metabolite. This is more than their inverse estrogenicity, so it has been theorized that NPEC could be the most relevant of the nonylphenolic compounds. Concentrations in secondary effluents were 2.1 to 7.6 times higher than in primary effluents, indicating that aerobic biological treatment results in their formation.[58]

AP and APEC are also sometimes found in raw sewage at low concentrations, despite not being found in commercial products. This implies that some partial degradation occurs in the sewage system before the influent reaches the STW.[65]

Nitrification during activated sludge can increase removal efficiencies.[58,71] Low-loading nitrifying conditions enable greater removal; almost complete removal for

TABLE 4.5
NPEOs and NP in a Mechanical-Biological Treatment Plant of Municipal Wastewater

	NP	NP_1EO	NP_2EO
Raw wastewater	14 µg l^{-1}	18 µg l^{-1}	18 µg l^{-1}
Treated wastewater	8 µg l^{-1}	49 µg l^{-1}	44 µg l^{-1}
Receiving water	3 µg l^{-1}	7 µg l^{-1}	10 µg l^{-1}
Activated sludge	128 mg kg^{-1}	76 mg kg^{-1}	61 mg kg^{-1}
Anaerobically digested sludge	1000 mg kg^{-1}	79 mg kg^{-1}	—
Anaerobic sludge effluent	467 µg l^{-1}	53 µg l^{-1}	6 µg l^{-1}

Source: From Ahel, M. and Giger, W., Determination of nonionic surfactants of the alkylphenol polyethoxylate type by HPLC, *Anal. Chem.*, 57, 2584, 1985. With permission.

TABLE 4.6
Influent and Effluent Concentrations for Alkylphenolic Compounds in STW

STW Location	Influent Concentrations (µg l^{-1})	Effluent Concentrations (µg l^{-1})
3 STW — Switzerland[174]		<10–35 NP
		24–133 NP_1EO
		<10–70 NP_2EO
1 STW (mechanical-biological) — Switzerland[173]	14 NP	8 NP
	18 NP_1EO	49 NP_1EO
	18 NP_2EO	44 NP_2EO
5 STW (mechanical-biological) — Switzerland[175]	844–2250 NPEO + NP	40–369 NPEO + NP
11 STW — Switzerland[58]	1090–2060 NP	240–760 NP
12 STW — U.K.[176]		<0.2–330 NP
1 STW (mechanical-biological) — Rome[177]	800 NPEO	8 NPEO
1 Municipal STW — U.S.[178]		143–272 NPEC

APEO with greater than 6 ethoxy units has been observed. The elimination rate decreases with decreasing chain length. In high loading, non-nitrifying conditions removal was reduced.

Temperature has a great influence on alkylphenolic surfactant removal in STW.[58,68,72–74] Removal efficiencies are greater in summer than winter due to increased temperatures. The metabolic rate of microorganisms slows in colder temperatures, so biodegradation decreases. The acclimation time required by bacteria also increases with a decrease in temperature.[74] The acclimation period has also been seen to increase with an increase in influent alkylphenolic compound concentration[75] and ethoxylate chain length.[76]

Three *Pseudomonas* species have been identified as degrading NP$_9$EO: *Pseudomonas putida* strain Fus1BI, *Pseudomonas* sp. strain SscB2, and *Xanthomonas* sp. strain SccB3. None of these species was able to biodegrade 4NP.[77] Another study[71] showed that *Pseudomonas* spp. only degrade the polyether moiety to produce NP$_2$EO. This study also backs up other observations that an attack on the ethoxylate chain at the terminal portion by microorganisms results in shorter mono substitutes and NP. APEOs with less than three ethoxy units are transformed into their corresponding APECs. Intermediate products will not be detected if APEOs are completely degraded by one microorganism, or if the growth rate of a microorganism using an intermediate product is greater than the APEO degrader. Microbial consortia are expected to be necessary due to the amphiphilic nature of the molecules, a consortia may operate synergistically or with a commensalistic relationship where one organism benefits from the breakdown while another is unaffected. Full-scale STW works generally provide greater removal efficiencies than smaller treatment works; this may be due to a more diverse microbial population and nutrient availability being present in larger works.[30]

Maki et al.[59] found that *Pseudomonas* sp. strain TR01 isolated from activated sludge degraded APEOs at an optimum temperature of 30°C and optimum pH 7. The bacteria were unable to mineralize NPEO, but were able to degrade ethoxy units exclusively. The dominant degradation product was NP$_2$EO and a small amount of NP$_2$EC without the presence of any other organisms. This species is expected to play an important role in STW activated sludge processes. A unique substrate assimilability was observed as it metabolizes the ethoxy chain only when the chain is linked to a large hydrophobic group. This work is supported by other studies that have observed shortened but unoxidized NPEO formed by mixed estuarine cultures.[78]

The biodegradation of NP has been observed by *Sphingomonas* sp.[79] *Pseudomonas* sp. was also present although it is thought it provided nutrients for the growth of *Sphingomonas* sp. rather than degrading 4NP itself. More than 95% of the NP was degraded within 10 days and no aromatic compounds were detected suggesting that the phenolic part was also degraded. The main degradation products were alcohols, the major one being nonanol. Different isomers of NP were used. This resulted in the formation of different isomers of nonanol, implying that the alcohols were derived from the alkyl group. *Candida maltosa* is a species of yeast that has been found to degrade 4NP to produce 4-acetylphenol.[80] The yeast was isolated from sludge at a textile industry treatment plant and used 4NP as its sole carbon source.

4.3.3 PESTICIDES

4.3.3.1 Triazine Herbicides

Tests in soils with triazines indicate that abiotic transformations occur[81] and that adsorption to soils results in an increase in half-life. In raw wastewater, triazine herbicides demonstrated negligible adsorption during bench scale primary sedimentation.[82] Changes in solids concentration and residence time had no effect on removal rates, and the compounds passed unchanged into secondary treatment processes. Triazines may partition into lipid structures of biological flocs. They may also chemically bind to bacterial proteins and nucleic acids in activated sludge, rather than adsorbing onto inorganic particulate matter. This may explain the negligible adsorption onto primary waste solids compared to the 40% removal during secondary treatment.[82] In batch test experiments,[83] a loss of 25% for atrazine and a loss of 33% simazine were observed. Primary degradation was thought to be the mechanism responsible. However, this is disputed in other tests[82] in which comparable losses were seen in live and dead activated sludge, making adsorption a more likely mechanism.

Atrazine has been seen to have a detrimental effect on STW at high concentrations.[84] At 20 mg l^{-1}, an increase in effluent chemical oxygen demand (COD) was observed after 4 days with a concomitant reduction in mixed liquor volatile suspended solids and viable (total) bacterial numbers.

4.3.3.2 Organochlorine Insecticides

Organochlorine insecticides behave in a similar manner to PCBs as they sorb to the solid phase in STW during primary and secondary treatment.[85,86] Because of this association with suspended solids, optimization of suspended solids removal should result in optimization of compound removal, but this has not been noted.[87] No obvious correlation between hydraulic and solids loading and pesticide or PCB sorption were observed.[11] It is possible that compounds associate with nonsettleable fine particles. Therefore, removal may not be related to suspended solids removal, since the particle size fraction with which the compounds may be associated will comprise a small portion of the total suspended solids.[87]

Lindane exhibited moderate sorption. Only 1 to 15% was removed through the sludge and was thought to only degrade anaerobically by reductive dehalogenation, although it has been seen to degrade aerobically by other workers.[88,89] After adaptation, substantial removal was observed through biodegradation. Biodegradation was optimal at intermediate and high sludge loadings and poor at high sludge age. At low SRT sorption became particularly important.[88] Removal of lindane by degradation was greatest at high sludge loading (70 to 80% at 0.3 to 0.8 mg BOD MLSS d^{-1}) and poor at high sludge ages (30 to 40% at 25 to 32 days). These results indicate that biodegradation by co-metabolism was the dominant process.[89]

4.3.3.3 Chlorophenoxy Acid Herbicides

Chlorophenoxy acid herbicides (CPHs) have a relatively high aqueous solubility and are less lipophilic and more polar than other pesticides and herbicides, which means that their association with solids is negligible. Partitioning tests showed that CPHs were poorly removed and that removal rates increased slightly with increasing suspended solids and lower flow rates.[90,91] Removal during primary sedimentation is minimal; however, removal during activated sludge treatment is greater since CPHs are reasonably biodegradable.[89] CPH structure appears to be important, indicating that biological mechanisms are involved.[92] Rate constants were highest at intermediate sludge loadings (0.16 to 0.17 mg BOD MLSS d^{-1}) and low at high sludge ages (25 to 32 days). This demonstrates co-metabolic rather than catabolic transformations are responsible.

2,4-dichlorophenoxyacetic acid (2,4-D) does not sorb to the solid phase in appreciable amounts, and biodegradation is a more important process.[89,93] Its transformation from an acid to a short chain ester is an important process. 2,4-D removal is a function of sludge age and sludge loading, respectively. Rate constants were highest at intermediate sludge loadings and lowest at high sludge ages, demonstrating that co-oxidation plays a role.

It has been reported that mecoprop, dichlorprop, and 2,4-D were degraded in activated sludge within 7 days,[94] when 86 to 98% elimination of dissolved organic carbon was also observed. During treatment in sequencing batch reactors, long acclimation periods of approximately 4 months were required before 2,4-D biodegradation was observed.[95] More than 99% removal was achieved after this period and was independent of HRT. The acclimation period required is linearly related to bacterial population density and the initial 2,4-D concentration.[96] 2,4-D was also degraded in anaerobic conditions with a first order rate constant after a lag phase. The main metabolite found after anaerobic degradation on 2,4-D was 2,4-dichlorophenol and small amounts of 4-chlorophenol. Nitschke et al.[97] observed 100% removal of mecoprop in laboratory activated sludge with prenitrification, although a long lag phase was required. Over the same 6-week period, 4% isoproturon and 8% terbutylazine were also removed by degradation.

4.3.3.4 Chlorophenols

Chlorophenol removal is influenced by sorption and degradation mechanisms. High removals have been seen in STW for dichlorophenol as it is easily degraded during activated sludge processes with a limited amount of sorption taking place.[89,98] Dichlorophenol is readily degraded. It is also produced as a biological breakdown product from CPHs and the chlorination of final effluents, resulting in an increase in concentrations found in effluents.

Negligible sorption of pentachlorophenol was found in laboratory scale tests but adsorption to activated sludge was found in STW.[88,99] Sorption could be a significant removal mechanism if the ratio of chemical oxygen demand to pentachlorophenol in the influent is very high. Longer sludge ages result in greater removal rates because pentachlorophenol is toxic to bacteria until acclimation has taken place.[89]

4.3.4 ORGANOTINS

Organotin degradation can involve the sequential removal of organic moieties to produce more toxic products, such as the formation of di- and monobutyltins from debutylation of tributyltin (TBT).[100] The high lipid solubility of organotins allows association with intracellular sites and degradation is known to take place in some bacteria, fungi, and algae. However, in STW, degradation of organotins, either aerobically or anaerobically, is insignificant and adsorption is the most important factor.[101–103] There were no observed differences in TBT degradation during aerobic, anaerobic, mesophilic, and thermophilic digestion.

In a Canadian STW,[103] monobutyltin was found in all influent samples but di- and tributyltin were found occasionally. The average reduction of monobutyltin was 40%; this significant reduction was due to biodegradation and sorption to sludge.

In the primary clarifier in a STW in Zurich,[101] 73% of total butyl tin was eliminated from wastewater due to adsorption, during secondary sedimentation an additional 17% was removed. Further removal during activated sludge treatment was minimal with biodegradation accounting for only 8% removal.

4.3.5 ORGANIC OXYGEN COMPOUNDS

4.3.5.1 Di-(2-ethyhexyl)phthalate

Di-(2-ethylhexyl)phthalate (DEHP) was detected in almost all sewage sludge samples collected by the Swedish Environmental Protection Agency up to 1991 in concentration ranges of 25 to 660 mg kg^{-1} dry weight.[25] Phthalates exhibit an 8 order of magnitude increase in K_{ow} and a 4 order of magnitude decrease in vapor pressure as alkyl chain length increases from 1 to 13 carbon atoms.[104] This tends to suggest that degradation processes will be less important as chain length increases and the compounds become more hydrophobic and less available for degradation.

In activated sludge, diethyl and dibutyl phthalate rapidly degrade by approximately 90% within 3 and 8 days, respectively. Dioctyl phthalate removal was slower (20% after 8 days), although all compounds demonstrated first order rate constants.[105] The degradation rate appears to be inversely proportional to alkyl side chain length.[106] Numerous other studies have demonstrated that a wide range of microorganisms in aerobic and anaerobic conditions degrade phthalate esters. Aerobic biodegradation tests with sewage inocula show that more than 50% ultimate degradation occurred within 28 days and rapid primary degradation occurred (90% within an week) for lower molecular mass compounds. For higher molecular mass phthalates, primary degradation took 12 days and an acclimation period was required.[104] Aerobic and anaerobic degradation initially starts with ester hydrolysis to form the monoester and alcohol (Figure 4.5[11]). In aerobic conditions, further enzymatic degradation of the monoester occurs via phthalic acid results. This is followed by ring cleavage, resulting in pyruvate or oxalacetate or acetyl CoA and succinate production.

Aerobic thermophilic treatment reduced the concentration of DEHP in tests.[107] A 70% reduction was seen within 96 hours at 70°C and an air flow rate of 16 m^3 m^{-3} hr^{-1}. Thermophilic treatment also has the advantage of removing pathogens.

4.3.5.2 Bisphenol A

Bisphenol A is easily removed during activated sludge treatment processes by biodegradation mechanisms.[108–110] The acclimation period required is short and more than 99% removal has been seen.[108] The two major metabolites produced are 2,2-*bis* (4-hydroxyphenyl) -1-propanol and 2,3-*bis* (4-hydroxyphenyl) -1,2-propanediol.[110,111]

ROOC-Ph-COOR → HOOC-Ph-COOR → HOOC-Ph-COOH → CH$_4$ + CO$_2$

Phthalate ester→ mono-phthalate ester → phthalate alcohol → methane + carbon dioxide

FIGURE 4.5 Theorized degradation pathway for phthalate esters.

4.3.6 POLYAROMATIC COMPOUNDS

4.3.6.1 Polychlorinated Biphenyls

PCBs are stable molecules with low aqueous solubilities and biological, chemical, and physical recalcitrance. As a result, they exhibit minimal degradation during wastewater treatment processes.[21,112] However, PCB degradation has been seen to occur to some extent both aerobically and anaerobically.[113] Nonbiological elimination mechanisms of PCB removal during activated sludge treatment, such as mass transfer to the atmosphere and chemical degradation, were dismissed due to the affinity of PCBs for suspended solids and their recalcitrance.[114]

PCB concentrations were 3 to 4 times higher in total atmosphere fallout than in wastewater. The influent concentrations of a sum of 7 PCB congeners were 15 to 26 ng l^{-1} in dry conditions and 31.5 to 53 ng l^{-1} in wet conditions. This indicates that PCBs in wastewater mainly originate from the atmosphere via rainwater washout.

PCB degradation in activated sludge decreases with an increasing number of chlorine atoms.[114–116] A chlorine content of greater than 42% prevents degradation. *Pseudomonas*, *Alcaligenes*, *Arthrobacter*, and *Acinetobacter* have all been found to be responsible for degradation processes. Initially, dioxygenation results in metabolites hydroxylated in the 2 and 3 positions; these are then degraded in turn by *meta* ring cleavage to ultimately give chlorinated benzoic acids.

The major mechanism for PCB removal is via adsorption to suspended matter and sludge flocs.[117] Direct correlation between the concentration of particulate matter in the raw wastewater and PCB removal has been observed,[21] and a relationship between PCBs and nonsettleable solids has been seen.[87,112] Primary sedimentation removed approximately 45% of the PCB load in a pilot plant study.[112] A 10- to 100-fold increase was seen from influent to primary sedimentation sludge, and final effluent loadings of 0.01 g total PCB day^{-1} were a 10-fold decrease on the influent.

PCB removal positively correlates with increasing sludge age.[114] A 9-day sludge age gave significantly lower effluent concentrations than at 4 days. Removal of 68% of PCBs was observed at a typical dry weather flow (DWF) (0.111 l s^{-1}). This decreased to 48% removal under higher flow conditions (0.333 l s^{-1}) and was 58% when flow varied in a typical daily manner (0.075 to 0.168 l s^{-1}).[118] PCB removal efficiencies were highest at DWF and lowest at 3 times DWF (Table 4.7), their removals being comparable with suspended solids removal. This is due to higher hydraulic loadings and surface loading rates associated with a retention time decrease.[118]

Air stripping has been noted as an important factor for compounds with an H_c greater than 100 Pa m^3 hr^{-1}. This makes less-chlorinated PCBs potentially susceptible,[119] although this is unlikely due to their high affinities for mixed liquor suspended solids (MLSS).[114] When compared to MLSS/effluent ratios, the most lipophilic compounds have the highest concentration ratios.

4.3.6.2 Polyaromatic Hydrocarbons

Polyaromatic hydrocarbon compounds (PAHs) are believed to be persistent in the environment, although there is some work that sees degradation within 12 to 18

TABLE 4.7
PCB Removal through STW with Varying DWF[118]

Flow	Average Concentrations (ng l⁻¹)		
	Raw Sewage	Settled Sewage	Primary Sludge
1 DWF	11.6	3.6	151.9
3 DWF	10.3	5.4	304.2
Variable	20.0	8.4	110.1

Source: From Garcia-Gutierrez, A.G., McIntyre, A.E., Perry, R., and Lester, J. N., The behaviour of PCBs in the primary sedimentation process of sewage treatment: a pilot plant study, *Sci. Total Environ.*, 22, 243, 1982. With permission.

hours.[120] In a conventional STW, degradation times could be as long as 80 to 600 hours, since the experiments were run in ideal conditions with temperatures of 20°C and with pre-adapted bacteria. PAH removal of 35.1 to 86.1% during one sedimentation process was observed.[121] During volatilization, significant removal was seen for 7 out of 13 PAHs, and during photodegradation 9 compounds demonstrated significant losses in settled sewage. Adsorption to suspended solids reduces the susceptibility of PAH to photodegradation.

The removal of these compounds during primary sedimentation is principally a function of physicochemical properties and unit process performance; their removal correlates with suspended solids removal.[121,122] Association with dissolved and colloidal matter in raw sewage dominated lower molecular weight compound removal. Low molecular weight PAH removal is dependant on suspended solids loading with mean removals of smaller compounds increased with increasing suspended solids loading. Removal of high molecular weight compounds is dependant on hydraulic loading. In Greece, low molecular weight compounds were removed efficiently but higher molecular weight compounds are resistant to biological degradation.[123,124] For example, during primary treatment naphthalene showed a 43% removal and acenaphthene showed 64% loss. Some compound concentrations were seen to increase because of supernatant return streams to influent prior to primary clarification. Good mass balances have been demonstrated during secondary treatment. Lower mass compounds showed losses of greater than 40% suggesting biodegradation or volatilization. Table 4.8 displays concentrations of PAH at various treatment stages.

PAHs' removal during primary sedimentation is a function of molecular weight and suspended solids removal efficiency, since they tend to partition onto the solid phase.[38,125–128] It has been estimated that 64% of total PAH sorb onto the solid phase during primary sedimentation. Because of their association with the solid phase, removal may be enhanced by more efficient physical solid-liquid separation techniques such as tertiary filters.[125]

Manoli and Samara[123] observed that sorption was the dominant process during primary treatment, particularly the higher mass compounds. Removal during secondary treatment is dependent on volatilization and biodegradation. Total mass balances showed that lower mass compounds were effectively removed but larger

TABLE 4.8
PAH Concentrations at Various Treatment Stages

	Raw Sewage		Primary Effluent		Secondary Effluent	
	Liquid	Solid	Liquid	Solid	Liquid	Solid
	$\mu g\ l^{-1}$	$\mu g\ g^{-1}$	$\mu g\ l^{-1}$	$\mu g\ g^{-1}$	$\mu g\ l^{-1}$	$\mu g\ g^{-1}$
Naphthalene	3.9	3.3	3.5	3.8	0.2	0.4
Acenaphthene	0.2	0.7	0.2	0.8	0	0
Fluorene	0.3	13.4	0.2	11.9	0	1.5
Pyrene	0.2	1.1	0.2	10	0	2.3

Source: From Melcer, H., Steel, P., and Bedford, W.K., Removal of polycyclic aromatic hydrocarbons and heterocyclic nitrogen compounds in a municipal treatment plant, *Water Environ. Res.*, 67, 926, 1995. With permission.

compounds were not. Fate modeling using Toxchem[125] demonstrated that increases in cross-sectional areas of primary and secondary clarification tanks would not improve PAH removal. Biomass also seemed unaffected by increases in SRT. This may be due to a reduction in effluent PAH concentration that no longer constitutes an effective substrate concentration for secondary metabolism. It may also be because of the difficulty in degrading PAH.

Atmospheric depositions were found not to be responsible for PAH distribution in STW in Paris.[113] Concentrations in STWs were higher after rainfall after a dry period. This indicates that PAHs enter STWs through ground leaching.

4.3.6.3 Polybrominated Flame Retardants

Very little has been reported on PBDEs flame retardants in STW. They have been detected near to site discharges, but their fate is still unknown. PBDEs are hydrophobic and are likely to sorb to the biomass in activated sludge treatment and be transported primarily by adsorption to particulate matter. Their K_{ow} values have been measured and support this theory (Table 4.2). Pre-1990 sewage sludge samples were analyzed for PCB and dichlorodiphenyltrichloroethanes alongside PBDEs and the levels of single congeners were comparable.[129] More recent samples had higher PBDE concentrations than PCB. Concentrations of BDE-47 (the tetra congener, 2,2',4,4',-TeBDE) and BDE-100 (the penta congener, 2,2',4,4',6-PeBDE) in 1988 were 14 to 110 ng g^{-1} and 3.3 to 28 ng g^{-1}, respectively. Approximately 10 years later, concentrations of 72 to 130 ng g^{-1} and 21 to 40 ng g^{-1} were detected.

Tetrabromobisphenol A (TBBPA) has been found in STWs. Concentrations in sludge from an STW that receives landfill leachate used by the plastics industry had levels of 100 ng g^{-1} ignition loss compared to 65 ng g^{-1} ignition loss for STW with no connection to the plastics industry.[130] TBBPA is removed during wastewater treatment, but whether this is due to physical or biological parameters is not known.[131]

For the breakdown of brominated compounds, microorganisms must be capable of cleavage of carbon-bromine bonds. A lag phase would be required for this to occur to allow the microorganisms to produce the hydrohalidase and dehalogenase enzymes necessary for dehalogenation. In anaerobic conditions, reductive dehalogenation may occur. This is the replacement of a bromine atom with hydrogen. Hydrolytic dehalogenation is the replacement of a bromine with a hydroxyl group.[132] One study[133] found a good correlation between the concentrations of polybrominated dibenzofurans and dibenzo dioxins. The study concluded that the main source of the organic compounds was from PBDEs.

4.3.7 ORGANO MERCURY COMPOUNDS

The fate of inorganic mercury in STW is well documented,[134,135] but less is known about possible methylation. It is expected that methylation and demethylation processes will occur in biological treatment systems, since many *Pseudomonas* spp. can methylate mercury.[136] Higher removal efficiencies have been reported during activated sludge treatment than during primary sedimentation.[137] Higher temperatures and higher sludge age are likely to increase methylation; however, in highly aerobic conditions demethylation may be the dominant mechanism. Association with settleable particulates has also been observed,[134,138] although removal purely through settled solids does not explain the full extent of removal. Physicochemical adsorption of soluble mercury in activated sludge has also been seen.[139] No detectable levels of methylmercury were found in the influent, indicating methylation occurred during treatment. In addition, no detectable levels were found in the final effluent suggesting all was adsorbed onto the biomass.[134]

4.3.8 CASE STUDY

The fate of nitrilotriacetic acid (NTA) during wastewater treatment, which has been well characterized, may provide an insight into the factors that could influence and the mechanisms that may control the fate and removal of the much less well understood EDCs. This compound has been considered for use as a detergent builder in place of phosphates, and although it is not implicated as an EDC, it is a cause for concern as it is able to sequest heavy metals and produce chelates that may cause mutagenic or teratogenic effects.[140] While NTA does degrade aerobically during biological wastewater treatment, with 90% removal reported,[114] the process is more sensitive to environmental perturbations than the removal of natural BOD, which is consistent with the degradation of other anthropogenic organic chemicals.[140]

During primary sedimentation NTA removal occurs through sorption to solids, with 16 to 40% removal.[142] During secondary treatment removal through biodegradation is more significant. However, it is subject to considerable variation and is influenced by temperature, water hardness, influent NTA concentrations, hydraulic loading and SRT.[143–150] Sorption onto sludge has been studied,[151] however, adsorption processes are thought to be only an intermediate process in the passage of the compound into bacterial cells, since catabolism of NTA is an intracellular process.[152]

The removal of NTA also demonstrates that temperature can play an important role in biotransformation processes (Figure 4.6). In temperate regions, temperatures

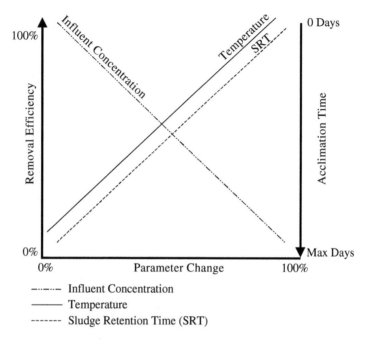

FIGURE 4.6 Effects of changing parameters on NTA acclimation time and removal efficiencies.

in STW frequently fall to approximately 7.5°C in winter,[153] and reductions in removal rates have been seen to occur at this level. Little effect on removal efficiency appears to occur above 10°C, but, at lower temperatures removal was reduced.[144,146,147,153,154] In studies using laboratory scale plant, NTA removal at 20°C at influent concentrations of 5 and 20 mg l^{-1} are essentially complete. When the temperature was reduced to 5°C, minimal degradation was observed; as the temperature was subsequently increased, removal rates again reached 95%.

The significance of variations in influent concentration is that there is frequently a time lag before degradation occurs. During this period the microbial population acclimatizes to the presence of a compound. This has been observed in the case of NTA, with increases in concentration leading to increases in acclimation time (Tables 4.9 and 4.10). At an influent concentration of 10 mg l^{-1}, an acclimation time of 23 days was required to achieve 95% removal. However, on doubling the concentration, removal rates decreased with effluent concentrations rising from 0 to 1.7 mg l^{-1} to 3.6 to 4.4 mg l^{-1} for a period of 10 days until 95% removal was again achieved.[150] Shorter-term variations in influent NTA concentration had less effect on its removal. However, the impact of a transient increase in influent concentration did cause a decrease in NTA removal in a pilot plant study, with adaptation occurring after 6 hours with corresponding improvement of removal efficiencies.[154] The importance of understanding the effects of variations in a combination of parameters on biological degradation mechanisms is illustrated by the impact of both transient increases in influent NTA concentration and a reduction in temperature. This temperature

TABLE 4.9
Acclimation Times Required for NTA Removal at Different Process Variable During Activated Sludge Treatment

Sludge Age (Days)	NTA Influent Concentration (mg l⁻¹)	MLSS (mg l⁻¹)	Acclimation Time (Days)
4	7.5	1514	15
4	15	—	20
9	7.5	3049	12
9	15	—	16
12	7.5	3536	6

Source: From Rossin, A.C., Lester, J. N., and Perry, R., Removal of nitrilotriacetic acid during primary sedimentation and its effects on metal removal, *Environ. Poll. (Series B)*, 4, 315, 1982. With permission.

TABLE 4.10
Effects of Parameter Changes on NTA Acclimation Time and Biodegradation Rates

	Parameter	Effect of Change
Acclimation	Increased temperature	Decreased acclimation
	Increased SRT	Decreased acclimation
	Increased influent concentration	Increased acclimation
Biodegradation	Increased temperature	Increased biodegradation
	Increased SRT	Increased biodegradation
	Increased influent concentration	Decreased biodegradation
	Increased HRT	Decreased biodegradation

reduction did cause significant decreases in NTA removal, indicating some synergistic effects (Figure 4.6). A further example of this type of effect is observed with NTA degradation: short-term increases in hydraulic loading in combination with a reduction in temperature also significantly reduced removal efficiencies.[145]

The impact of sludge age on removal of organic compounds is also demonstrated by its impact on acclimation time and removal efficiencies of NTA. Increasing sludge age can lead to a reduction in acclimation time required for NTA removal (Figure 4.6). In an activated sludge pilot plant with low metal concentrations, the acclimation period was 20 days with a 4 day sludge age. At a 9 day sludge age, the acclimation time reduced to 15 days (Table 4.9).[142] Plants operating at higher sludge ages have higher MLSS concentration with a low rate of sludge wastage. This has been shown to reduce acclimation time[152] and results in greater removal rates.[143] Once the system is acclimatized, removal has been shown to be dependent on MLSS concentration and sludge age. Although NTA removal has been seen to correlate with hydraulic conditions, no correlation with suspended solids or COD removal was observed.[148]

For use as a detergent builder, NTA is used to complex with calcium and magnesium. However, other metal complexes, such as cadmium, copper, lead, nickel, and zinc have greater stability constants and will therefore be preferentially formed during wastewater treatment.[152] The presence of metals, resulting in formation of different NTA-metal complexes, has been demonstrated to increase acclimatization time from 10 to 15 days in a pilot scale activated sludge plant.[152] Water hardness also affects NTA removal efficiencies; no degradation occurred over 60 days in soft water compared to an acclimation time of 16 to 31 days in the same test with hard water, as the calcium complex is more readily biodegradable.[149] A range of organic compounds form complexes with metals and these effects are unlikely to be limited to NTA. Moderately strong metal-complexing ligands, such as activated sludge biopolymers and humic substances, are responsible for the complexation of some metals.[155] They are also responsible for complexation with synthetic compounds such as the booster biocide pyrithione[156] that forms zinc-complexes.

The fate of NTA in anaerobic digestion of sludge is also worth consideration. In laboratory experiments with sludges previously unexposed to NTA, no anaerobic degradation was observed to occur.[157] However, subsequent work utilizing waste activated sludge from a pilot plant which had been acclimatized to NTA demonstrated an ability to subsequently degrade NTA in anaerobic digesters.[158] The waste activated sludge will be essentially free of NTA (or of compounds that are readily degraded in aerobic treatment), as aerobic degradation will have occurred. However, when this sludge was co-digested with primary sludges, which contain sorbed contaminants, the previous acclimatization in aerobic conditions allows for subsequent anaerobic degradation. Aerobically acclimatized sludge at addition of 15% to primary sludge was able to reduce NTA concentrations from 28 mg l^{-1} to <0.5 mg l^{-1} over 10 days with a lag time of 4 days.[158]

This case study illustrates that the removal of synthetic organic compounds from STWs effluent is dependent to a certain extent on parameters, such as:

1. Temperature
2. Sludge age
3. Treatment type
4. Influent concentrations
5. Concentration of co-metabolites
6. HRT

It is therefore possible that these parameters will also affect the removal of some EDCs. Continued research into this area is required in order to elucidate EDC removal mechanisms, the parameters affecting removal rates, and the production of metabolites that may be endocrine disrupting.

The foregoing is only intended to provide possible mechanisms by which EDCs may biodegrade or the factors that may influence the processes. It is highly probable that co-metabolism will be important at the much lower concentrations at which EDCs exist (μg l^{-1} or ng l^{-1}) in raw wastewater as compared to those that NTA would exist in (mg l^{-1}). In addition, the biochemical activity of the bacteria cells is not confined to viable cells alone. It also includes non-viable intact cells which are biochemically active [159.] These cells have been termed moribund or senescent, and such cells may also have an important function that is still unrecognized. Moreover, these cells are a transitory state prior to their lysis and release of intracellular enzymes into the mixed liquor where they continue to be metabolically active for some time.[10] However, under such extracellular conditions, it is highly unlikely in the absence of cellular constraints that the specificity of these active enzymes is retained.

It is in activated sludge with the highest sludge age that the highest concentrations of transitory moribund and senescent cells exist. The highest concentrations of extracellular enzymes exist as well, which under conditions of more rapid growth would be intracellular. High sludge ages characterize activated sludge plants that are producing fully nitrified effluents and converting ammonium ions to nitrate ions. The nitrifying genera of bacteria *Nitrosomonas* and *Nitrobacter* are autotrophic, very slow growing, and require long SRTs (i.e., high sludge age) to avoid washout. To convert ammonium to nitrite then nitrate, they utilize high concentrations of monooxygenase enzymes. If these are released into solution in the mixed liquor in the absence of cellular control they will oxidize other compounds in addition to ammonium and nitrite ions. These other compounds could include some EDCs and explain the observations of Vader et al.[50]

Physical factors such as temperature, dissolved oxygen content, and pH can all influence biodegradation by affecting the growth of the aerobes. Sludge pH is important because it affects the bioavailability of compounds due to their changing solubility, sorption potential, and aerobic growth. Most activated sludge plants operate at a neutral pH, although some aerobe species may be more efficient in acidic or alkaline conditions.[160,161]

Temperature influences the activity of the microbial community and metabolic pathways; generally mesophilic conditions are used. A reduction in temperature will reduce the effluent quality as the metabolic rate of the organisms slows. The effluent COD and effluent suspended solids would be likely to increase during periods of low metabolic activity. In one study looking at NTA removal,[162] a reduction in temperature resulted in a denser liquid fraction in the aerator that prevented good floc settling and may have caused an increase in effluent suspended solids; recovery of the system occurred when the temperature returned to normal. STWs are operated at ambient temperatures that are fluctuating rather than in temperature controlled environments; therefore, temperature effects are inevitable.

Further research using pilot plants to assess treatability is the next important research step. Information on the fate of some EDCs during wastewater treatment is limited particularly for PBDEs, for example, and requires further research. Wastewater treatment standards and consent limits differ greatly between different countries, resulting in varying effluent and sludge quality.

This chapter highlights the importance of treatment conditions as well as compound characteristics in controlling EDC distribution throughout the STW. The concentration of some EDCs in sewage sludge is greatly influenced by the wastewater treatment process. One of the most important factors observed is the sludge age during activated sludge treatment. A higher sludge age results in a faster acclimation period followed by more complete degradation. It also increases the sorption potential because of higher suspended solids concentration in the mixed liquor. The literature available on APEOs illustrates this particularly well. [58,72.]

The K_{ow}, K_{oc}, and H_c can be used to determine the effectiveness of sorption, degradation, and volatilization during treatment. The removal of the majority EDCs from wastewater streams in efficient STW is generally not a major concern. The problem arises from their hydrophobic nature and tendency to sorb to the solid phase. This results in amplified concentrations in sludge making disposal of sewage sludge the major concern. EDCs applied to land in sludge have the potential to enter the food chain or watercourses.

REFERENCES

1. Brown, J.F., Bedard, D.L., Brennan, M.J., Carnahan, J.C., Feng, H., and Wagner, R.E., Polychlorinated biphenyl dechlorination in aquatic sediments, *Science*, 236, 709, 1987.
2. Brown, J.F., Wagner, R.E., Feng, H., Bedard, D.L., Brennan, M.J., Carnahan, J.C., and May, R.J., Environmental dechlorination of PCBs, *Environ. Toxicol. Chem.*, 6, 579, 1987.
3. Genthner, B.R., Sharak, W.A., Price, I.I., and Pritchard, P.H., Anaerobic degradation of chloroaromatic compounds in aquatic sediments under a variety of enrichment conditions, *Appl. Environ. Microbiol.*, 55, 1466, 1989.
4. Stangroom, S.J., Collins, C.D., and Lester, J.N., Abiotic behaviour of organic, micro pollutants in soils and the aquatic environment: a review. II. Transformations, *Environ. Technol.*, 21, 865, 2000.
5. Lester, J.N. and Edge, D., Sewage and sewage sludge treatment, in *Pollution Causes, Effects and Control*, 4th ed., Harrison, R.M., Ed., The Royal Society of Chemistry, Cambridge, 2000.
6. Hamer, G., Microbial consortia for multiple pollutant biodegradation, *Pure Appl. Chem.*, 69 (11), 2343–2356, 1997.
7. Fronteau, C., Bauwens, W., and Vanrolleghem, P.A., Integrated modelling: comparison of state variables, processes and parameters in sewer and wastewater treatment plant models, *Water Sci. Technol.*, 36, 373, 1997.
8. Hahn, H., Hoffmann, E., and Schafer, M., Managing residuals from wastewater treatment for priority pollutants, *Eur. Water Manage.*, 2, 49, 1999.
9. European Commission, Pollutants in Urban Wastewater and Sewage Sludge, London, 2001.
10. Lester, J.N. and Birkett, J.W., *Microbiology and Chemistry for Environmental Scientists and Engineers*, 2nd ed., E. and F.N. Spon, London, 1999.
11. Rogers, H.R., Sources, behaviour and fate of organic contaminants during sewage treatment and in sewage sludges, *Sci. Total Environ.*, 185, 3, 1996.
12. Dobbs, R.A., Wang, L., and Govind, R., Sorption of toxic organic compounds on waste water solids: correlation with fundamental properties, *Environ. Sci. Technol.*, 23, 1092, 1989.
13. Byrns, G., The fate of xenobiotic organic compounds in wastewater treatment plants, *Water Res.*, 35, 2523, 2001.
14. Danielsson, L.G. and Zhang, Y.H., Methods for determining n-octanol-water partition coefficient, *Trac–Trends Anal. Chem.*, 25, 188, 1996.

15. Furhacker, M., Breithofer, A., and Jungbauer, A., 17β-estradiol: behaviour during waste water treatment, *Chemosphere*, 39, 1903, 1999.
16. Bell, J.P. and Tsezos, M., Removal of hazardous organic pollutants by biomass adsorption, *J. Water Poll. Cont. Fed.*, 59, 191, 1987.
17. Byrns G, The fate of xenobiotic organic compounds in wastewater treatment plants, *Water Res.*, 35, 2523, 2001.
18. Grathwohl, P., Influence of organic matter from soils and sediments from various origins on the sorption of some chlorinated aliphatic hydrocarbons, implications on K_{oc} correlations, *Environ. Sci. Technol.*, 24, 1687, 1990.
19. De Paolis, F. and Kukkonen, J., Binding of organic pollutants to humic and fulvic acids: influence of pH and the structure of humic material, *Chemosphere*, 34, 1693, 1997.
20. Bedding, N.D., McIntyre, A.E., Perry, R., and Lester, J.N., Organic contaminants in the aquatic environment. 11. Behaviour and the fate in the hydrological cycle, *Sci. Total Environ.*, 26, 255, 1983.
21. Chevreuil, M., Granier, L., Chesterikoff, A., and Letolle, R., Polychlorinated biphenyls partitioning in waters from river filtration plant and wastewater plant: the case for Paris, France, *Water Res.*, 24, 1325, 1991.
22. Karickhoff, S.W., Brown, D.S., and Scott, T. A., Sorption of hydrophobic pollutants on natural sediments, *Water Res.*, 13, 241, 1979.
23. Jaffe, R., Fate of hydrophobic organic pollutants in the aquatic environment: a review, *Environ. Poll.*, 69, 237, 1991.
24. Jacobsen, B.N., Nyholm, N., Pedersen, B.M., Poulsen, O., and Ostfeldt, P., Removal of organic micropollutants in lab activated sludge reactors under various operating conditions: sorption, *Water Res.*, 27, 1505, 1993.
25. Alcock, R.E., Sweetman, A., and Jones, K.C., Assessment of organic contaminant fate in waste water treatment plants. I. Selected compounds and physicochemical properties, *Chemosphere*, 38, 2247, 1999.
26. Meakins, N.C., Bubb, J. M., and Lester, J. N., The fate and behaviour of organic micro pollutants during waste water treatment, *Intern. J. Environ. Poll.*, 4, 27, 1994.
27. Meakins, N.C., Bubb, J.M., and Lester, J.N., The fate and behaviour of organic micro pollutants during wastewater treatment, *Intern. J. Environ. Poll.*, 4, 27, 1994.
28. Okutman, D.O.S. and Orhon, D., Hydrolysis of settleable substrate in domestic sewage, *Biotechnol. Lett.*, 23, 1907, 2001.
29. Johnson, A.C., Belfroid, A., and Di Corcia, A., Estimating steroid oestrogen inputs into activated sludge treatment works and observations on their removal from the effluent, *Sci. Total Environ.*, 256, 163, 2000.
30. Maguire, R.J., Review of the persistence of nonylphenol and nonylphenol ethoxylates in aquatic environments, *Water Qual. Res. J. Canada*, 34, 37, 1999.
31. Janke, D. and Fritsche, W., Nature and significance of microbial cometabolism of xenobiotics, *J. Basic Microbiol.*, 25, 603, 1985.
32. Knackmuss, H.J., Basic knowledge and perspectives of bioelimination of xenobiotic compounds, *J. Biotechnol.*, 51, 287, 1996.
33. Rittmann, B.E., Jackson, D.E., and Storck, S. L., *Potential for Treatment of Hazardous Organic Chemicals with Biological Processes*, CRC Press, Boca Raton, FL, 1988.
34. Tabak, H.H., Bloomhuff, R.N., and Bunch, R.L., Steroid hormones as wastewater pollutants II. Studies on the persistence and stability of natural urinary and synthetic ovulation-inhibiting hormones in untreated and treated wastewaters, *Devel. Indust. Microbiol.*, 22, 497, 1981.

35. Ahel, M. and Giger, W., Partitioning of alkylphenols and alkylphenol polyethoxylates between water and organic solvents, *Chemosphere*, 26, 1471, 1993b.

36. Duarte-Davidson, R. and Jones, K.C., Screening the environmental fate of organic contaminants in sewage sludge applied to agricultural soils. II. The potential for transfers to plants and grazing animals, *Sci. Total Environ.*, 185, 59, 1996.

37. Zitomer, D.H.S. and Speece, R.E., Sequential environments for enhanced biotransformation of aqueous contaminants, *Environ. Sci. Technol.*, 27, 227, 1993.

38. Petrasek, A.C., Austern, B.M., and Neiheisel, T. W., Removal and partitioning of volatile organic priority pollutants in wastewater treatment, in 9th US-Japan Conference on Sewage Treatment Technology, Tokyo, Japan, 1983.

39. Melcer, H., Bell, J.P., and Thompson, D., Predicting the fate of volatile organic compounds in municipal wastewater treatment plants, *Water Sci. Technol.*, 25, 383, 1992.

40. Desbrow, C., Routledge, E.J., Brighty, G.C., Sumpter, J.P., and Waldock, M., Identification of estrogenic chemicals in STW effluent. 1. Chemical fractionation and *in vitro* biological screening, *Environ. Sci. Technol.*, 32, 1549, 1998.

41. Baronti, C., Curini, R., D'Ascenzo, G., Di Corcia, A., Gentili, A., and Samperi, R., Monitoring natural and synthetic estrogens at activated sludge sewage treatment plants and in a receiving river water, *Environ. Sci. Technol.*, 34, 5059, 2000.

42. Jobling, S., Nolan, M., Tyler, C.R., Brighty, G., and Sumpter, J.P., Widespread sexual disruption in wild fish, *Environ. Sci. Technol.*, 32, 2498, 1998.

43. Routledge, E.J., Sheahan, D., Desbrow, C., Brighty, G.C., Waldock, M., and Sumpter, J. P., Identification of estrogenic chemical in STW effluent. 2. *In vivo* responses in trout and roach, *Environ. Sci. Technol.*, 32, 1559, 1998.

44. Belfroid, A.C., Van der Horst, A., Vethaak, A.D., Schafer, A.J., Rijs, G.B.J., Wegener, J., and Cofino, W.P., Analysis and occurrence of estrogenic hormones and their glucuronides in surface water and wastewater in the Netherlands, *Sci. Total Environ.*, 225, 101, 1999.

45. Johnson, A.C. and Sumpter, J.P., Removal of endocrine-disrupting chemicals in activated sludge treatment works, *Environ. Sci. Technol.*, 35, 4697, 2001.

46. Ternes, T.A., Kreckel, P., and Mueller, J., Behaviour and occurrence of estrogens in municipal sewage treatment plants. II. Aerobic batch experiments with activated sludge, *Sci. Total Environ.*, 225, 91, 1999.

47. Ternes, T.A., Stumpf, M., Mueller, J., Haberer, K., Wilken, R.D., and Servos, M., Behaviour and occurrence of estrogens in municipal sewage treatment plants. I. Investigations in Germany, Canada and Brazil, *Sci. Total Environ.*, 225, 81, 1999.

48. Turan, A., Excretion of natural and synthetic estrogens and their metabolites: occurrence and behaviour, in *Endocrinically Active Chemicals in the Environment*, German Federal Environment Agency, Berlin, 1995.

49. Matsui, S., Takigami, H., Matsuda, T., Taniguchi, N., Adachi, J., Kawami, H., and Shimizu, Y., Estrogen and estrogen mimics contamination in water and the role of sewage treatment, *Water Sci. Technol.*, 42, 173, 2000.

50. Vader, J.S., van Ginkel, C. G., Sperling, F. M. G. M., de Jong, J., de Boer, W., de Graaf, J. S., van der Most, M., and Stokman, P.G.W., Degradation of ethinyl estradiol by nitrifying activated sludge, *Chemosphere*, 241, 1239, 2000.

51. Korner, W., Bolz, U., Submuth, W., Hiller, G., Schuller, W., Hanf, V., and Hagenmaier, H., Input/output balance of estrogenic active compounds in a major municipal sewage plant in Germany, *Chemosphere*, 40, 1131, 2000.

52. Johnson, K., The Partitioning of Natural and Synthetic Estrogens between Aqueous and Solid phases, MSc. thesis, Imperial College of Science Technology and Medicine, London, 1999.

53. Lai, K.M., Johnson, K.L., Scrimshaw, M.D., and Lester, J.N., Binding of waterborne steroid estrogens to solid phases in river and estuarine systems, *Environ. Sci. Technol.*, 34, 3890, 2000.

54. Takigami, H., Taniguchi, N., Matsuda, T., Yamada, M., Shimizu, Y., and Matsui, S., The fate and behaviour of human estrogens in night soil treatment processes, *Water Sci. Technol.*, 42, 45, 2000.

55. Layton, A.C., Gregory, B.W., Seward, J.R., Schuktz, T.W., and Sayler, G.S., Mineralization of steroidal hormones by biosolids in wastewater treatment systems in Tennessee, USA, *Environ. Sci. Technol.*, 34, 3925, 2000.

56. Routledge, E.J. and Sumpter, P., Structural features of alkylphenolic chemicals associated with estrogenic activity, *J. Biol. Chem.*, 272, 3280, 1997.

57. Jobling, S., Reynolds, T., White, R., Parker, M.G., and Sumpter, J.P., A variety of environmentally persistent chemicals, including some phthalate plasticizers, are weakly estrogenic, *Environ. Health Pers.*, 103, 582, 1995.

58. Ahel, M., Giger, W., and Koch, M., Behaviour of alkylphenol ethoxylates surfactants in the aquatic environment. 1. Occurrence and transformation in sewage treatment, *Water Res.*, 28, 1131, 1994.

59. Maki, H., Masuda, N., Furiwara, Y., Ike, M., and Fujita, M., Degradation of alkylphenol ethoxylates by *Pseudomonas* sp. strain TR01, *Appl. Environ. Microbiol.*, 60, 2265, 1994.

60. Carvalho, G., Paul, E., Novais, J. M., and Pinheiro, H.M., Studies on activated sludge response to variations in the composition of a synthetic surfactant-containing feed effluent, *Water Sci. Technol.*, 42, 135, 2000.

61. Nimrod, A.C. and Benson, W.H., Environmental estrogenic effects of alkylphenol ethoxylates, *Crit. Rev. Toxicol.*, 26, 335, 1996.

62. Kvestak, R., Terzic, S., and Ahel, M., Input and distribution of alkylphenol polyethoxylates in a stratified estuary, *Marine Chem.*, 46, 89, 1994.

63. Ejlertsson, J., Nilsson, M.L., Kylin, H., Bergman, A., Karlson, L., Oquist, M., and Svensson, B. H., Anaerobic degradation of nonylphenol mono- and diethoxylates in digester sludge, landfilled municipal solid waste, and landfilled sludge, *Environ. Sci. Technol.*, 33, 301, 1999.

64. John, D.M., House, W.A., and White, G.F., Environmental fate of nonylphenol ethoxylates: Differential adsorption of homologs to components of river sediment, *Environ. Toxicol. Chem.*, 19, 293, 2000.

65. Lee, H.B. and Peart, T.E., Occurrence and elimination of nonylphenol ethoxylates and metabolites in municipal wastewater and effluents, *Water Qual. Res. J. Canada*, 33, 389, 1998.

66. Isobe, T., Nishiyama, H., Nakashima, A., and Takada, H., Distribution and behavior of nonylphenol, octylphenol, and nonylphenol monoethoxylate in Tokyo metropolitan area: their association with aquatic particles and sedimentary distributions, *Environ. Sci Technol.*, 35, 1041, 2001.

67. Tanghe, T., Devriese, G., and Verstraete, W., Nonylphenol degradation in lab scale activated sludge units is temperature dependent, *Water Res.*, 32, 2889, 1998.

68. Ahel, M., Molnar, E., and Giger, W., Estrogenic metabolites of alkylphenol polyethoxylates in secondary sewage effluents and rivers, *Water Sci. Technol.*, 42, 15, 2000.

69. Lee, H. B., Weng, J., Peart, T.E., and Maguire, R.J., Occurrence of alkylphenoxyacetic acids in Canadian sewage treatment plant effluents, *Water Qual. Res. J. Canada*, 33, 19, 1998.

70. Solé, M., de Alda, L.M.J., Castillo, M., Porte, C., Ladegaard-Pedersen, K., and Barceló, D., Estrogenicity determination in sewage treatment plants and surface waters from the Catalonian area (NE Spain), *Environ. Sci. Technol.*, 34, 5076, 2000.

71. van Ginkel, C.G., Complete degradation of xenobiotic surfactants by consortia of aerobic microorganisms, *Biodegradation*, 7, 151, 1996.

72. Birch, R., Prediction of the fate of detergent chemicals during sewage treatment, *J. Chem. Technol. Biotechnol.*, 50, 411, 1991.

73. Maruyama, K., Yuan, M., and Otsu, A., Seasonal changes in ethylene oxide chain length of poly(oxyethylene)alkylphenyl ether nonionic surfactants in three main rivers in Tokyo, *Environ. Sci. Technol.*, 34, 343, 2000.

74. Naylor, C.G., Environmental fate and safety on nonylphenol ethoxylates, *Alkyl. Alkyl. Ethoxy. Rev.*, 1, 23, 1998.

75. Figueroa, L.A., Miller, J., and Dawson, H.E., Biodegradation of two ethoxylated nonionic surfactants in sequencing batch reactors, *Water Environ. Res.*, 69, 1282, 1997.

76. Ball, H., Reinhard, M., and McCarty, P., Biotransformation of halogenated and nonhalogenated octylphenol polyethoxylate residues under aerobic and anaerobic conditions, *Environ. Sci. Technol.*, 23, 951, 1989.

77. Frassinetti, S., Isoppo, A.L., Corti, A., and Allini, G.V., Bacterial attack of nonionic aromatic surfactants: Comparison of degradative capabilities of new isolates from nonylphenol polyethoxylate polluted waste waters, *Environ. Technol.*, 17, 1999, 1996.

78. Kvestak, R. and Ahel, M., Biotransformation of nonylphenol ethoxylates surfactants by estuarine mixed bacterial cultures, *Arch. Environ. Contam. Toxicol.*, 29, 551, 1995.

79. Fujii, K., Urano, N., Ushio, H., Satomi, M., Iida, H., Ushio-Sata, N., and Kimura, S., Profile of a nonylphenol-degrading microflora and its potential for bioremedial applications, *J. Biochem.*, 128, 909, 2000.

80. Corti, A., Frassinetti, S., Vallini, G., D'Antone, S., Fichi, C., and Solaro, R., Biodegradation of nonionic surfactants. I. Biotransformation of 4-(1-Nonyl)phenol by a *Candida maltosa* isolate, *Environ Poll.*, 90, 83, 1995.

81. Burkhard, N. and Guth, J.A., Chemical hydrolysis of 2-chloro-4,6-bis(alkylamino)-1,3,5-triazine herbicides and their breakdown in soil under the influence of adsorption, *Pest. Sci.*, 12, 45, 1981.

82. Meakins, N.C., Bubb, L.M., and Lester, J.N., The fate and behaviour of s-triazine herbicides, atrazine and simazine, in primary and secondary wastewater treatment processes, *Chemosphere*, 155, 61, 1994.

83. Leoni, V., Cremisini, C., Giovinazzo, R., Puccetti, G., and Vitali, M., Activated sludge biodegradation test as a screening method to evaluate persistence of pesticides in soil, *Sci. Total Environ.*, 123/124, 279, 1992.

84. Nsabimana, E., Bohatier, J., Belan, A., Pepin, D., and Charles, L., Effects of the herbicide atrazine on the activated sludge process: microbiology and functional views, *Chemosphere*, 33, 479, 1996.

85. Hannah, S.A., Austern, B.M., Eralp, A.E., and Dobbs, R.A., Comparative removal of toxic pollutants by six wastewater treatment processes, *J. Water Poll. Cont. Fed.*, 58, 27, 1986.

86. Hannah, S.A., Austern, B.M., Eralp, A.E., and Dobbs, R.A., Removal of organic toxic pollutants by trickling filter and activated sludge, *J. Water Poll. Cont. Fed.*, 50, 1281, 1988.

87. Garcia-Gutierrez, A.G., McIntyre, A.E., Perry, R., and Lester, J.N., Behaviour of persistent organochlorine micropollutants during primary sedimentation of waste water, *Sci. Total Environ.*, 39, 27, 1984.
88. Jacobsen, B.N., Nyholm, N., Pedersen, B.M., Poulsen, O., and Ostfeldt, P., Microbial degradation of pentachlorophenol and lindane in laboratory scale activated sludge reactors, *Water Sci. Technol.*, 23, 349, 1991.
89. Nyholm, N., Jacobsen, B.O., Pedersen, B.M., Poulsen, O., Damborg, A., and Schultz, B., Removal of organic micropollutants at ppb levels in lab activated sludge reactors under various operating conditions: biodegradation, *Water Res.*, 26, 339, 1992.
90. Hill, N.P., McIntyre, A.E., Perry, R., and Lester, J.N., Development of a method for the analysis of chlorophenoxy herbicides in wastewaters and wastewater sludges, *Int. J. Environ. Anal. Chem.*, 15, 107, 1983.
91. Hill, N.P., McIntyre, A.E., Perry, R., and Lester, J.N., Behaviour of chlorophenoxy herbicides during primary sedimentation, *J. Water Poll. Cont. Fed.*, 57, 60, 1985.
92. Hill, N.P., McIntyre, A.E., Perry, R., and Lester, J.N., Behaviour of chlorophenoxy herbicides during the activated sludge treatment of municipal wastewater, *Water Res.*, 20, 45, 1986.
93. Saleh, F.L., Lee, G.F., and Wolf, H.W., Selected organic pesticides, occurrence, transformation and removal form domestic wastewater, *J. Water Poll. Cont. Fed.*, 52, 19, 1980.
94. Zipper, C., Bolliger, C., Fleischmann, T., Suter, E.M.J., Angst, W., Muller, M.D., and Kohler, H-P.E., Fate of the herbicides mecoprop, dichloroprop, and 2,4-D in aerobic and anaerobic sewage sludges as determined by laboratory batch studies and enantiomer-specific analysis, *Biodegradation*, 10, 271, 1999.
95. Mangat, S.S. and Elefsiniotis, P., Biodegradation of the herbicide 2,4-dichlorophenoxyacetic acid (2,4-D) in sequencing batch reactors, *Water Res.*, 33, 861, 1999.
96. Greer, C.W., Hawari, J., and Samson, S., Influence of environmental factors on 2,4-dichlorophenoxyacetic acid degradation by *Pseudomonas cepacia* isolated from peat, *Arch. Microbiol.*, 154, 317, 1990.
97. Nitschke, L., Wilk, A., Schüssler, W., Metzner, G., and Lind, G., Biodegradation in laboratory activated sludge plants and aquatic toxicity of herbicides, *Chemosphere*, 39, 2313, 1999.
98. Chudoba, J., Albokava, J., Lentage, B., and Kummel, R., Biodegradation of 2,4-dichlorophenol by activated sludge microorganisms, *Water Res.*, 23, 538, 1989.
99. Tsezos, M. and Bell, J.P., Comparison of the bisorption and desorption of hazardous organic pollutants by live and dead biomass, *Water Res.*, 23, 561, 1989.
100. Gadd, G.M., Microbial interactions with tributyltin compounds: detoxification, accumulation, and environmental fate, *Sci. Total Environ.*, 258, 119, 2000.
101. Fent, K., Organotin compounds in municipal wastewater and sewage sludge: contamination, fate in treatment processes and ecotoxicological consequences, *Sci. Total Environ.*, 185, 151, 1996.
102. Fent, K., *Organotin in Municipal Wastewater and Sewage Sludge*, 1st ed., Chapman and Hall, London, 1996.
103. Chau, Y.K., Zhang, S., and Maguire, R.J., Occurrence of butyltin species in sewage and sludge in Canada, *Sci. Total Environ.*, 121, 271, 1992.
104. Staples, C.A., Peterson, D.R., Parkerton, T.F., and Adams, W.J., The environmental fate of phthalate esters: a literature review, *Chemosphere*, 35, 667, 1997.
105. Jianlong, W., Ping, L., and Yi, Q., Biodegradation of phthalic esters by acclimated activated sludge, *Environ. Int.*, 22, 737, 1996.
106. O'Grady, D.P., Howard, P.H., and Werner, A.F., Activated sludge biodegradation of 12 commercial phthalate esters, *Appl. Environ. Microbiol.*, 49, 443, 1985.

107. Banat, F.A., Prechtl, S., and Bischof, F., Experimental assessment of bio-reduction of di-2-ethylhexyl phthalate (DEHP) under aerobic thermophilic conditions, *Chemosphere*, 39, 2097, 1999.

108. Matsui, S., Murakami, T., Sasaki, T., Hirose, Y., and Iguma, Y., Activated sludge degradability of organic substances in the wastewater of the Kashima petroleum and petrochemical industrial complex in Japan, *Prog. Water Technol.*, 7, 645, 1975.

109. Matsui, S., Okawa, Y., and Ota, R., Experience of 16 years operation and maintenance of the Fukashiba industrial wastewater treatment plant of the Kashima petrochemical complex. II. Biodegradation of 37 organic substances and 28 process wastewaters, *Water Sci. Technol.*, 20, 201, 1988.

110. Staples, C.A., Dorn, P.B., Klecka, G.M., O'Block, S.T., and Harris, L.R., A review of the environmental fate, effects, and exposures of bisphenol A, *Chemosphere*, 36, 2149, 1998.

111. Spivack, J., Leib, T.K., and Lobos, J.H., Novel pathway for bacterial metabolism of bisphenol A. Rearrangement and stilbene cleavage in bisphenol A metabolism, *J. Biol. Chem.*, 269, 7323, 1994.

112. Morris, S.L. and Lester, J.N., Behaviour and fate of polychlorinated biphenyls in a pilot wastewater treatment plant, *Water Res.*, 28, 1553, 1994.

113. Blanchard, M., Teil, M.J., Ollivon, D., Garban, B., Chesterikoff, C., and Chevreuil, M., Origin and distribution of polyaromatic hydrocarbons and polychlorinated biphenyls in urban effluents to wastewater treatment plans of the Paris area (France), *Water Res.*, 35, 3679, 2001.

114. Buisson, R.S.K., Kirk, P.W.W., and Lester, J.N., The behaviour of selected chlorinated organic micropollutants in the activated sludge process: a pilot plant study, *Water Air Soil Poll.*, 37, 419, 1988.

115. Havel, J., Reineke, W., Total degradation of various chlorobiphenyls by co-cultures and in-vivo constructed hybrid pseudomonads, *FEMS Microbiol. Lett.*, 78, 163, 1991.

116. van Haelst, A.G., Bakboord, J., Parsons, J.R., and Govers, H.A.J., Biodegradability of tetrachlorobenzyltoluenes and polychlorinated biphenyls in activated sludge and in cultures of *Alcaligenes* sp. jB1: a preliminary study, *Chemosphere*, 31, 3799, 1995.

117. McIntyre, A.P. and Lester, J.N., The behaviour of polychlorinated biphenyls and organochlorine insecticides in primary mechanical wastewater treatment, *Environ. Poll. (Series B)*, 2, 223, 1981.

118. Garcia-Gutierrez, A.G., McIntyre, A.E., Perry, R., and Lester, J. N., The behaviour of PCBs in the primary sedimentation process of sewage treatment: a pilot plant study, *Sci. Total Environ.*, 22, 243, 1982.

119. Wild, S.R., Jones, K.C., The effect of sludge treatment on the organic contaminant content of sewage sludge, *Chemosphere*, 19, 1765, 1989.

120. McNally, D.L., Mihelcic, J.R., and Lueking, D.R., Biodegradation of three and four ring polycyclic aromatic hydrocarbons under aerobic and denitrifying conditions, *Environ. Sci. Technol.*, 32, 2633, 1998.

121. Bedding, N.D., Taylor, P.N., and Lester, J.N., Physicochemical behaviour of polynuclear aromatic hydrocarbons in primary sedimentation. 1. Batch studies, *Environ. Technol.*, 16, 801, 1995.

122. Bedding, N.D., Taylor, P.N., and Lester, J.N., Physicochemical behaviour of polynuclear aromatic hydrocarbons in primary sedimentation. 2. Pilot-scale studies, *Environ. Technol.*, 16, 813, 1995.

123. Manoli, E. and Samara, C., Occurrence and mass balance of polycyclic aromatic hydrocarbons in the Thessaloniki sewage treatment plant, *J. Environ. Qual.*, 28, 176, 1999.

124. Samara, C., Lintelmann, J., and Kettrup, A., Determination of selected polynuclear aromatic hydrocarbons in wastewater and sludge samples by HPLC with fluorescence detection, *Toxicol. Environ. Chem.*, 48, 89, 1995.

125. Melcer, H., Steel, P., and Bedford, W.K., Removal of polycyclic aromatic hydrocarbons and heterocyclic nitrogen compounds in a municipal treatment plant, *Water Environ. Res.*, 67, 926, 1995.

126. Hegeman, M.J.M., Van der Weijden, C., and Loch, F.G., Sorption of benzo(a)pyrene and phenanthrene on suspended harbour sediment as a function of sediment concentration and salinity, a laboratory study using solvent partition coefficient, *Environ. Sci. Technol.*, 29, 363, 1995.

127. Fu, G.M., Kan, A.T., and Tomsen, M., Adsorption and desorption hysteresis of PAHs in surface sediment, *Environ. Toxicol. Chem.*, 13, 1559, 1994.

128. Chiou, C.T., McGroddy, S.E., and Kile, D.E., Partition characteristics of polycyclic aromatic hydrocarbons on soils and sediments, *Environ. Sci Technol.*, 32, 264, 1998.

129. Sellstrom U., Determination of Some Polybrominated Flame Retardants in Biota, Sediment and Sewage Sludge, PhD dissertation, Stockholm University, Sweden, 1999.

130. Sellström, U. and Jansson, B., Analysis of tetrabromobisphenol A in a product and environmental samples, *Chemosphere*, 31, 3085, 1995.

131. Ronen, Z. and Abeliovich, A., Anaerobic-aerobic process for microbial degradation of tetrabromobisphenol A, *Appl. Environ. Microbiol.*, 66, 2372, 2000.

132. Bitton, G., *Wastewater Microbiology*, 2nd ed., Wiley-Liss Inc., New York, 1994.

133. Hagenmaier, H., She, J., Benz, T., Dawidowsky, N., Dusterhoft, L., and Lindig, C., Analysis of sewage sludge for polyhalogenated dibenzo-p-dioxins, dibenzofurans and diphenylethers, *Chemosphere*, 25, 1457, 1992.

134. Goldstone, M.E., Atkinson, C., Kirk, P.W.W., and Lester, L.N., The behaviour of heavy metals during wastewater treatment. 3. Mercury arsenic, *Sci. Total Environ.*, 95, 271, 1990.

135. Goto, M., Shibakawa, T., Arita, T., and Ishii, D., Continuous monitoring of total and inorganic mercury in wastewater and other waters, *Anal. Chim. Acta*, 140, 179, 1982.

136. Compeau, G.C. and Bartha, R., Sulphate reducing bacteria: principal methylators of mercury in anoxic estuarine sediment, *Appl. Environ. Microbiol.*, 50, 498, 1985.

137. Lester, J.N., Significance and behaviour of heavy metals in wastewater treatment processes. 1. Sewage treatment and effluent discharge, *Sci. Total Environ.*, 30, 1, 1983.

138. Chen, K.Y., Young, C.S., and Rohatgi, N., Trace metals in wastewater effluents, *J. Water Poll. Cont. Fed.*, 46, 2663, 1974.

139. Wu, J.S. and Hilger, H., Chemodynamic behaviour of mercury in activated sludge processes, *Amer. Inst. Chem. Eng.*, 81, 109, 1985.

140. Perry, R., Kirk, P.W.W., Stephenson, T., and Lester, J.N., Environmental aspects of the use of NTA as a detergent builder, *Water Res.*, 18, 255, 1984.

141. Brouwer, N.M. and Terpstra, P.M.J., Ecological and toxicological properties of nitrilotriacetic acid (NTA) as a detergent builder, *Tenside Surfac. Deter.*, 32, 225, 1995.

142. Rossin, A.C., Lester, J. N., and Perry, R., Removal of nitrilotriacetic acid during primary sedimentation and its effects on metal removal, *Environ. Poll. (Series B)*, 4, 315, 1982.

143. Obeng, L.L., Lester, J.N. Perry, R., Effect of mixed liquor suspended solids concentration on the biodegradation of nitrilotriacetic acid in the activated sludge process, *Chemosphere*, 10, 1005, 1981.

144. Obeng, L.P., Perry, R., and Lester, J.N., The influence of transient temperature changes on the biodegradation of nitrilotriacetic acid in the activated sludge process, *Environ. Poll. (Series A)*, 28, 149, 1982.

145. Hunter, M., Stephenson, T., Lester, J.N., and Perry, R., The influence of transient phenomena on the biodegradation of nitrilotriacetic acid in the activated sludge process. 1. Variations in hydraulic loading and sewage strength, *Water Air Soil Poll.*, 25, 415, 1984.

146. Wei, N., Stickney, R., Crescuolo, P., and LeClair, B.P., Impact of nitrilotriacetic acid (NTA) on an activated sludge plant: a field study, Environment Canada Research Report, No. 91, Project No. 71-3-3, Ontario, 1979.

147. Eden, G.E., Culley, G.E., and Rootham, R.C., Effect of temperature on the removal of NTA (nitrilotriacetic acid) during sewage treatment, *Water Res.*, 6, 877, 1972.

148. Rossin, A.C., Perry, R., and Lester, J.N., The removal of nitrilotriacetic acid and its effect on metal removal during biological sewage treatment. 1. Adsorption and acclimatization, *Environ. Poll. (Series A)*, 29, 271, 1982.

149. Stoveland, S., Lester, J.N., and Perry, R., The influence of nitrilotriacetic acid on heavy metal transfer in the activated sludge process. 1. At constant loading, *Water Res.*, 13, 949, 1979.

150. Stoveland, S., Perry, R., and Lester, J.N., The influence of nitrilotriacetic acid on heavy metal transfer in the activated sludge process. II. At varying and shock loadings, *Water Res.*, 13, 1043, 1979.

151. Fischer, K., Sorption of chelating-agents (HEDP and NTA) onto mineral phases and sediments in aquatic model systems. 2. Sorption onto sediments and sewage sludges, *Chemosphere*, 24, 51, 1992.

152. Stephenson, T.L., Lester, J.N., and Perry, R., Acclimatisation to nitrilotriacetic acid in the activated sludge process, *Chemosphere*, 13, 1033, 1984.

153. Stephenson, T., Lester, J.N., and Perry, R., The influence of transient temperature changes on the biodegradation of nitrilotriacetic acid in the activated sludge process: a pilot plant study, *Environ. Poll., (Series A)*, 32, 1, 1983.

154. Stephenson, T., Perry, R., and Lester, J.N., The influence of transient phenomena on the biodegradation of nitrilotriacetic acid in the activated sludge process. II. Variations in influent metal and NTA concentrations, *Water Air Soil Poll.*, 25, 431, 1985.

155. Sedlak, D.L., Phinney, J.T., and Bedsworth, W.W., Strongly complexed Cu and Ni in wastewater effluents and surface runoff, *Environ. Sci Technol.*, 31, 3010, 1997.

156. Thomas, K.V., Determination of the antifouling agent zinc pyrithione in water samples by copper chelate formation and high-performance liquid chromatography-atmospheric pressure chemical ionisation mass spectrometry, *J. Chrom. A.*, 833, 105, 1999.

157. Kirk, P.W.W., Perry, R., and Lester, J. N., The behaviour of nitrilotriacetic acid during the anaerobic digestion of sewage sludge, *Water Res.*, 16, 1223, 1982.

158. Stephenson, T., Perry, R., and Lester, J. N., The behaviour of nitrilotriacetic acid during anaerobic digestion of cosettled sewage sludges, *Water Res.*, 17, 1337, 1983.

159. Wooldridge, W.R. and Standfast, A.F.B., The biochemical oxidation of sewage, *J. Biochem.*, 27, 183, 1933.

160. Burgess, J.E., Quarmby, J., and Stephenson, T., Role of micronutrients in activated sludge-based biotreatment of industrial effluents, *Biotechnol. Adv.*, 17, 49, 1999.

161. Singleton, I., Microbial metabolism of xenobiotics: fundamental and applied research, *J. Chem. Technol. Eng.*, 59, 9, 1994.

162. Hunter, M., Stephenson, T., Lester, J.N., and Perry, R., The influences of transient phenomena on the biodegradation of nitrilotriacetic acid on the activated sludge process. 1. Variations in hydraulic loading and sewage strength, *Water, Air, Soil Pollut.*, 25, 415, 1984.

163. Huang, C.-H., Sedlak, D.L., Analysis of estrogenic hormones in municipal wastewater effluent and surface water using enzyme-linked immunosorbent assay and gas chromatography/tandem mass spectrometry, *Environ. Toxicol. Chem.*, 20, 133, 2001.

164. Lai, K.M., Scrimshaw, M.D., and Lester, J.N., Prediction of the bioaccumulation factors and body burden of natural and synthetic estrogens in aquatic organisms in the river systems, *Sci. Total Environ.*, 289, 159, 2002.

165. Ahel, M. and Giger, W., Aqueous solubility of alkylphenols and alkylphenol polyethoxylates, *Chemosphere*, 26, 1461, 1993a.

166. Ahel, M. and Giger, W., Aqueous solubility of alkylphenols and alkylphenol polyethoxylates, *Chemosphere*, 26, 1461, 1993.

167. Ahel, M. and Giger, W., Partitioning of alkylphenols and alkylphenol polyethoxylates between water and organic solvents, *Chemosphere*, 26, 1471, 1993.

168. Sellstrom, U., Determination of Some Polybrominated Flame Retardants in Biota, Sediment and Sewage Sludge, PhD dissertation, Stockholm University, Sweden, 1999.

169. de Wit, C.A., An overview of brominated flame retardants in the environment, *Chemosphere*, 46, 583, 2002.

170. World Health Organization/International Carnivorous Plant Society, Environmental Health Criteria 172: Tetrabromobisphenol A, Geneva, 1995.

171. Fent, K. and Muller M.D., Occurrence of organotins in municipal waste water and sewage sludge and behaviour in a treatment plant, *Environ. Sci. Technol.*, 25, 489, 1991.

172. Wang, J., Liu, P., and Qian, Y., Biodegradation of phthalic acid esters by acclimated activated sludge, *Environ. Int.*, 22, 737, 1996.

173. Ahel, M. and Giger, W., Determination of nonionic surfactants of the alkylphenol polyethoxylate type by HPLC, *Anal. Chem.*, 57, 2584, 1985.

174. Stephanou, E. and Giger, W., Persistent organic chemicals in sewage effluents. 2. Quantitative determinations of nonylphenols and nonylphenol ethoxylates by glass capillary gas chromatography, *Environ. Sci. Technol.*, 16, 800, 1982.

175. Ahel, M. and Giger, W., Determination of alkylphenol and alkylphenol mono- and diethoxylates in environmental samples by HPLC, *Anal. Chem.*, 57, 1577, 1985.

176. Blackburn, M. and Waldock, M., Concentrations of alkylphenols in rivers and estuaries in England and Wales, *Water Res.*, 29, 861, 1995.

177. Crescenzi, C., Di Corcia, A., and Samperi, R., Determination of nonionic polyethoxylate surfactants in environmental waters by liquid chromatography/electrospray mass spectrometry, *Anal. Chem.*, 67, 1797, 1995.

178. Field, J. A. and Reed, R. L., Nonylphenol polyethoxy carboxylate metabolites of nonionic surfactants in U.S. paper mill effluents, municipal sewage treatment plant effluents, and river waters, *Environ. Sci. Technol.*, 30, 3544, 1996.

5 Fate and Behavior of Endocrine Disrupters in Sludge Treatment and Disposal

M.D. Scrimshaw and J.N. Lester

CONTENTS

0-56670-601-7/03/$0.00+$1.50
© 2003 by CRC Press LLC

5.1 INTRODUCTION

The fate of organic compounds during wastewater treatment processes is controlled by physical, chemical, and biological processes. The presence of compounds in the sludge will be determined predominantly by their partitioning to the solid phase during earlier stages of the wastewater treatment processes. They will occur in the sludge either through partitioning during primary or secondary treatment and possibly through active uptake into the biomass. It is likely that compounds that occur in the sludge are recalcitrant and not readily degraded through aerobic metabolic pathways. They are also chemically stable in terms of oxidation state and hydrolysis.

Sludge treatment processes have a number of objectives that are aimed at altering bulk properties of the material to convert it to a form more suitable for subsequent reuse.[1] The objectives can be described as:

1. Render it less offensive and reduce associated health hazards.
2. Reduce the volume of material.

The first objective is achieved through biological, thermal, or chemical (lime) treatment or a combination of these processes. Proposed standards for sludge treatment processes within the European Union (EU) specify that the effectiveness of the process should be evaluated by the reduction in *Escherichia coli*, with conventional treatments achieving a 2 $\log_{10}$ reduction in number and advanced treatment (hygienization) resulting in a 6 $\log_{10}$ reduction.[2] Similar standards, applying to Class A or Class B sludges apply in the United States, with certain site restrictions applied to Class B sludges under the federal regulations.[3] The treatment processes for hygienization effectively produce a pasteurized product, with a significantly reduced pathogen population in comparison to the raw sludge; the choice between this and conventional treatment will determine, or be determined by, the ultimate disposal route of the sludge. The second objective is achieved through physical treatment, such as thickening, filtration, or centrifugation, of digested sludge material or during the composting process through drainage and evaporation of water.

The treatment and disposal of sewage sludge within the EU has become increasingly important over the last decade as a result of legislation. For example, the Urban Wastewater Treatment Directive has resulted in a large increase in the volume of sludge requiring disposal.[4] The major options for the disposal of sludge within the EU are presently either recycling to agricultural land or through disposal to landfill either directly or as ash following incineration. However, more innovative techniques are also under investigation.[5] Recycling of sludge to agricultural land is in most cases the least expensive option for disposal, and also results in a degree of recycling of carbon, nitrogen, phosphorus, and other minerals. However, sewage sludge may also contain a range of organic and inorganic contaminants, including many potential endocrine disrupting chemicals (EDCs). Within the EU, the quality of sludge used on land is controlled by the sludge in agriculture directive,[4] although new standards are being developed.[2] The focus is on sludge and soil quality criteria based on protection of health throughout the food chain by considering both pathogen reduction and concentrations of heavy metals and nutrients. The system in Europe differs from that used in the United States, where the U.S Environmental Protection Agency (EPA) (503) regulations cover all sludge treatment and disposal options.[4] The regulations represent the most risk assessed produced by the agency.[6] However, EPA's position is that further research on land application is not needed. This opinion has been criticized as being inconsistent with the arguments of using insufficient data in eliminating many contaminants from regulation.[3]

Sewage sludge (or biosolids) produced as a result of wastewater treatment are known to contain a range of organic micropollutants. The use, or reuse, and recycling of sewage sludges within agriculture or for other purposes, is based on the premise that they are "safe" or fit for such purposes. Within the EC, sludge has been specifically excluded for the hazardous waste directive,[4] which would have had implications for most currently utilized disposal routes. There are however, concerns about the presence of trace organic compounds, in particular those that exhibit or are suspected of having effects on the endocrine system.

5.1.1 Compounds Associated with Sludges

It is possible to identify the compounds of greatest concern within sewage sludges through field sampling and analysis and models that predict the fate of compounds within sewage treatment works (STWs).[7] In general, the more hydrophobic a chemical the greater the amount that will pass through to the sludge.[8] Such transfer has been described as an inevitable consequence of the removal of contaminants from wastewater streams.[9] The polyaromatic hydrocarbons (PAHs) are a significant group of hydrophobic compounds, which may occur at up to 2000 mg kg^{-1} (dry weight [dw]) in sludge.[10] The concentrations of compounds implicated as EDCs found in sewage sludges vary through almost 10 orders of magnitude from part per trillion to percentage levels (Figure 5.1).

Some groups of organic compounds are degraded during the aerobic treatment processes to more hydrophobic compounds. Of particular concern are the alkylphenol ethoxylates (APEs), which include the nonylphenol ethoxylates. These compounds degrade to short (1 to 2 carbon) chain ethoxylates or to the parent alkylphenol (AP), which both increases lipophilicity and enhances their estrogenic activity.[11,12] However, during anaerobic digestion, further removal of residual ethoxylate groups occurs, and nonylphenol persists in digested sludges[13] with similar behavior reported for the octylphenols.[14] Natural and synthetic estrogens are also removed during activated sludge processes,[13,15–17] with losses of 20 to 90% occurring; the mechanisms, which could be biotransformation or binding to the solids, have not yet been elucidated.

One of the major factors that determines the effectiveness of sludge treatment is the availability of the organic material for further degradation. In the case of primary sludges, the organic material is readily available for degradation. In secondary sludges, much of the organic matter is contained within intact cells and not readily available. One of the objectives of sludge pretreatment is to free the organic material for subsequent digestion by anaerobic bacteria.[21] It is also therefore likely that such processes may have an impact on the availability of contaminants bound to sludges and their subsequent fate. However, it has been proposed that many

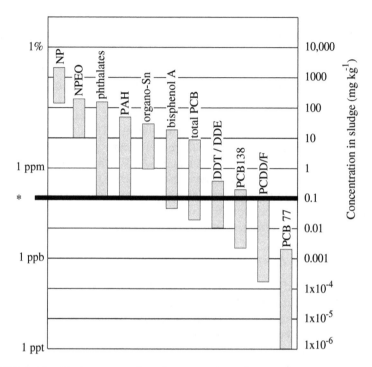

FIGURE 5.1 Graphical representation of concentrations of some compounds associated with sewage sludges. (*Represents the concentration of 17β-estradiol reported by Takigami, H., et al., *Water Sci. Technol.*, 42, 45, 2000. Adapted after Klopffer, W., *Chemosphere*, 33, 1067, 1996, and Giger, W., *EAWAG News,* 1989.)

contaminants rapidly adsorb back to the solid phase.[22] Before considering the impact of sludge stabilization processes, some discussion of pretreatment techniques is appropriate, since they are likely to have an impact on the ultimate fate of EDCs associated with sludges.

5.2   PREDIGESTION TREATMENT TECHNIQUES

There are a range of pretreatment options available to facilitate the stabilization of sewage sludges. The first step involved in any disposal or treatment process may be the removal of excess moisture; the dry-solids content (DS) is a key factor in the determination of capital and operational costs of treatment processes.[5] Operational costs for transport (if sludge is not treated on-site), the size of plant required, and energy costs involved in processing material with a high moisture content are significant factors when treatment techniques are selected. As a result of these considerations, the initial process in almost all sludge treatment involves a dewatering step, or thickening, to reduce the volume of the sludge requiring further processing. Thickening results in the production of a sludge of higher DS content and a liquid that is recirculated to the head of the treatment works. Removal of water will occur at the site of production or transport costs would be significant. To facilitate removal of water from the primary or secondary sludges, chemical or physical treatments may be applied prior to thickening. However, in general terms, pretreatment techniques do not have a significant impact on the final concentration of any recalcitrant compounds present, with mechanical treatment, biological, and freeze-thaw giving some improvement in terms of reduction of hard chemical oxygen demand (COD).[23]

5.2.1 SLUDGE THICKENING

Wastewater treatment plants produce both primary sludges from initial separation of solids from influent streams (primary sludge), and biological sludges from the aerobic treatment of wastewater, secondary sludge, or humus sludge.

There are a range of techniques available to thicken sludges, and the most common processes are gravity or mechanical thickening. Before thickening, sludges may be conditioned to make the process more effective, the addition of polymeric or inorganic chemicals or some type of thermal action is required. However, in the context of the removal of estrogenic substances, these will affect their fate in terms of the effectiveness of the water separation from the bulk solids. The predominant factor controlling the fate of estrogenic organic compounds in any thickening process will be their physicochemical properties. For example, the octanol-water partition coefficient may be a useful predictor as to which compounds will remain in the sludge solids and which will be recycled within the treatment works.

5.2.2 CONDITIONING OF SLUDGES

A number of pretreatments (other than or in addition to dewatering) are available for the conditioning of sludges prior to digestion. Normally, the surplus activated sludge (SAS) is more difficult to digest than primary sludge because it consists of microorganisms that are difficult to degrade.[24] The objectives of sludge pretreatment are to disrupt the cellular structure and to increase the amount of material in the dissolved phase prior to treatment.[23] However, dewatering of sludge prior to stabilization also reduces the volume of material to be treated, with liquid being returned for treatment to the head of the works. Implications of pretreatment for EDCs associated with the sludge include being recycled back into the aerobic treatment process as part of returned liquors from thickening before anaerobic digestion. An additional implication is that through lysis of cells EDCs may be more readily bioavailable during the subsequent sludge treatment process.

Coagulants and polyelectrolytes, such as lime, ferric chloride, ferric sulfate, and aluminum chloride are frequently added to sludges to aid removal of water. Dewatering raw sludge with the polyelectrolyte Zetag 94S before removing water with filter presses had no effect on the partitioning of the hydrophobic PCBs to the solids; however, there was some indication that more soluble compounds (in this case dieldrin) may exhibit some degree of removal in the aqueous phase.[25] The use of ultrasonic pre-treatment has been demonstrated to significantly increase the amount of soluble chemical oxygen demand (SCOD), with up to 90% of the total COD solubilized in SAS.[26] Clark and Nujjoo assumed that the increase in SCOD in SAS was due to cavitation-induced cell lysis and such an effect may also significantly increase the availability of cell-bound contaminants for degradation during the digestion process. Of significance was that the use of ultrasonic energy also increased SCOD in primary sludges. Primary sludges contain less biomass than SAS, and an increase in SCOD in this material is more likely to be a result of particle size reduction and solubilization. The overall effect of an increase in SCOD is likely to result in increased bioavailability of any potential endocrine disrupters, and if this is a limiting factor in their subsequent biodegradation, it would result in further transformation. However, if the compounds remain unchanged, though in solution, they would probably be returned to the head of the works in recycled liquors.

5.2.3 SLUDGE PASTEURIZATION

There is a lack of any specific information on the impact of pasteurization as a pretreatment. Thermal treatment between 60 to 180°C destroys cell walls and will make the cell contents more accessible to subsequent biological degradation.[23] Specifications within Annex I of the EU Working Document on Sludge specify temperatures of 70°C for 30 minutes (prior to mesophilic anaerobic digestion) for liquid sludge, or 80°C for the first hour in thermal drying.[2] Most compounds identified as EDCs exhibit a degree of thermal stability. However, some loss of the more volatile compounds may occur during such treatment processes, and simple fugacity modeling using physicochemical data may be of more value than experimentally derived data.

5.3 SLUDGE STABILIZATION TECHNIQUES

These treatment processes lead to the formation of a stable, inoffensive end product in which the number of viable pathogens has been significantly reduced. The pretreatment process will have been based upon the requirements of this process that may also have been selected for the production of a product suitable for a particular final disposal route, such as use on agricultural land, incineration, or other combustion process. These techniques are microbiological (aerobic or anaerobic) or chemical. If the process has not resulted in a significant loss of water (e.g., through evaporation in composting) the end product will again be dewatered prior to final disposal. Contaminants within raw sludges may be recalcitrant and remain bound to solids. If so, then as the total bulk of sludge is reduced during treatment, concentrations within the final product will be greater than in the raw material.

Anaerobic digestion is the most widely used sludge treatment process,[1,27,28] and the fate of EDCs during this process has received attention over the years. This attention is not only from the aspect of wastewater treatment, but also from the view that to some extent the process represents an accelerated view of the long-term fate of contaminants in anaerobic sediments. A summary of the reported removal for a range of xenobiotic EDCs is given in Table 5.1 These processes and more extensive data on the reported concentrations of the compounds present in sewage sludges are discussed in further detail.

TABLE 5.1
Summary of Treatment Processes and Impact on Contaminant Concentrations

Treatment Process	Compounds	Effect
Anaerobic digestion	Chlorophenoxy herbicides	2,4-D degraded[29,30]; (to chlorophenols)[31–33]
		2,4,5-T degraded[29,30,33]
		2,4,5-TP removed 90% removal over 4–32 days[34]
		Dichlorprop persistent[31]
		MCPA degraded[29,30]
		MCPP not removed[30,31]
		MCPB 60–88% removal over 32 days[34]
		4-chlorphenoxy acetic acid (4-CPA) removed in 8 days[34]
	Chlorophenols	2-chlorophenol 90% reduction[30]; 80% removal (16 days)[34,35]; 30% removal (56 days)[36]; removal[32]; degraded (no lag)[37]
		3-chlorophenol degraded (lag period)[37]
		4-chlorophenol 30% removal (56 days)[36]; 20–80% loss over 70 days[33]; degraded (lag period)[37]
		4-chloro-3-methyl phenol 80% removal over 16 days[34,35]
		4-chloro-2-methyl phenol removed in 2 days[30]
		dichlorophenols 3,4- and 3,5- persistent[37]
		2,4-dichlorophenol persistent[29,30,36]; degraded[32,38]; also 2,3-, 2,5- and 2,6- degraded[37]
		2,4,5-trichlorophenol removed[30,33,38]
		2,4,6-trichlorophenol removed[30,38]
		2,3,4,6-tetrachlorophenol removed[30,38]
	PCP	28% loss (adapted consortium 100%)[39]; 60% removal[40] removal[38,41,42]
	Chlorobenzenes	Up to 80% removal[43,44]
	Organochlorine pesticides	γ-HCH removed (abiotic)[29,30,35]
		Dieldrin persistent[29,30]; 24% loss[35]
		DDE persistent[29,30,40]; 6% loss[35]
		Toxaphene dechlorination observed some recalcitrant metabolites[45]
	Polychlorinated biphenyls	Aroclor 1260 persistent[29,30]
		8% loss over 32 days[35]
		Aroclor 1242 degraded[46,47]
	Alkylphenol polyethoxylates	removal of ethoxylate groups to form parent alkylphenol[48,49]
	Other insecticides	Permethrins removal (30 days)[43]
	Organotins	Not degraded in sludges[50]

(continued)

TABLE 5.1 (continued)
Summary of Treatment Processes and Impact on Contaminant
Concentrations

Treatment Process	Compounds	Effect
Anaerobic digestion	PAH	Anthracene persistent[51]; fluoranthene persistent[51]; pyrene persistent[51]; benzo(c)phenanthrene persistent[51]; benzo(b)fluoranthene persistent[51]; benzo(a) anthracene persistent[51]; chrysene persistent[51]; benzo(b)naphtho(2,1-d)thiophene persistent[51]; benzo(e)pyrene persistent[51]; benzo(k)fluoranthene persistent[51]; benzo(a)pyrene persistent[51]; benzo(ghi)perylene removed[51]; indeno(1,2,3-cd)pyrene removed[51]; coronene removed[51]
	Triazines	
	Phthalates	Dimethyl degraded (4 days)[52]; 30% loss (56 days)[36]; 82% (70 days)[53]; 90% removed[54]
		Diethyl degraded (4 days)[52]; 90% (70 days)[53]
		Dibutyl degraded (4 days)[52]; 30% loss (56 days)[36]; 80% (70 days)[53]; 90% removed[54]
		Butyl-benzyl phthalate degraded (4 days)[52]; (48 days)[53]; 52% in 63 days[55]
		DEHP persistent[52]
		DOP persistent[52,54]; 30% (70 days)[53]
Aerobic thermophilic composting	Phthalates	70% reduction of di-2-ethylhexyl phthalate[56]
	Steroid estrogens	50% reduction of estradiol; 90% for testosterone[57]
	PAH	57–73% reduction[58]

5.3.1 ANAEROBIC DIGESTION

Anaerobic digestion involves the conversion of organic matter into methane and carbon dioxide by anaerobic bacteria.[59] The bacteria present also utilize nitrates and nitrites with the generation of ammonia. The use of anaerobic conditions induces degradation processes for some organic pollutants that were previously recalcitrant under aerobic conditions. These processes include reductive dehalogenation, nitroreduction, and reduction of sulfoxides.[59] Such processes may lead to the production of metabolites that are subsequently more susceptible to further aerobic degradation.[60] During the last decade of the 20th century, the use of anaerobic sludge digestion increased significantly in Western Europe as a result of the 1991 Urban Wastewater Treatment Directive.[4] A diagram summarizing the fate of compounds during sludge digestion is presented in Figure 5.2.

Although it is known that wastewater treatment processes, in particular the activated sludge process, do reduce concentrations of the steroid estrogens, their concentrations within sludges have not been determined.[15,16,61,62] It has been inferred

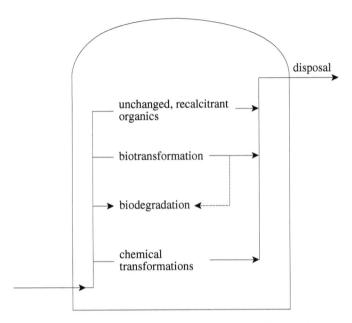

FIGURE 5.2 Possible removal processes for EDCs in the anaerobic digestion process. (From Meakins, N.C., Bubb, J.M., and Lester, J.N., The fate and behaviour of organic micropollutants during wastewater treatment processes: a review, *Int. J. Environ. Pollut.*, 4, 27, 1994. With permission.)

that up to 20% of inputs of these compounds to STW were bound to the particulate phase in final effluents.[61] From this evidence it would appear to be probable that some proportion of these compounds will be present in sludges. Evidence from work on suspended sediments indicates that binding to solids of the steroid estrogens, estrone, estradiol, estriol, and the synthetic hormone, ethinylestradiol, is similar to a prediction from a level 1 fugacity model.[63] Analysis of sludges from a night soil treatment process has been undertaken by an enzyme-linked immunosorbent assay technique; concentrations of 17β-estradiol (E2) were reported as 101 μg kg^{-1} in sludge at the end of the treatment process,[18] compared to 303 and 274 μg kg^{-1} in the untreated raw night soil and septic tank sludges, respectively. The liquor from anaerobic digestion tanks has been demonstrated to contain estrogens (estradiol and estrone) as determined by a radioimmunoassay at a concentration of 143 pmol l^{-1}, which was 61% of the concentration (234 pmol l^{-1}) present in the activated sludge liquors.[64] However, there are no available data for the concentrations of these important estrogenic compounds in sludges from wastewater treatment plans or their fate during treatment processes This chapter will focus on xenoestrogenic compounds.

One of the most widely studied mechanisms of anaerobic biodegradation is reductive dehalogenation of chlorinated organic compounds, which are frequently found in sewage sludges (Table 5.2). This is defined as the removal of a halogen substituent from a molecule with concurrent addition of electrons. In examples of biologically catalyzed reductive dehalogenation, the halogen atoms are released as halide ions.[65] This transformation pathway has been observed to occur with chlorophenoxy herbicides (CPHs), chlorophenols, polychlorinated biphenyls (PCBs), and a range of insecticides such as dichlorodiphenyltrichloroethane (DDT). The process involves either hydrogenolysis, the substitution of chlorine atoms with hydrogen, or vicinal reduction, the removal of two halogens from adjacent carbon atoms with the formation of a C=C bond. To some extent, this process is dependent on the position of the halogen (usually a chlorine) on the aromatic ring, the degree of chlorination, and on the acclimation of the microbial population to the compounds, with the rate of dechlorination generally increasing with sludge age and degree of chlorination.[66] For certain systems, it has been noted that orthochlorine atoms were most readily removed from chlorophenols.[67] Reductive dechlorination also depends on the presence of electron receptors, and these are frequently a limiting resource for anaerobic communities.[65] However, this may be more of an issue within the environment than in wastewater treatment systems.

5.3.1.1 Chlorophenoxy Herbicides

CPHs are known to degrade during anaerobic digestion although some are more recalcitrant than others, with mecoprop (MCPP) being shown to be persistent in a number of studies (Table 5.1). The degradation of 2,4-dichlorophenol (2,4-D), 2,4,5-trichlorophenol (2,4,5-T), 2-methyl-4-chlorophenoxyacetic acid (MCPA) and MCPP in laboratory scale, semi-continuous digestion utilizing both primary and mixed sludge digesters was studied by Buisson et al.[30] In this study, MCPP demonstrated the greatest recalcitrance at concentrations of ~ 100 μg l^{-1}. However, batch studies indicated that when degradation did occur, it was very rapid.[29] MCPP's recalcitrance was also exhibited by dichlorprop. Dichlorprop also has a methyl group in the 2–position on the alkanoic side chain that appears to inhibit cleavage of the ether link.[31]

In the case of the 2,4-D, where the ether link can be cleaved, this has been demonstrated to degrade through the formation of chlorophenols.[31–33] The acclimatization period required for degradation to begin was dependent upon the initial concentration.[32] The ability to cleave ether links from substituted benzene rings (O-CH$_3$) has also been demonstrated in the case of p-methoxyphenol.[36] It would appear that the major factors controlling the degradation of the CPH compounds in anaerobic conditions are the type of ether group and the position of the chlorine substitutions on the aromatic ring.

5.3.1.2 Chlorophenols and Chlorobenzenes

The degradation of chlorophenols, including pentachlorophenol (PCP) has been investigated by a number of workers,[29,31,36,40,68] and a range of microbial species capable of degrading PCP have been studied.[69] Sludge samples taken from anaerobic digesters have been demonstrated to remove 28% of parent PCP over 96 hours,[39] but this work did not indicate the age of the sludge, and sludge ages over 8 days have been noted to be required for dechlorination of PCP.[70] Both the source of the sludge tested[33] and the concentration of chlorophenols used in tests may influence the degradation rate, with inhibition occurring as concentration increases. Concentrations of PCP above 6.3 mg l^{-1} have been observed to inhibit degradation.[69] Concentrations of 2,4-D above

70 mg l[-1] added to sewage sludge inhibited gas production by 50% compared to controls[36] and in samples spiked at 10 µg l[-1] each with a range of chlorophenols; recalcitrance was demonstrated for 2,4-D.[29,31] However, other work has demonstrated degradation of this compound with loss of up to 30%.[36]

The persistence of 2,4-D may be related to the positioning of the chlorine atoms on the ring structure. Pentachlorophenol degradation occurs by initial dehalogenation at the para-position,[40] and in contaminated soils, the use of anaerobic digestion has been demonstrated to remove 95% of PCP.[71] The degradation pathway appears to then produce 3,5-chlorophenol and 3-chlorophenol as observed by Tartakovsky et al.[41] The degradation of 2,4-D has been demonstrated to occur with the loss of the chlorine in the 2- position with the preferential formation of 4-chlorophenol; both 2- and 4-chlorophenol were subsequently observed to degrade to phenol.[32] The preferential formation of 4-chlorophenol may be expected, as substitution in this position has been shown to be most difficult to remove in methanogenic sludges, with the ortho-substitution being least recalcitrant.[33,38] However not all observations support this.[41]

Once particular species responsible for transformation have been identified, the opportunity to study the kinetics of the degradation processes more closely arises. The kinetics for the degradation of PCP by *Mycobacterium chlorophenolicus comb nov.* in laboratory simulations have been described and modeled.[69] However, the success of any microbial species in the environment is related to its competitiveness and growth rate in any given situation. Inoculation with an adapted consortium of bacteria has been demonstrated to completely degrade PCP after a lag period of 24 hours.[39] It has also been suggested that inoculation of upflow anaerobic sludge blanket (UASB) reactors may also introduce the ability to successfully dechlorinate PCP in systems where it is not inherent.[42] Such reactors have been well-studied for their ability to degrade the chlorophenols, and there have been significant removal rates. Although such systems are anaerobic and may be indicative of processes occurring during anaerobic digestion, they are not at present suitable for treatment of sludges. Conclusions made from observations of degradation of chlorophenols in such systems indicate that:

1. Recovery from toxic loadings is possible
2. Degradation of substituted groups occurs in the order ortho >meta >para
3. Phenol does not accumulate as an end product[37,67]

The chlorobenzene compounds consist of a single benzene ring substituted with 1 to 6 chlorine atoms. Hexachlorobenzene (HCB) has found widespread use as an insecticide for fumigation and other members of the family are utilized as process solvents and precursors for a range of organic compounds.[43] It has recently been found to be present in sludges in Switzerland (Table 5.2), and reductive dechlorination of HCB has been observed with accumulation of 1,3,5-trichlorobenzene.[44]

5.3.1.3 Organochlorine Pesticides

Many of the chlorinated pesticides follow similar degradation processes to the chlorophenols in that reductive dehalogenation is the major pathway for initial

TABLE 5.2
Concentration (Range or Mean) of Chlorinated Compounds Reported in Sewage Sludges

Compound	Concentration (mg kg⁻¹), Location, and References
Chlorophenols	0.4 (total) UK[79]; 9.8–60.5 (total),[26] (2,4-dichlorophenol) UK[80]
Pentachlorophenol	0.1–2.0 UK[80]
Organochlorine pesticides	
HCB	ND–0.013 Switzerland[78]
γHCH	0.01–70 U.K.[81]; ND–0.057 Switzerland[78]
Aldrin	0.01–0.21 U.K.[81]; ND–0.029 Switzerland[78]
Endrin	0.01–0.71 U.K.[81]; ND–0.022 Switzerland[78]
Dieldrin	0.01–52.9 U.K.[81]; ND–0.043 Switzerland[78]
p,p'-DDE	0.5 Germany[81]; 0.036–0.097 Switzerland[78]
p,p'-DDD	ND–0.048 Switzerland[78]
p,p'-DDT	ND–0.017 Switzerland[78]
Heptachlor	ND–0.004 Switzerland[78]
Heptachlor epoxide	ND–0.165 Switzerland[78]
PCB (total)	0.15–3.6 USA[a]; 0.02–0.46, 0.01–21.5, 0.3 U.K.[73,74,81]; 0.13–1.63 Canada[a]; 0.5–8.0, 0.36–7.6 Switzerland[a]; 0.6–6.6 Holland[a]; 0.52–15.2, 1.38–6.65 Germany[a]
PCB28	0.041, 0.054 Germany[a]; 0.007, 0.001–0.009 Switzerland[78,83]; 0.086 France[84]
PCB52	0.028, 0.022 Germany[a]; 0.023, 0.004–0.0272 Switzerland[78,83]; 0.049–0.086 France[84]
PCB 77 (µg kg⁻¹)	0.540–4.27 U.K.[75]; 0.234–5.30 Switzerland[83]; 0.34 Germany[85]; 0.8–1.1 Spain[86]; 0.54–1.00 Sweden[87]; 0.03–0.66 Finland[88]
PCB101	0.052, 0.061 Germany[a]; 0.053, 0.006–0.030 Switzerland[78,83]; 0.082, 0.086 France[84]
PCB118	0.040, 0.006–0.028 Switzerland[78,83]; 0.060 0.086 France[84]
PCB 126 (µg kg⁻¹)	ND–0.280 UK75; 0.04–1.59 Switzerland[83]; 0.03–0.06 Spain[86]
PCB138	0.082, 0.093 Germany[a]; 0.056, 0.006–0.029 Switzerland[78,83]; 0.117, 0.086 France[84]
PCB153	0.084, 0.1 Germany[a]; 0.060, 0.009–0.031 Switzerland[78,83]; 0.174, 0.086 France[84]
PCB 169 (µg kg⁻¹)	ND–0.55 U.K.[75]; 0.02–0.23 Switzerland[83]; 0.01 Spain[86]
PCB180	0.053, 0.064 Germany[a]; 0.021, 0.005–0.016 Switzerland[78,83]; 0.129, 0.086 France[84]
PCDD/PCDF (µg kg⁻¹)	Up to 63 U.K.[75]; 9.1 (OCDD) Sweden[89]; 32.9 (OCDD) Sweden[90]; 7.5 (OCDD) Sweden[91]

ND = not detected

[a] Cited in Alcock, R.E. and Jones, K.C., *Chemosphere*, 26, 2199, 1993.

removal of chlorine atoms. This makes the ring structure amenable to further hydroxylation and cleavage in any subsequent exposure to aerobic conditions. Their occurrence in sewage sludges is well-documented (Table 5.2). Some compounds, however, have been demonstrated to be removed via abiotic pathways. This includes γHCH (which has a saturated, rather than aromatic ring structure), with removal being observed in both batch and semicontinuous operation.[29,30]

The use of bacteria from anaerobic sludge in sequencing aerobic/anaerobic reactors demonstrated the recalcitrance of DDE[40] that agreed with data from both semicontinuous and batch anaerobic digestion, which indicated that neither DDE nor dieldrin is amenable to anaerobic degradation.[29,30] Other compounds are rapidly degraded in anaerobic sludges. Toxaphene, a complex mixture of chlorobornanes and other bicyclic compounds has the general composition of $C_{10}H_{18-n}Cl_n$. $C_{10}H_{16-n}Cl_n$ was observed to be degraded with the greater rates for the more highly chlorinated compounds.[45]

5.3.1.4 Polychlorinated Biphenyls, Dioxins, and Difurans

PCBs, Polychlorinated dibenzo-p-dioxins (PCDDs), and polychlorinated dibenzo-furans (PCDFs) compounds represent a challenge in terms of analysis. (There are 207 possible congeners for the PCB alone). Some consideration may be given to the relative toxicity of each compound/congener in terms of toxic equivalence factors (TEF).[72] It is also important when considering comparison of data that historically workers may have quantified samples using commercial mixtures of PCB as standards. However, more recent (post-1990) data are more likely to have been generated through quantification of individual congeners. The presence of these compounds in sludges and digested sludges has been well-documented for over a quarter of a century and reports of their occurrence have been included in Table 5.2. Data from the United Kingdom demonstrate the changes in analytical techniques and priorities of analysis. In 1982, McIntyre and Lester[73] reported values based on Arochlor 1260. However, by 1993, a range of 29 specific congeners were quantified,[74] and in 2001, the nonortho-(planar) congeners which are deemed to be more toxic have been quantified.[75,76] Data from the United States in the late 1970s showed concentrations of PCB (as Arochlors) in dry sludge ranging from 238 to 1700 mg kg^{-1}.[77] More recent analysis of samples of sludge from Switzerland has indicated that concentrations of PCB do appear to be decreasing.[78]

When laboratory scale batch anaerobic digesters were spiked with 40 μg l^{-1} of Aroclor 1260, no degradation was observed,[29] which was subsequently reflected in semi-continuous digesters where native PCBs were quantified using the same standard.[30] However, it is well documented that anaerobic cultures are able to reduce the level of chlorination of PCBs through reductive dehalogenation, although the process does not lead to complete degradation.[65] Complete dehalogenation (to biphenyl) of Aroclor 1242 has been observed in a single-stage coupled aerobic/anaerobic bioreactor[47] and in a continuously operated upflow anaerobic sludge bed (UASB) reactor inoculated with granular anaerobic sludge.[46]

Despite the restrictions on the use of PCBs, their presence in digested sewage sludges continues to be of concern. Improved analytical techniques now allow for determination of the non-*ortho* congeners and their concentrations in relation to those of the PCDDs and PCDFs, with results often reported in terms of TEF. Increasing concern about the possible transfer of these compounds into the food chain through sludge use on agricultural land continues to drive research into their occurrence in sludges. Recent work has demonstrated that the toxic equivalents (TEQs) in 14 samples of digested sludge from the United Kingdom were predominantly due to PCDD/PCDF. However, at one site, the contribution from PCB77 increased the TEQ from 30 to ~680 ng kg^{-1}.[76]

5.3.1.5 Alkylphenol Ethoxylates

Alkylphenol ethoxylates, which are considered significant in terms of occurrence in wastewater and estrogenic activity, are liable to both aerobic and anaerobic degradation in STW. Greater concentrations of the degradation products, in particular nonylphenol (NP), nonylphenol ethoxylate (NP$_1$EO), and nonylphenol diethoxylate (NP$_2$EO), have been reported to occur in digested sludge than the parent compounds.[92,93] Concentrations of NPs in sludge of up to 3.6 mg kg^{-1} and octylphenol at 0.2 mg kg^{-1} have been reported in raw sludges.[94] However, concentrations in digested sludges are frequently greater than in raw[48] and the concentration of NP in digested sludge has been reported as high as 250 mg kg^{-1}.[95] Data on occurrence of these compounds in sludges are summarized in Table 5.3. In fact, anaerobic treatment of sludges containing residues of alkylphenol polyethoxylates (AP$_n$EO) results in the removal of residual ethoxylate groups resulting in the formation of NP, which

TABLE 5.3
Concentrations of Alkylphenols and Degradation Products and Derivatives in Raw and Digested Sludges

Compound	Matrix	Concentration	References
Octylphenol	Digested sludge	7.5 mg kg^{-1}	Spain[101]
OPEO	Digested sludge	12 mg kg^{-1}	Spain[101]
Nonylphenol	Raw sludge	3.7 mg kg^{-1}	Germany[94]
		137–470 mg kg^{-1}	Canada[105]
		4.6 mg kg^{-1}	Germany[99]
	Digested sludge	250 mg kg^{-1}	Taiwan[95]
		78 mg l^{-1}	Switzerland[11]
		450–2530 mg kg^{-1}	Switzerland[106]
		638 and 326 mg kg^{-1}	U.K.[107]
		172 mg kg^{-1}	Spain[101]
		80–120 mg kg^{-1}	Germany[82]
NP$_1$EO	Digested sludge	51–304 mg kg^{-1}	Canada[93]
NP$_2$EO	Digested sludge	4–118 mg kg^{-1}	Canada[93]
NP$_n$EO	Digested sludge	133 mg kg^{-1}	Spain[101]
		9–169 mg kg^{-1}	Canada[93]
NP$_1$EC	Digested sludge	<0.5–25 mg kg^{-1}	Canada[93]
NP$_2$EC	Digested sludge	<0.5–38 mg kg^{-1}	Canada[93]

is more recalcitrant and hydrophobic than the parent compounds.[48] These character-istics make the degradation products more toxic and their accumulation in aquatic organisms has been demonstrated.[96,97]

Up to 96% of NP produced during wastewater treatment is likely to be associated with the digested sludge. The long-term significance of NP residues in sludges may be controlled by restrictions on the use of these compounds. Studies have indicated that concentrations in sludges from Germany, where restrictions have applied since 1992, have declined from an average of 128 to 4.6 mg kg^{-1} (dry matter) between the periods 1987–89 and 1994–95.[98,99] Analysis of sludges from STW in the Cata-lonian region of Spain demonstrated the occurrence of NP$_{1+2}$EO, carboxylic acid metabolites (NP$_{1+2}$EC) and NP.[100] Both NP$_1$EO and NP$_2$EO occurred from 16 to 86 mg kg^{-1} in the sludges, and it was inferred that these compounds were produced faster than they degraded during sludge digestion. The carboxylic acid metabolites, nonylphenoxyacetic acid (NP$_1$EC) and nonylphenoxyeyhoxyacetic acid (NP$_2$EC), were not detected in all samples; however, when detected, concentrations ranged from 10 to 48 mg kg^{-1} and NP was ubiquitous from 14 to 74 mg kg^{-1}. Other work supports the finding that carboxylated degradation products, formed during aerobic treatment, are less likely to occur in sludges. Whether this is due to further trans-formation in the anaerobic conditions or a result of their solubility is not clear.[101]

The ability of anaerobic bacteria to degrade NP$_1$EO and NP$_2$EO to NP appears to be relatively ubiquitous in sludge from anaerobic digesters, landfilled sludge and municipal waste.[102] Furthermore, the results from work involving ^{14}C labeled mate-rial indicated that more complete degradation or mineralization of the NP did not occur in any samples, which demonstrates the recalcitrance of the AP products. Because of their recalcitrance, the AP concentration in digested sludges is likely to reach relatively high concentrations (Table 5.3). It has been noted that such concen-trations may become inhibitory for the production of methane, although acclimatized cultures are less likely to be affected.[102,103]

APs may also be affected through chemical processes. Halogenated derivatives can be formed during disinfection with chlorine in the presence of the bromide ion. However, although they were found at certain stages of drinking water treatment, they were removed during flocculation and not found in sewage influent or sludge samples.[101]

5.3.1.6 Polyaromatic Hydrocarbons

PAHs are frequently monitored for, and observed in, sewage sludge. This is because they have a high persistence since they are lipophilic with low biode-gradability.[104] The concentrations of these compounds found in sewage sludges and a range of other nonchlorinated contaminants are summarized in Table 5.4. The data for total PAH need to be treated with some caution, as the value will be influenced by the number of PAH determined; for example, figures from Switzerland were based on the sum of 16 PAH[78] and data from Spain[104] based on 40 PAH (Table 5.4). Studies on their degradation have been undertaken by a number of workers and there is evidence that they are recalcitrant and removal processes may be abiotic. The use of sodium azide to arrest biological activity

TABLE 5.4
Concentration (Range or Mean) of Compounds Reported in Sewage Sludges

Compound	Concentration (mg kg⁻¹), Location, and References
PAH (total)	257, 1.13–5.52 Spain[104,127]; 4.7–22.6 Switzerland[78]; 3–5 Denmark[128]
Fluorene	0.5, 2.4 U.K.[129,130]; 7.1, 0.03–0.91 Spain[104,127]; 0–6 Poland[10]; 0.02–0.63 Switzerland[78]
Phenanthrene	4.7, 3.0 U.K.[129,130]; 7.2, 0.25–2.03 Spain[104,127]; 0.3–4.4 Poland[10]; 0.11–1.72 Switzerland[78]
Anthracene	ND–1.6 USA[131]; 0.6, 0.3 U.K.[129,130]; 1.2, 0.09–0.29 Spain[104,127]; ND–7 Poland[10]; 0.01–0.22 Switzerland[78]
Benzo(a)anthracene	ND–0.7 USA[131]; 0.005–0.3 U.K.[132]; 0.03–0.18 Spain[127]; 2–16 Poland[10]; 0.05–1.7 Switzerland[78]
Chrysene	ND–11.2 USA[131]; 0.05–0.31 Spain[127]
Fluoranthene	<0.01–10.4 USA[131]; 16–400, 8.1, 1.5 UK[129,130]; 3, 4, 0.06–0.69 Spain[104,127]; 2.2 France[133]; 2–19 Poland[10]; 0.14–3.76 Switzerland[78]
Benzo(k)fluoranthene	Switzerland[78]
Pyrene	0.005–0.16 U.K.[132]; ND–0.29 Spain[127]; 0.06–1.27 Switzerland[78]
Benzo(a)pyrene	<0.01–10.5 USA[131]; 6.8, 2.0 U.K.[129,130]; 2.8, 0.11–0.71 Spain[104,127]; 2.4 France[133]; 2–15 Poland[10]; 0.13–3.36 Switzerland[78]
Benzo(e)pyrene	ND–11 USA[131]; 0.01–0.32 U.K.[132]; 0.02–0.52 Spain[127]; 0.7 France[133]; 3–25 Poland[10]; 0.08–2.09 Switzerland[78]
Benzo(ghi)perylene	3–31 Poland[10]
Indeno(1,2,3-cd)pyrene	0.005–0.32, 10.5 2.5, 0.7 U.K.[129,130,132]; 0.7, ND–0.59 Spain[104,127]; 1–2 Poland[10]; 0.08–0.56 Switzerland[78]
Coronene	ND U.S.[131]; 0.01–0.24 U.K.[132]; ND–0.46 Spain[127]; ND–20 Poland[10]; 0.07–1.83 Switzerland[78] 0.6. U.K.[129]; 0.5 Spain[104]
Phthalates	20–110 Germany[82]
Di(ethylhexyl)phthalate	20–100 Denmark[128]; 0.21–8.44 Germany[115]
Bisphenol A	0.070–0.770 Germany[94]; 0.1–36 Canada[114]
Biocides	
Tributyl tin	1.1, 1.0, 3.4 Switzerland[50,134,135]; 0.1[136]; 1–10 Germany[82]
Dibutyl tin	1.5, 1.0, 4.8 Switzerland[50,134,135]; 0.04[136]
Monobutyl tin	0.5, 0.8, 3.2 Switzerland[50,134,135]; 0.02[136]
Triphenyl tin	0.5, 0.1, 2.2 Switzerland[50,134,135]

ND = not detected

in laboratory batch anaerobic digestion indicated that these compounds were stable during the anaerobic process, with the exception of the six- and seven-ring compounds, benzo(ghi)perylene, indeno(1,2,3-cd)pyrene, and coronene.[51] The four- to five-ring PAH have been identified as representing the major proportion of PAH present in sewage sludges,[10] with benzo(a)pyrene, a significant carcinogen, representing 8% of the total 32 PAH determined. The usefulness of benzo(a)pyrene as an indicator of total contamination is that it makes 10% of the total contamination, as observed by Berset et al.,[78] who quantified 16 PAH. When degradation and transformation products have been determined, the concentrations of oxygenated and nitrogenated derivatives of PAH have been observed to be above those of nonsubstituted, parent, compounds by an order of magnitude with carbonyl groups most prevalent.[10]

The concentrations of PAH in co-composted sewage sludge and wood chips (20.8 mg kg^{-1}) in samples from Ohio have been shown to be higher than in tree/shrub/grass (16.0 mg kg^{-1}) or leaf (14.4 mg kg^{-1}) compost samples which in all cases were greater than remediation goals.[108]

5.3.1.7 Other Compounds Implicated as EDCs

5.3.1.7.1 Herbicides and Insecticides

Degradation of the triazine herbicide, atrazine, has been observed during studies into anaerobic treatment of wastewaters with a high COD demand in hybrid reactors. Atrazine reduction after 5 days by the microbial mass was 43.8, 40, and 33.2% with initial concentrations of 0.5, 1.0, and 2.0 mg l^{-1}, respectively.[109] Both *cis*- and *trans*-permethrin have also been noted to degrade during sludge treatment (Table 5.1) through what would appear to be chemical, rather than biological, pathways. It would appear that the organophosphorus pesticides are less likely to be present in sludges than the organochlorine compounds. This is probably due to their chemical structure rendering them susceptible to hydrolysis. A survey of sludges from 12 U.K. works in 1981 reported negative results for the presence of 16 compounds including diazinon, malathion, and parathion.[110]

5.3.1.7.2 Phthalates

Phthalates do exhibit a degree of degradation during anaerobic digestion (Table 5.1). Added to sludge at concentrations of 50 mg (C) l^{-1}, up to 30% of dimethyl- and dibutyl phthalate were observed to degrade over 56 days.[36] Other work supports the observation that some phthalates are degraded; however, di-*n*-octyl phthalate and di-(2-ethylhexyl)phthalate have both been reported to be recalcitrant by a number of workers[52,53,111] with the length of the alkyl side chain influencing amenability to degradation.[54] A review of the fate of phthalate esters indicates that degradation under anaerobic conditions is initiated through ester hydrolysis, followed by breakdown of the monoester to phthalic acid.[112] Further degradation of the acid follows the same pathway as for benzoate. Degradation rates tend to be slower in anaerobic conditions than in aerobic processes. The large variability in results for such rates may be indicate that the source of the sludge or inoculum used is significant in influencing results.

5.3.1.7.3 Biocides

Biocides are of particular concern as they are intended to be toxic to a broad spectrum of organisms, but they also find use in a range of other applications. The organotin compounds have been well-studied because of their use as antifoulants. However, the use of di- and tri-substituted organotins for thermal and ultraviolet stabilization of rigid and semirigid PVC accounts for 70% of production.[50] During wastewater treatment, the organotins readily associate with particulate matter, and they are not subsequently degraded during anaerobic treatment.[50] Sludges contaminated with TBT at up to 10 mg kg^{-1} demonstrated fungotoxic effects, but the significance of the organotin compounds in this was not evaluated.[82]

5.3.1.7.4 Bisphenol A

Identified as estrogenic, bisphenol A has been reported to degrade in activated sludge microcosms, although over a longer time than the usual hydraulic retention time of a WWTW. The production of refractory metabolites, 2,3-bis(4-hydroxyphenyl)-1,2-propanediol and *p*-hydroxyphenacyl alcohol may also be of concern.[113] Residues of bisphenol A have been observed in sludges[94,114] with concentrations ranging from 0.004 to 1.363 mg kg^{-1} dw.[115] Comments during these observations were that "since the decomposition has not been studied to date" (presumably in sludge digestion), then further research on this issue was required. Estrogenic activity in sewage sludge has also been related to the removal of bisphenol A from influent, although this activity was not specifically related to bisphenol A concentrations in the sludges.[116]

5.3.1.7.5 Flame Retardants

The presence of organophosphorus-based products in sludges at concentrations up to 6 mg l^{-1} was noted by McIntyre et al. when investigating the occurrence of organophosphorus pesticides.[110] The use of such compounds has been superseded by brominated organic compounds, in particular the polybrominated diphenyl ethers (PBDEs).[117] There is little information available on their fate in sludges, although reductive debromination has been reported in sediments.[118] Concentrations of both PBDE and tetrabromobisphenol A (TBBPA) in sludges from Sweden of up to 450 and 220 µg kg^{-1} have been reported, with wide variation in concentrations between STW.[119] Both TBBPA at 31 to 56 µg kg^{-1} and its dimethylated degradation product were also found in samples of sewage from sites with industrial related and non-identifiable point sources of inputs for the target compounds.[120]

5.3.1.7.6 Pharmaceuticals

The occurrence, fate, and effects of pharmaceuticals in the environment have recently been reviewed.[121,122] The available evidence would indicate that they are unlikely to occur in sewage sludges. Recent modeling data undertaken with the STPWIN fugacity model[123] demonstrated that of the 25 most commonly prescribed pharmaceuticals in the United Kingdom, only 4 demonstrated partitioning of >10% to sludges.[124] Of these compounds, ibuprofen has also been identified by other workers as being likely to accumulate in sludge; predicted concentrations range from 0.1 to 34.2 mg kg^{-1} and the values are strongly influenced by the sludge/water partition coefficient (K_ds) used.[125] Utilizing an experimentally derived K_ds, the predicted concentration of ibuprofen in sludge was 20.3 mg kg^{-1}.

5.3.2 THERMOPHILIC AEROBIC DIGESTION

Thermophilic aerobic digestion involves aerating the sludge in a closed reactor with bacterial activity generating heat. In terms of the degradation processes acting on recalcitrant organic compounds, this process is likely to be similar to the activated sludge process; however, the higher temperatures involved will result in a different population of bacteria. Also, processes generally function at 45 to 70°C,[56] although Hudson (1996) indicates the process should reach 55°C for at least 4 hours with an overall time of 7 days. The system has been applied to the stabilization of industrial SAS.[126] Although there is no extensive literature on the fate of EDCs in such processes, they have been demonstrated to enhance the degradation of diethylhexyl phthalate in laboratory simulations using spiked sludge.[56]

5.3.3 COMPOSTING

Composting is an aerobic process that utilizes sludge or manure, usually in combination with other waste materials such as domestic, municipal, agricultural, or horticultural materials. The larger size of such material allows for ingress of oxygen into the windrows used for this process. Temperatures within the waste material of >60°C are achieved. In studies investigating the possibility of using composting to clean up contaminated land, concentrations of PAH were observed to fall by up to 73% with the larger five- and six-ring compounds being more stable.[58] Degradation appeared to be relatively rapid over the first 2 weeks of composting. Concentrations then remained constant indicating that bioavailability, either through binding or stress-induced changes in bacterial membrane permeability, was limiting transformation.

The steroid estrogen, E2, and testosterone have been noted to "degrade" during the composting of chicken manure with hay and other amendments.[57] Over a period of 85 days, the concentration of E2 decreased from a starting concentration of 95 to 42 ng g^{-1}. However, the project monitored only for the parent compounds and not for any other related compounds or metabolites. No evidence of mineralization or on the pathways of degradation was presented. Mineralization of E2, estrone, and testosterone has been observed to occur in biosolids taken from municipal wastewater treatment plants, but biosolids from an industrial works did not mineralize E2 or estrone. This indicates that the influent composition affected the ability of biosolids to degrade the compounds.[137] Such studies would indicate that there is potential for the degradation of steroid estrogens during composting processes. However, the source of biosolids and adaptation of the microbial population is likely to have a significant impact of the removal of steroid estrogens, and factors controlling degradation are not fully understood.

5.3.4 LIME STABILIZATION

Lime (CaO) may be added to either liquid sludge or sludge cakes (dewatered sludge). Addition of lime raises the pH, however, as the reaction between lime and moisture in the sludge is exothermic, heat is also created. Over time, the calcium hydroxide produced in the initial reaction reacts with atmospheric CO_2, resulting in a subsequent fall in pH.[27] Due to the chemical and thermal stability of many recalcitrant organic compounds, it is unlikely that the stabilization of sludge with lime would significantly affect residual concentrations. However, there is little available information within the literature on their fate. The treatment of spiked raw sludge with lime at 20% w/w dry solids in laboratory conditions has been demonstrated to result in rapid degradation of the organophosphorus pesticides diazinon and malathion, over 36 hours, with parathion being more stable and still present after 48 hours.[138] As lime stabilization is relatively insignificant as a process in terms of volume of sludge treated (4% within the EU[4]), it is not likely to be a significant source of EDCs to the environment.

5.3.5 HEAT TREATMENT

Heating of sludges is another option for treatment before dewatering processes. This treatment ruptures cells, releasing their contents, and prevents clogging of presses and filters. The impact on concentrations of PAH released to liquors after treatment was observed to be <20% for treatment of mixed sludges followed by vacuum filtration or belt pressing.[139] In an extension of this work using a pilot plant study, concentrations of most PAH were not statistically different after heat treatment at between 180 to 200°C. Initial concentrations in the sludges ranged from 0.3 to 14 $\mu g\ l^{-1}$. Most PAH demonstrated some reduction in concentration, with benzo(k)fluoranthene exhibiting a 38 and 40% reduction in primary and activated sludge, respectively.[140]

Overall, therefore, the treatment processes applied to sewage sludges may have a significant impact on the concentrations of contaminants present in the end products. Some of this effect will be due to the biological or chemical transformation of the parent compounds, with some of the transformation products, such as nonylphenol, being more recalcitrant in anaerobic conditions than the parent compounds. In other cases, such as with chlorinated compounds, dehalogenation will have produced less hydrophobic compounds. While relatively stable in anaerobic conditions, these compounds are more amenable to subsequent breakdown should aerobic conditions be encountered again.

5.4 POSTDIGESTION TREATMENT

One of the major objectives of treating the raw sludges is to make them more amenable to dewatering, as any removal of water results in less material to dispose. Sludges that have been treated through the above processes are frequently subjected to a final dewatering process. This is because transport costs associated with removal of the final product are of significance, and any subsequent processing would also benefit in costs terms from a reduction in volume of raw material. Although a range of techniques are available (e.g., belt presses, centrifugation, and membrane technology), the fate of EDCs will be controlled by partitioning between the liquid and solid phases, with any EDCs in the liquid phase recycled to the head of the treatment works. However, it is possible that a range of degradation products will have formed as a result of the sludge treatment process, which if products of metabolic activity, are likely to be more water soluble than the parent compounds. An exception to such a generalization is the formation of APs during anaerobic digestion, which are more hydrophobic than the parent ethoxylates.

The effect of vacuum filtration after addition of aluminum chlorohydrate as a coagulant of the fate of PCB, DDE, and dieldrin indicates that the more water soluble compounds (in this case dieldrin) are more likely to be removed during final dewatering. This reflects the effect of dewatering prior to digestion.[25] Laboratory partitioning studies, using centrifugation at $600 \times$ gravity demonstrated little change in the absolute concentration of PCB in centrifuged solids (due to removal of water) from 2.55 to 2.67 mg kg^{-1} (dw).[35] More water soluble compounds exhibited a fall in concentrations associated with the solids, by 53% for dieldrin and of >80% for the herbicides 2,4-D and 2,4,5-T. The recycling of liquors extracted from sludges back to the head of the works will result in a subsequent aerobic treatment phase. Any highly chlorinated compounds, which have undergone reductive dehalogenation during anaerobic digestion, may be further degraded as described by Tartakovsky et al.[47] However, the recycling of high concentrations of compounds to the head of STW may have subsequent impacts on the performance of the works if bacterial activity is affected.

5.5   SLUDGE DISPOSAL OPTIONS AND THEIR IMPACT

All sludge treatment processes produce an end product that will be utilized in some form (e.g., agriculture or thermal processing) or disposed of in some way (e.g., landfill or incineration). Sewage sludge is by far the largest of the by-products resulting from wastewater treatment. Production is expected to increase to around 10 million tons (dw) within the EU by 2005,[141] while the United States reported over 7 million tons in 1990 and Japan produced 4.5 million tons in 1991.[142] The disposal and utilization of this material will have an impact on the environment, and the presence and fate of any EDCs within the product will be of potential concern. If final treatment is to involve a high temperature process, such as incineration, then it is unlikely that the presence of many EDCs will be of concern, since they will be oxidized to CO_2 and H_2O by the process. However, if the material is to be recycled, then it is likely that the parent compounds present, or their degradation and transformation products will be of concern. Although the presence of EDCs will have been controlled through sorption processes, transformation could have resulted in changes in their solubility. This may result in increased concentrations in any liquors produced during post-treatment thickening. Parent compounds that have remained unchanged will predominantly remain associated with the solid phase. Increasingly, the fate of the contaminants within sludges is coming under scrutiny. Procedures based on evaluating risks through understanding of transfer pathways are being used to formulate policy and research priorities.[143]

5.5.1 SLUDGE TO LAND

The use of treated municipal sewage sludge for soil amendment in agriculture has historically been considered more desirable than incineration or other disposal routes. However, there is demonstrated potential for environmental impact through:

1. Run-off of contaminants into surface waters
2. Percolation through to groundwater
3. Possible impact on human health by uptake through crops or grazing
4. Possible ecotoxicological impact on the soils[82]

Because of concern about the possible impact of contaminants on human health, regulations restricting the use of contaminated sludges are in force or being considered. The lack of knowledge concerning the toxicity of the organic compounds, compounded with difficulties in analytical methods, led to them being omitted from the regulations (40 CFR Part 503) of the U.S. Clean Water Act 1993.[133] However, recent amendments to the regulations (40 CFR 503) prohibit land application of sewage sludge containing greater than 0.0003 mg TEQ kg^{-1} (dw) sewage sludge.[144] Some U.S. states have incorporated PCB into regulations for sludges in some situations, with both Texas and New York having limits of 1 mg kg^{-1} (dw) in composts.[3]

Regulations in some European countries do state maximum allowable concentrations of contaminants in sludges for agricultural use. For example, Germany, Switzerland, and the Netherlands set values of 200 µg kg^{-1} for individual PCB congeners and 100 ng kg^{-1} toxic equivalents for PCDD/PCDF.[127] The European Community (EC) is presently drafting regulations specifying maximum concentrations for a range of compounds in sludges to be used in agriculture (Table 5.5).[2]

The sorption capacity of amended soils for triazines, including atrazine, has been noted to be the same in soils before and after addition of sludge.[145] However, in a field trial, transport of atrazine was noted to increase after amendment that was attributed to complexation with more mobile colloidal and dissolved organic matter.[146]

Some compounds, such as DDE, which are not degraded during the (aerobic) wastewater treatment processes or subsequent anaerobic digestion of sludges, have been shown to be particularly persistent in soils amended with sludges.[40,147] Other chlorinated organic compounds have been shown to decline in concentration after initial application of sludge to land. Concentrations of both PCBs and chlorinated phenols declined to values observed in control soils over a period of 260 days that attributed to volatilization and biodegradation processes. PCDD/PCDF and the non-ortho PCB 77 appeared more persistent.[148]

Other compounds with strong potential for causing problems related to endocrine disruption are known to be degraded in aerobic conditions in soils. 4NP, which is produced during anaerobic treatment, has been shown to degrade within 38 days in aerobic conditions in treated soils.[149] During laboratory tests 62% was lost over 28 days, but a lag period was observed.[150] The compound persisted in larger sludge aggregates where oxygen transport was limited. However, due to its hydrophobic nature, it did not leach from the soil, and extrapolation of data indicated that the substance would persist for 1 year in a 2 cm sludge aggregate.

TABLE 5.5
Limit Values (Dry Matter) for Concentrations of Organic Compounds and Dioxins in Sludge for Use on Land

Organic Compounds	Limit Values (mg kg^{-1})
AOX[a]	500
LAS[b]	2600
DEHP[c]	100
NPE[d]	50
PAH[e]	6
PCB[f]	0.8
PCDD/PCDF[g]	100 (ng TE kg^{-1})

[a] Sum of halogenated organic compounds.

[b] Linear alkylbenzene sulfonates.

[c] Di(2-ethylhexyl)phthalate.

[d] It comprises the substances nonylphenol and nonylphenolethoxylates with 1 or 2 ethoxy groups.

[e] Sum of the following polycyclic aromatic hydrocarbons: acenaphthene, phenanthrene, fluorene, fluoranthene, pyrene, benzo(b+j+k)fluoranthene, benzo(a)pyrene, benzo(ghi)perylene, indeno(1, 2, 3-c,d)pyrene.

[f] Sum of the polychlorinated biphenyls components number 28, 52, 101, 118, 138, 153, 180.

[g] Polychlorinated dibenzodioxins/polychlorinated dibenzofuranes.

5.5.2 THERMAL PROCESSING OF SLUDGE

There are a range of thermal processing options available resulting in the production of a combination of water, oil, gas, and char. These processes are used to a greater or lesser extent to generate energy that results in the final production of a char residue of approximately 10 to 20% of the original volume of sludge.[151] It has been stated that the primary objective of thermal recovery processes for sewage sludge is the safe disposal of hazardous substances.[152] A further aim is to generate energy or residual materials that can be introduced into recovery or recycling systems. It is expected that by 2005, 40% of sludge produced in the EC will be disposed of by a thermal route.[27] Thermal processing can be categorized into incineration, gasification/pyrolysis, and the co-fuelling of cement kilns and power stations, with several technologies available for thermal processing.[142]

Compounds of concern associated with combustion processes are typically chlorinated aromatics, such as PCDDs PCDFs, and the PAH. The emission of PCDD/PCDF during laboratory scale combustion and co-combustion has been demonstrated by Samaras et al.[141] Emissions ranged from 0.5 to 300 ng kg^{-1} of fuel in terms of toxic equivalent quantities, with high copper and chlorine concentrations within the sludge resulting in formation of the products during combustion. Other systems have been demonstrated to comply with strict limits on emissions of PCDD/PCDF of <0.01 ng m^{-3} when dealing with sludge samples containing up to 770 ng kg^{-1} PCDD/PCDF.[152] The system also produced gas consisting of approximately 36% CO and 23% H$_2$. This could be utilized for a range of power applications (fuel cells, gas turbines or boilers), sulphur in usable quantities, and vitrified slag suitable for use in the construction industry. Leachability studies indicated that there was no risk of contaminants being released from the vitrified slag.

REFERENCES

1. Meakins, N.C., Bubb, J.M., and Lester, J.N., The fate and behaviour of organic micropollutants during wastewater treatment processes: a review, *Int. J. Environ. Pollut.*, 4, 27, 1994.
2. CEC, Working Document on Sludge 3rd Draft, Brussels, Apr. 27, 2000.
3. Harrison, E.Z., McBride, M.B., and Bouldin, D.R., Land application of sewage sludges: an appraisal of the US regulations, *Int. J. Environ. Pollut.*, 11, 1, 1999.
4. Hall, J.E., Sewage-sludge production, treatment and disposal in the European Union, *J. Chart. Inst. Water. Environ. Manage.*, 9, 335, 1995.
5. Hudson, J.A. and Lowe, P., Current technologies for sludge treatment and disposal, *J. Chart. Inst. Water. Environ. Manage.*, 10, 436, 1996.
6. O'Dette, R.G., US EPA's biosolids regulations: 40 CFR part 503, *Environ. Prog.*, 17, F3, 1998.
7. Mikkelsen, J., Nyholm, N., Neergaard Jacobsen, B., and Fredenslund, F.C., Evaluation and modification of the simpletreat chemical fate model for activated sludge sewage treatment plants, *Water Sci. Technol.*, 33, 279, 1996.
8. Byrns, G., The fate of xenobiotic organic compounds in wastewater treatment plants, *Water Res.*, 35, 2523, 2001.
9. Wild, S.R. and Jones, K.C., The effect of sludge treatment on the organic contaminant content of sewage sludges, *Chemosphere*, 19, 1765, 1989.
10. Bodzek, D., Janoszka, B., Dobosz, C., Warzecha, L., and Bodzek, M., Determination of polycyclic aromatic compounds and heavy metals in sludges from biological sewage treatment plants, *J. Chromatogr.*, A 774, 177, 1997.
11. Brunner, P.H., Capri, S., Marcomini, A., and Giger, W., Occurrence and behavior of linear alkylbenzenesulfonates, nonylphenol, nonylphenol monophenol and nonylphenol diethoxylates in sewage and sewage-sludge treatment, Water Res., 22, 1465, 1988.
12. Swisher, R.D., *Surfactant Biodegradation*, Marcel Dekker, New York, 1987.
13. Johnson, A.C. and Sumpter, J.P., Removal of endocrine-disrupting chemicals in activated sludge treatment works, *Environ. Sci. Technol.*, 35, 4697, 2001.
14. Ball, H.A., Reinhard, M., and McCarty, P.L., Biotransformation of halogenated and nonhalogenated octylphenol polyethoxylate residues under aerobic and anaerobic conditions, *Environ. Sci. Technol.*, 23, 951, 1989.
15. Ternes, T.A., Stumpf, M., Mueller, J., Haberer, K., Wilken, R.D., and Servos, M., Behavior and occurrence of estrogens in municipal sewage treatment plants. I. Investigations in Germany, Canada and Brazil, *Sci. Total Environ.*, 225, 81, 1999.
16. Ternes, T.A., Kreckel, P., and Mueller, J., Behaviour and occurrence of estrogens in municipal sewage treatment plants. II. Aerobic batch experiments with activated sludge, *Sci. Total Environ.*, 225, 91, 1999.
17. Belfroid, A.C., Van der Horst, A., Vethaak, A.D., Schafer, A.J., Rijs, G.B. J., Wegener, J., and Cofino, W.P., Analysis and occurrence of estrogenic hormones and their glucuronides in surface water and waste water in The Netherlands, *Sci. Total Environ.*, 225, 101, 1999.
18. Takiugami, H., Taniguchi, N., Matsuda, T., Yamada, M., Shimizu, Y., and Matsui, S., The fate and behaviour of human estrogens in a night soil treatment process, *Water Sci. Technol.*, 42, 45, 2000.
19. Klopffer, W., Environmental hazard assessment of chemicals and products. Part V. Anthropogenic chemicals in sewage sludge, *Chemosphere*, 33, 1067, 1996.
20. Giger, W., Organische Verunreinigungen im Klarschlamm: Herunft und Verhalten in der Umwelt, *EAWAG News* Dubendorf, Switzerland, 1989, p. 8.

21. Weemaes, M.P.J. and Verstraete, W.H., Evaluation of current wet sludge disintegration techniques, *J. Chem. Technol. Biotechnol.*, 73, 83, 1998.
22. Muller, J.A., Pretreatment processes for the recycling and reuse of sewage sludge, *Water Sci. Technol.*, 42, 167, 2000.
23. Muller, J.A., Prospects and problems of sludge pre-treatment processes, *Water Sci. Technol.*, 44, 121, 2001.
24. Ghosh, S., Buoy, K., Dressel, L., Miller, T., Wilcox, G., and Loos, D., Pilot-scale and full-scale 2-phase anaerobic-digestion of municipal sludge, *Water Environ. Res.*, 67, 206, 1995.
25. McIntyre, A.E., Lester, J.N., and Perry, R., The influence of chemical conditioning and de-watering on the distribution of polychlorinated biphenyls and organochlorine insecticides in sewage sludges, *Environ. Pollut. (Series B)*, 2, 309, 1981.
26. Clark, P.B. and Nujjoo, I., Ultrasonic sludge pretreatment for enhanced sludge digestion, *J. Chart. Inst. Water. Environ. Manage., 14, 66, 2000.*
27. Lester, J.N. and Edge, D., Sewage and sewage sludge treatment, in *Pollution Causes, Effects and Control*, 4th ed., Harrison, R.M., Ed., The Royal Society of Chemistry, Cambridge, 2000, p. 113.
28. Weiland, P., Anaerobic waste digestion in Germany: status and recent developments, *Biodegradation*, 11, 415, 2000.
29. Buisson, R.S.K., Kirk, P.W.W., Lester, J.N., and Campbell, J. A., Behavior of selected chlorinated organic micropollutants during batch anaerobic-digestion, *Water Pollut. Control*, 85, 387, 1986.
30. Buisson, R.S.K., Kirk, P.W.W., and Lester, J.N., Fate of selected chlorinated organic compounds during semi-continuous anaerobic sludge digestion, *Arch. Environ. Contam. Toxicol.*, 19, 428, 1990.
31. Zipper, C., Bolliger, C., Fleischmann, T., Suter, M.J.F., Angst, W., Muller, M.D., and Kohler, H.P. E., Fate of the herbicides mecoprop, dichlorprop, and 2,4-D in aerobic and anaerobic sewage sludge as determined by laboratory batch studies and enantiomer-specific analysis, *Biodegradation*, 10, 271, 1999.
32. Berestovskaya, Y.Y., Ignatov, V.V., Markina, L.N., Kamenev, A.A., and Makarov, O.E., Degradation of ortho-chlorophenol, para-chlorophenol, and 2,4-dichlorophenoxyacetic acid by the bacterial community of anaerobic sludge, *Microbiology*, 69, 397, 2000.
33. Mikesell, M.D. and Boyd, S.A., Reductive dechlorination of the pesticides 2,4-d, 2,4,5-t, and pentachlorophenol in anaerobic sludges, *J. Environ. Qual.*, 14, 337, 1985.
34. Kirk, P.W.W. and Lester, J.N., Degradation of phenol, selected chlorophenols and chlorophenoxy herbicides during anaerobic sludge-digestion, *Environ. Technol. Lett.*, 10, 405, 1989.
35. Kirk, P.W.W. and Lester, J.N., The behavior of chlorinated organics during activated-sludge treatment and anaerobic-digestion, *Water Sci. Technol.*, 20, 353, 1988.
36. Madsen, T. and Rasmussen, H.B., Anaerobic biodegradation potentials in digested sludge, a freshwater swamp and a marine sediment, *Chemosphere*, 31, 4243, 1995.
37. Boyd, S.A. and Shelton, D.R., Anaerobic biodegradation of chlorophenols in fresh and acclimated sludge, *Appl. Environ. Microbiol.*, 47, 272, 1984.
38. Takeuchi, R., Suwa, Y., Yamagishi, T., and Yonezawa, Y., Anaerobic transformation of chlorophenols in methanogenic sludge unexposed to chlorophenols, *Chemosphere*, 41, 1457, 2000.
39. Chang, B.V., Chiang, C.W., and Yuan, S.Y., Dechlorination of pentachlorophenol in anaerobic sewage sludge, *Chemosphere*, 36, 537, 1998.

40. Strompl, C. and Thiele, J.H., Comparative fate of 1,1-diphenylethylene (DPE), 1,1-dichloro- 2,2-bis(4-chlorophenyl)-ethylene (DDE), and pentachlorophenol (PCP) under alternating aerobic and anaerobic conditions, *Arch. Environ. Contam. Toxicol.*, 33, 350, 1997.

41. Tartakovsky, B., Manuel, M.F., Beaumier, D., Greer, C.W., and Guiot, S.R., Enhanced selection of an anaerobic pentachlorophenol-degrading consortium, *Biotechnol. Bioeng.*, 73, 476, 2001.

42. Christiansen, N., Hendriksen, H.V., Jarvinen, K.T., and Ahring, B.K., Degradation of chlorinated aromatic-compounds in UASB reactors, *Water Sci. Technol.*, 31, 249, 1995.

43. Kirk, P.W.W., Rogers, H.R., and Lester, J.N., The Fate of chlorobenzenes and per-methrins during anaerobic sewage-sludge digestion, *Chemosphere*, 18, 1771, 1989.

44. Fathepure, B.Z., Tiedje, J.M., and Boyd, S.A., Reductive dechlorination of hexachlorobenzene to trichlorobenzenes and dichlorobenzenes in anaerobic sewage-sludge, *Appl. Environ. Microbiol.*, 54, 327, 1988.

45. Buser, H.R., Haglund, P., Muller, M.D., Poiger, T., and Rappe, C., Rapid anaerobic degradation of toxaphene in sewage sludge, *Chemosphere*, 40, 1213, 2000.

46. Tartakovsky, B., Hawari, J., and Guiot, S.R., Enhanced dechlorination of aroclor 1242 in an anaerobic continuous bioreactor, *Water Res.*, 34, 85, 2000.

47. Tartakovsky, B., Michotte, A., Cadieux, J.C.A., Lau, P.C.K., Hawari, J., and Guiot, S.R., Degradation of aroclor 1242 in a single-stage coupled anaerobic/aerobic bioreactor, *Water Res.*, 35, 4323, 2001.

48. Ahel, M., Giger, W., and Koch, M., Behavior of alkylphenol polyethoxylate surfactants in the aquatic environment. 1. Occurrence and transformation in sewage-treatment, *Water Res.*, 28, 1131, 1994.

49. Wahlberg, C., Renberg, L., and Wideqvist, U., Determination of nonylphenol and nonylphenol ethoxylates as their pentafluorobenzoates in water, sewage-sludge and biota, *Chemosphere*, 20, 179, 1990.

50. Fent, K., Organotin compounds in municipal wastewater and sewage sludge: contamination, fate in treatment process and ecotoxicological consequences, *Sci. Total Environ.*, 185, 151, 1996.

51. Kirk, P.W.W. and Lester, J.N., The fate of polycyclic aromatic-hydrocarbons during sewage-sludge digestion, *Environ. Technol.*, 12, 13, 1991.

52. Ziogou, K., Kirk, P.W.W., and Lester, J.N., Behavior of phthalic-acid esters during batch anaerobic-digestion of sludge, *Water Res.*, 23, 743, 1989.

53. Shelton, D.R., Boyd, S.A., and Tiedje, J.M., Anaerobic degradation of phthalic acid esters in sludge, *Environ. Sci. Technol.*, 18, 93, 1984.

54. Wang, J.L., Chen, L.J., Shi, H.C., and Qian, Y., Microbial degradation of phthalic acid esters under anaerobic digestion of sludge, *Chemosphere*, 41, 1245, 2000.

55. Painter, S.E. and Jones, W.J., Anaerobic bioconversion of phthalic acid esters by natural inocula, *Environ. Technol.*, 11, 1015, 1990.

56. Banat, F.A., Prechtl, S., and Bischof, F., Experimental assessment of big-reduction of Di-2-thylhexyl phthalate (DEHP) under aerobic thermophilic conditions, *Chemosphere*, 39, 2097, 1999.

57. Hakk, H., Millner, P., and Larsen, G., Fate of the Endogenous Hormones 17β-Estradiol and Testosterone in Composted Poultry Manure, in *2nd International Conference on Pharmaceuticals and Endocrine Disrupting Chemicals in Water*, National Groundwater Association, Minneapolis, MN, 2001, p. 128.

58. Potter, C.L., Glaser, J.A., Chang, L.W., Meier, J.R., Dosani, M.A., and Herrmann, R.F., Degradation of polynuclear aromatic hydrocarbons under bench-scale compost conditions, *Environ. Sci. Technol.*, 33, 1717, 1999.
59. Kobayashi, H. and Rittmann, B.E., Microbial removal of hazardous organic-compounds, *Environ. Sci. Technol.*, 16, A170, 1982.
60. Grady, C.P.L., Biodegradation of toxic organics: status and potential, *J. Environ. Eng.-ASCE*, 116, 805, 1990.
61. Johnson, A.C., Belfroid, A., and Di Corcia, A., Estimating steroid oestrogen inputs into activated sludge treatment works and observations their removal from the effluent, *Sci. Total Environ.*, 256, 163, 2000.
62. Vader, J.S., van Ginkel, C.G., Sperling, F.M.G.M., de Jong, G., de Boer, W., de Graaf, J.S., van der Most, M., and Stokman, P.G.W., Degradation of ethinyl estradiol by nitrifying activated sludge, *Chemosphere*, 41, 1239, 2000.
63. Lai, K.M., Johnson, K.L., Scrimshaw, M.D., and Lester, J.N., Binding of waterborne steroid estrogens to solid phases in river and estuarine systems, *Environ. Sci. Technol.*, 34, 3890, 2000.
64. Shore, L.S., Gurevitz, M., and Shemesh, M., Estrogen as an environmental-pollutant, *Bull. Environ. Contam. Toxicol.*, 51, 361, 1993.
65. Mohn, W.W. and Tiedje, J.M., Microbial reductive dehalogenation, *Microbiol. Rev.*, 56, 482, 1992.
66. van Eekert, M.H. A. and Schraa, G., The potential of anaerobic bacteria to degrade chlorinated compounds, *Water Sci. Technol.*, 44, 49, 2001.
67. Droste, R.L., Kennedy, K.J., Lu, J.G., and Lentz, M., Removal of chlorinated phenols in upflow anaerobic sludge blanket reactors, *Water Sci. Technol.*, 38, 359, 1998.
68. Cass, Q.B., Freitas, L.G., Foresti, E., and Damianovic, M., Development of HPLC method for the analysis of chlorophenols in samples from anaerobic reactors for wastewater treatment, *J. Liq. Chromatogr. Relat. Technol.*, 23, 1089, 2000.
69. Jacobsen, B.N. and Arvin, E., Biodegradation kinetics and fate modelling of pentachlorophenol in bioaugmented activated sludge reactors, *Water Res.*, 30, 1184, 1996.
70. Nyholm, N., Jacobsen, B.N., Pedersen, B.M., Poulsen, O., Damborg, A., and Schultz, B., Removal of organic micropollutants at ppb levels in laboratory activated-sludge reactors under various operating-conditions: biodegradation, *Water Res.*, 26, 339, 1992.
71. Chen, S.T. and Berthouex, P.M., Treating an aged pentachlorophenol- (PCP-) contaminated soil through three sludge handling processes, anaerobic sludge digestion, post-sludge digestion and sludge land application, *Water Sci. Technol.*, 44, 149, 2001.
72. Safe, S., Polychlorinated biphenyls (PCBs), dibenzofurans (PCDFs) and related compounds: environmental and mechanistic considerations which support the development of toxic equivalency factors (TEFs), *Crit. Rev. Toxicol.*, 21, 51, 1990.
73. McIntyre, A.E. and Lester, J.N., Polychlorinated biphenyl and organochlorine insecticide concentrations in forty sewage sludges in England, *Environ. Pollut. (Series B)*, 3, 225, 1982.
74. Alcock, R.E. and Jones, K.C., Polychlorinated biphenyls in digested UK sewage sludges, *Chemosphere*, 26, 2199, 1993.
75. Sewart, A.P., Harrad, S.J., McLachlan, M.S., McGrath, S.P., and Jones, K.C., PCDD/PCDFs and non-o-PCBs in digested U.K. sewage sludges, *Chemosphere*, 30, 51, 1995.
76. Stevens, J., Green, N.J.L., and Jones, K.C., Survey of PCDD/PCDFs and non-ortho PCBs in UK sewage sludges, *Chemosphere*, 44, 1455, 2001.

77. Bergh, A.K. and Peoples, R.S., Distribution of polychlorinated biphenyls in a municipal wastewater treatment plant and environs, *Sci. Total Environ.*, 8, 197, 1977.

78. Berset, J.D. and Holzer, R., Quantitative determination of polycyclic aromatic hydrocarbons, polychlorinated biphenyls and organochlorine pesticides in sewage sludges using supercritical fluid extraction and mass spectrometric detection, *J. Chromatogr.*, A 852, 545, 1999.

79. Buisson, R.S. K., Kirk, P.W.W., and Lester, J.N., Determination of chlorinated phenols in water, wastewater, and wastewater-sludge by capillary GC/ECD, *J. Chromatogr. Sci.*, 22, 339, 1984.

80. Wild, S.R., Harrad, S.J., and Jones, K.C., Chlorophenols in digested UK sewage sludges, *Water Res.*, 27, 1527, 1993.

81. McIntyre, A.E. and Lester, J.N., Occurrence and distribution of persistent organochlorine compounds in U.K. sewage sludges, *Water, Air Soil Pollut.*, 23, 379, 1984.

82. Schnaak, W., Kuchler, T., Kujawa, M., Henschel, K.P., Sussenbach, D., and Donau, R., Organic contaminants in sewage sludge and their ecotoxicological significance in the agricultural utilization of sewage sludge, *Chemosphere*, 35, 5, 1997.

83. Berset, J.D. and Holzer, R., Determination of coplanar and ortho substituted PCBs in some sewage sludges of Switzerland using HRGC/ECD and HRGC/MSD, *Chemosphere*, 32, 2317, 1996.

84. Dupont, G., Delteil, C., Camel, V., and Bermond, A., The determination of polychlorinated biphenyls in municipal sewage sludges using microwave-assisted extraction and gas chromatography mass spectrometry, *Analyst*, 124, 453, 1999.

85. Steinwandter, H., Research in Environmental-pollution. 8. Identification of non- *o,o'*-Cl and mono-*o,o'*-Cl substituted PCB congeners in Hessian sewage sludges, *Fresenius J. Anal. Chem.*, 344, 66, 1992.

86. Molina, L., Cabes, M., Diaz-Ferrero, J., Coll, M., Marti, R., Broto-Puig, F., Comellas, L., and Rodriguez-Larena, M.C., Separation of non-ortho polychlorinated biphenyl congeners on per-packed carbon tubes: application to analysis in sewage sludge and soil samples, *Chemosphere*, 40, 921, 2000.

87. Nylund, K., Asplund, L., Jansson, B., Jonsson, P., Litzen, K., and Sellstrom, U., Analysis of Some Polyhalogenated Organic Pollutants in Sediment and Sewage-Sludge, *Chemosphere*, 24, 1721, 1992.

88. Koistinen, J., Alkyl polychlorobibenzyls and planar aromatic chlorocompounds in pulp-mill products, effluents, sludges and exposed biota, *Chemosphere*, 24, 559, 1992.

89. Rappe, C., Kjeller, L.O., and Andersson, R., Analyses of PCDDs and PCDFs in sludge and water samples, *Chemosphere*, 19, 13, 1989.

90. Broman, D., Naf, C., Rolff, C., and Zebuhr, Y., Analysis of polychlorinated dibenzo-para-dioxins (PCDD) and polychlorinated dibenzofurans (PCDF) in soil and digested sewage-sludge from Stockholm, Sweden, *Chemosphere*, 21, 1213, 1990.

91. Naf, C., Broman, D., Ishaq, R., and Zebuhr, Y., PCDDs and PCDFs in water, sludge and air samples from various levels in a waste-water treatment-plant with respect to composition changes and total flux, *Chemosphere*, 20, 1503, 1990.

92. Bennie, D.T., Review of the environmental occurrence of alkylphenols and alkylphenol ethoxylates, *Water Qual. Res. J. Canada*, 34, 79, 1999.

93. Lee, H.B., Peart, T.E., Bennie, D.T., and Maguire, R.J., Determination of nonylphenol polyethoxylates and their carboxylic acid metabolites in sewage treatment plant sludge by supercritical carbon dioxide extraction, *J. Chromatogr.*, A 785, 385, 1997.

94. Bolz, U., Hagenmaier, H., and Korner, W., Phenolic xenoestrogens in surface water, sediments, and sewage sludge from Baden-Wurttemberg, south-west Germany, *Environ. Pollut.*, 115, 291, 2001.

95. Lin, J.G., Arunkumar, R., and Liu, C.H., Efficiency of supercritical fluid extraction for determining 4- nonylphenol in municipal sewage sludge, *J. Chromatogr.*, A 840, 71, 1999.

96. Ahel, M., McEvoy, J., and Giger, W., Bioaccumulation of the lipophilic metabolites of nonionic surfactants in fresh-water organisms, *Environ. Pollut.*, 79, 243, 1993.

97. Ekelund, R., Bergman, A., Granmo, A., and Berggren, M., Bioaccumulation of 4-nonylphenol in marine animals: a reevaluation, *Environ. Pollut.*, 64, 107, 1990.

98. Jobst, H., Chlorophenols and nonylphenols in sewage sludges. 1. Occurrence in sewage sludges of western German treatment plants from 1987 to 1989, *Acta Hydrochim. Hydrobiol.*, 23, 20, 1995.

99. Jobst, H., Chlorophenols and nonylphenols in sewage sludges. Part II. Did contents of pentachlorophenol and nonylphenols reduce?, *Acta Hydrochim. Hydrobiol.*, 26, 344, 1998.

100. Castillo, M., Martínez, E., Ginebreda, A., Tirapu, A., and Barceló, D., Determination of non-ionic surfactants and polar degradation products in influent and effluent water samples and sludges of sewage treatment plants by a generic solid-phase extraction protocol, *Analyst*, 125, 1733 2000.

101. Petrovic, M., Diaz, A., Ventura, F., and Barcelo, D., Simultaneous determination of halogenated derivatives of alkylphenol ethoxylates and their metabolites in sludges, river sediments, and surface, drinking, and wastewaters by liquid chromatography-mass spectrometry, *Anal. Chem.*, 73, 5886, 2001.

102. Ejlertsson, J., Nilsson, M.L., Kylin, H., Bergman, A., Karlson, L., Oquist, M., and Svensson, B.H., Anaerobic degradation of nonylphenol mono- and diethoxylates in digestor sludge, landfilled municipal solid waste, and landfilled sludge, *Environ. Sci. Technol.*, 33, 301, 1999.

103. Salanitro, J.P. and Diaz, L.A., Anaerobic biodegradability testing of surfactants, *Chemosphere*, 30, 813, 1995.

104. Moreda, J.M., Arranz, A., De Betono, S.F., Cid, A., and Arranz, J.F., Chromatographic determination of aliphatic hydrocarbons and polyaromatic hydrocarbons (PAHs) in a sewage sludge, *Sci. Total Environ.*, 220, 33, 1998.

105. Lee, H.B. and Peart, T.E., Determination of 4-nonylphenol in effluent and sludge from sewage-treatment plants, *Anal. Chem.*, 67, 1976, 1995.

106. Giger, W., Brunner, P.H., and Schaffner, C., 4-Nonylphenol in sewage-sludge: accumulation of toxic metabolites from nonionic surfactants, *Science*, 225, 623, 1984.

107. Sweetman, A.J., Development and application of a multi-residue analytical method for the determination of n-alkanes, linear alkylbenzenes, polynuclear aromatic-hydrocarbons and 4-nonylphenol in digested sewage sludges, *Water Res.*, 28, 343, 1994.

108. McGowin, A.E., Adom, K.K., and Obubuafo, A.K., Screening of compost for PAHs and pesticides using static subcritical water extraction, *Chemosphere*, 45, 857, 2001.

109. Ghosh, P.K., Philip, L., and Bandyopadhyay, M., Anaerobic treatment of atrazine bearing wastewater, *J. Environ. Sci. Health Part B-Pestic. Contam. Agric. Wastes*, 36, 301, 2001.

110. McIntyre, A.E., Perry, R., and Lester, J.N., Analysis and incidence of organo-phosphorus compounds in sewage sludges, *Bull. Environ. Contam. Toxicol.*, 26, 116, 1981.

111. Horowitz, A., Shelton, D.R., Connell, C.P., and Tiedje, J.M., Anaerobic degradation of aromatic compounds in sediments and digested sludges, *Devel. Ind. Microbiol.*, 23, 435, 1982.

112. Staples, C.A., Peterson, D.R., Parkerton, T.F., and Adams, W.J., The environmental fate of phthalate esters: a literature review, *Chemosphere*, 35, 667, 1997.

113. Ike, M., Jin, C.S., and Fujita, M., Biodegradation of bisphenol A in the aquatic environment, *Water Sci. Technol.*, 42, 31, 2000.

114. Lee, H.B. and Peart, T.E., Determination of bisphenol A in sewage effluent and sludge by solid-phase and supercritical fluid extraction and gas chromatography/mass spectrometry, *J. AOAC Int.*, 83, 290, 2000.

115. Fromme, H., Kuchler, T., Otto, T., Pilz, K., Muller, J., and Wenzel, A., Occurrence of phthalates and bisphenol A and F in the environment, *Water Res.*, 36, 1429, 2002.

116. Körner, W., Bolz, U., Sussmuth, W., Hiller, G., Schuller, W., Hanf, V., and Hagenmaier, H., Input/output balance of estrogenic active compounds in a major municipal sewage plant in Germany, *Chemosphere*, 40, 1131, 2000.

117. Rahman, F., Langford, K.H., Scrimshaw, M.D., and Lester, J.N., Polybrominated diphenyl ether (PBDE) flame retardants, *Sci. Total Environ.*, 275, 1, 2001.

118. Morris, P.J., Quensen, J.F., Tiedje, J.M., and Boyd, S.A., An assessment of the reductive debromination of polybrominated biphenyls in the Pine River reservoir, *Environ. Sci. Technol.*, 27, 1580, 1993.

119. Oberg, K., Warman, K., and Oberg, T., Distribution and levels of brominated flame retardants in sewage sludge, *Chemosphere*, 48, 805, 2002.

120. Sellstrom, U. and Jansson, B., Analysis of tetrabromobisphenol a in a product and environmental-samples, *Chemosphere*, 31, 3085, 1995.

121. Jones, O.A.H., Voulvoulis, N., and Lester, J.N., Human pharmaceuticals in the aquatic environment: a review, *Environ. Technol.*, 22, 1383, 2001.

122. Halling-Sorensen, B., Nielsen, S.N., Lanzky, P.F., Ingerslev, F., Lutzhoft, H.C.H., and Jorgensen, S.E., Occurrence, fate and effects of pharmaceutical substances in the environment: a review, *Chemosphere*, 36, 357, 1998.

123. U.S. Environmental Protection Agency, Estimation Program Interface (EPI) Suite, version 3.10, 2001.

124. Jones, O.A.H., Voulvoulis, N., and Lester, J.N., Aquatic environmental risk assessment for the top 25 English prescription pharmaceuticals, *Water Res.*, 36, 5013, 2002.

125. Stuer-Lauridsen, F., Birkved, M., Hansen, L.P., Holten Lutzhoft, H.C., and Halling-Sorensen, B., Environmental risk assessment of human pharmaceuticals in Denmark after normal therapeutic use, *Chemosphere*, 40, 783, 2000.

126. Kim, Y.K., Eom, Y.S., Oh, B.K., Lee, W.H., and Choi, J.W., Application of a thermophilic aerobic digestion process to industrial waste activated sludge treatment, *J. Microbiol. Biotechnol.*, 11, 570, 2001.

127. Perez, S., Guillamon, M., and Barcelo, D., Quantitative analysis of polycyclic aromatic hydrocarbons in sewage sludge from wastewater treatment plants, *J. Chromatogr.*, A 938, 57, 2001.

128. Knudsen, L., Kristensen, G.H., Jorgensen, P.E., and Jepsen, S.E., Reduction of the content of organic micropollutants in digested sludge by a post-aeration process: a full-scale demonstration, *Water Sci. Technol.*, 42, 111, 2000.

129. Wild, S.R., Waterhouse, K.S., McGrath, S.P., and Jones, K.C., Organic contaminants in an agricultural soil with a known history of sewage-sludge amendments: polynuclear aromatic-hydrocarbons, *Environ. Sci. Technol.*, 24, 1706, 1990.

130. Wild, S.R. and Jones, K.C., Organic-chemicals entering agricultural soils in sewage sludges: screening for their potential to transfer to crop plants and livestock, *Sci. Total Environ.*, 119, 85, 1992.

131. U.S. Environmental Protection Agency, Fate of Priority Pollutants in Publicly Owned Treatment Works, EPA 440/1–83/303, Washington, DC, 1982.

132. McIntyre, A.E., Perry, R., and Lester, J.N., Analysis of polynuclear aromatic hydrocarbons in sewage sludges, *Anal. Lett.*, 14, 291, 1981.

133. Miege, C., Dugay, J., and Hennion, M.C., Optimization and validation of solvent and supercritical-fluid extractions for the trace-determination of polycyclic aromatic hydrocarbons in sewage sludges by liquid chromatography coupled to diode-array and fluorescence detection, *J. Chromatogr.*, A 823, 219, 1998.

134. Fent, K. and Muller, M.D., Occurrence of organotins in municipal waste-water and sewage-sludge and behavior in a treatment-plant, *Environ. Sci. Technol.*, 25, 489, 1991.

135. Muller, M.D., Comprehensive trace level determination of organotin compounds in environmental-samples using high-resolution gas-chromatography with flame photometric detection, *Anal. Chem.*, 59, 617, 1987.

136. Chau, Y.K., Zhang, S.Z., and Maguire, R.J., Occurrence of Butyltin species in sewage and sludge in Canada, *Sci. Total Environ.*, 121, 271, 1992.

137. Layton, A.C., Gregory, B.W., Seward, J.R., Schultz, T.W., and Sayler, G.S., Mineralization of steroidal hormones by biosolids in wastewater treatment systems in Tennessee USA, *Environ. Sci. Technol.*, 34, 3925, 2000.

138. McIntyre, A.E., Lester, J.N., and Perry, R., Persistence of organo-phosphorus insecticides in sewage sludges, *Environ. Technol. Lett.*, 2, 111, 1981.

139. Nicholls, T.P., Lester, J.N., and Perry, R., The influence of heat treatment on the metallic and polycyclic aromatic hydrocarbon content of sewage sludge, *Sci. Total Environ.*, 12, 137, 1979.

140. Nicholls, T.P., Lester, J.N., and Perry, R., The influence of heat treatment on the metallic and polycyclic aromatic hydrocarbon content of sewage sludge: a pilot plant study, *Sci. Total Environ.*, 14, 19, 1980.

141. Samaras, P., Blumenstock, M., Schramm, K.W., and Kettrup, A., Emissions of chlorinated aromatics during sludge combustion, *Water Sci. Technol.*, 42, 251, 2000.

142. Werther, J. and Ogada, T., Sewage sludge combustion, *Prog. Energy Combust. Sci.*, 25, 55, 1999.

143. Chaney, R.L., Ryan, J.A., and O'Connor, G.A., Organic contaminants in municipal biosolids: risk assessment, quantitative pathways analysis, and current research priorities, *Sci. Total Environ.*, 185, 187, 1996.

144. U.S. Environmental Protection Agency, Sewage sludge; use or disposal standards: dioxin and dioxin-like compounds; numeric concentration limits, *Federal Register 64*, 246, 72045, Dec. 23, 1999.

145. Sluszny, C., Graber, E.R., and Gerstl, Z., Sorption of s-triazine herbicides in organic matter amended soils: fresh and incubated systems, *Water Air Soil Pollut.*, 115, 395, 1999.

146. Graber, E.R., Dror, I., Bercovich, F.C., and Rosner, M., Enhanced transport of pesticides in a field trial with treated sewage sludge, *Chemosphere*, 44, 805, 2001.

147. Boul, H.L., Garnham, M.L., Hucker, D., Baird, D., and Aislabie, J., The influence of agricultural practices on the levels of DDT and its residues in soil, *Environ. Sci. Technol.*, 28, 1397, 1994.

148. Wilson, S.C., Alcock, R.E., Sewart, A.P., and Jones, K.C., Persistence of organic contaminants in sewage sludge-amended soil: a field experiment, *J. Environ. Qual.*, 26, 1467, 1997.

149. Hesselsoe, M., Jensen, D., Skals, K., Olesen, T., Moldrup, P., Roslev, P., Mortensen, G.K., and Henriksen, K., Degradation of 4-nonylphenol in homogeneous and nonhomogeneous mixtures of soil and sewage sludge, *Environ. Sci. Technol.*, 35, 3695, 2001.

150. Staples, C.A., Williams, J.B., Blessing, R.L., and Varineau, P.T., Measuring the biodegradability of nonylphenol ether carboxylates, octylphenol ether carboxylates, and nonylphenol, *Chemosphere*, 38, 2029, 1999.

151. Furness, D.T., Hoggett, L.A., and Judd, S.J., Thermochemical treatment of sewage sludge, *J. Chart. Inst. Water. Environ. Manage.*, 14, 57, 2000.
152. Jaeger, M. and Mayer, M., The Noell Conversion Process: a gasification process for the pollutant-free disposal of sewage sludge and the recovery of energy and materials, *Water Sci. Technol.*, 41, 37, 2000.

6 Endocrine Disrupters in Receiving Waters

R.L. Gomes and J.N. Lester

CONTENTS

6.1 INTRODUCTION

Sewage treatment processes clean wastewater to a level fit for reintegration into the aquatic environment. Before releasing into receiving waters, effluent must reach certain standards. The chemical revolution, which commenced in the 1930s, has led to the creation of thousands of synthetic organic chemicals (SOCs). The advent of the 1970s with increased organic chemical use and water reuse, gave rise to an interest into SOCs that may be present in sewage effluent.[1,2]

0-56670-601-7/03/$0.00+$1.50
© 2003 by CRC Press LLC

Sewage treatment works (STWs) are able to remove endocrine disrupting chemicals (EDCs) throughout the treatment process. However, depending on the level of treatment, varying concentrations of EDCs have been identified in sewage effluent and sludge (see Chapters 4 and 5). Consequently, wastewater effluent has been shown to contain concentrations of EDCs capable of inducing adverse reactions in biota, which are present in surface waters.[3,4] When industrial discharges contribute to wastewater, an even wider range of EDCs may be found in the effluent. Although EDCs may be present in effluent at trace concentrations, adverse effects have been observed in wildlife and may have health implications for humans.[5,6]

Sewage effluent has been determined as the point source for the endocrine disruption observed in biota, present in surface waters.[7] Diffuse sources of EDCs occur from agricultural runoff, from crops sprayed with EDCs or manure from grazing farm animals.[8] Sludge application to land is another route where the presence of EDCs may be transferred to surface waters via runoff (see Chapter 5). Additionally, percolation from these and other sources into groundwater have caused contamination with EDCs.[9]

6.2 ENDOCRINE DISRUPTERS IN THE RECEIVING AQUATIC ENVIRONMENT

The receiving aquatic environment for effluent discharge can be groundwater or more commonly surface waters (streams, rivers, estuaries, and marine). EDCs have been identified as responsible for estrogenic effects observed in fish present in surface waters. The main causative EDCs have been determined as the steroid estrogens, estrone (E1), 17β-estradiol (E2), 17α-ethinylestradiol (EE2),[10–12] and to a lesser extent alkylphenols (APs) and their ethoxylates (APEOs).[13–15] The presence of these compounds in the aqueous environment has been primarily attributed to incomplete removal during the sewage treatment process.[7,11,16,17] The fate of EDCs present within the effluent is partly dependent on the type of receiving water as well as the compounds' own physicochemical properties. Receiving waters are also receptacles for EDCs from various other sources.

6.2.1 FATE AND BEHAVIOR IN RECEIVING WATERS

Estrogenic effects are reduced after discharge into receiving waters because of dilution, degradation, and sorption processes.[18] In many developed countries, the percentage of sewage effluent in receiving waters can be in the order of 50%. In periods of low flow when rainfall is low and demand is at its highest, the percentage can be up to 90%.[19] A 1:3 dilution of strongly estrogenic STW effluent has been shown to be enough to remove the capacity to induce vitellogenin (VTG) in male trout present in the receiving surface waters.[20,21]

Regarding discharge of EDCs to surface waters, partitioning to the solid phase (suspended solids or whole sediment) and biota will decrease their presence in the water column. Seasonal variation can affect the presence of EDCs in surface waters by the parameters, microbial activity, temperature, and rainfall.[22] Figure 6.1 illustrates several fate processes associated with receiving surface waters.

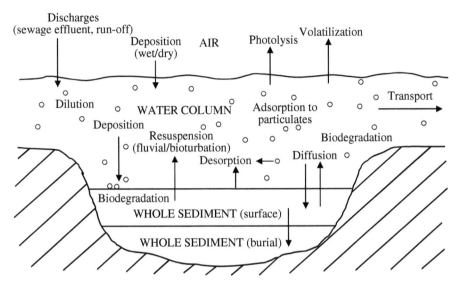

FIGURE 6.1 Sources of EDCs and several fate processes in the water column and sediment phase or receiving surface waters (O = particulate).

Entrance of EDCs to groundwater sources is by percolation or effluent recharge. The hydrophobic and low solubility nature of most EDCs will decrease percolation through to underlying groundwater due to sorption to solid organic matter. For EDCs introduced via effluent recharge, behavior and transport are dependent on advection, dispersion, sorption and biodegradation.[23] Compared to surface waters, the more closed system of groundwater is not influenced to the same extent by seasonal variability, and many transformation processes are either compromised or unable to take place.

6.2.1.1 Partitioning and Transport

Sediments constitute an important compartment of aquatic ecosystems. The environmental consequences resulting from EDCs partitioning into the suspended (particulate) or whole sediment phase are:

1. There is a prolonged persistence of the EDCs in the aquatic environment.
2. Fluvial sediment phase is not static and can therefore act as an effective transport system, introducing contamination to previously unaffected areas.
3. Transformation processes are affected, such as the degree of hydrolysis or EDC availability for biodegradation.
4. The bioavailability of EDCs to benthic organisms and demersal fish results in endocrine disrupting effects to the organism and bioaccumulation in the food chain.

The main processes that determine the fate of sediment-bound EDCs are sorption, degradation (see Section 6.2.1.2) and transport. Most EDCs favor sorption to solid surfaces. Therefore, the fate and behavior of EDCs are dependent to a large extent on the fate and behavior of the suspended particles that constitute a large proportion of the total area of the solid surface in the aqueous environment. Adsorption to organic particles is correlated with the organic carbon partition coefficient (K_{oc}) and inversely correlated to the water solubility of the EDC. Natural organic matter (NOM) plays a major role in the degree of adsorption of low solubility hydrophobic EDCs. Comparatively, adsorption to clay particles is weaker and not as favorable, with the exception of ionic pesticides.[23]

The exchange equilibrium that exists between adsorbed and dissolved EDCs, greatly favors the adsorbed state because of the compound's hydrophobic nature. Unless interactions between the EDC and particle are very strong, the adsorption process is reversible. However, the extent of desorption is limited because of the equilibrium favoring adsorption.

The sediment phase is a dynamic process and can act as an effective transport system redistributing the bound EDCs great distances. EDCs adsorbed to suspended particulate sediments are able to undergo transport depending on the hydrological conditions, organic components in the system and the structure of the EDC. The estrogenic potential of sewage effluent persists for several kilometers downstream of the discharge point and declines as the distance from the source increases.[4,21] However, studies on the River Aire in the U.K. showed that maximum potency remained for at least 5 km from the point effluent source.[24]

Depending on fluvial conditions, sedimentation of particulates does occur to some extent. This explains why EDCs are often identified in sediment at higher concentrations than in the aqueous phase. The role of sediments extends beyond the traditional view of a permanent sink for EDCs present in the aquatic environment. Physical or chemical changes in environmental conditions can alter the fate of the bound chemical, causing release into the water column. Resuspension from turbulence in the water column and chemical processes, such as molecular diffusion and desorption, will affect sediment distribution. Biological processes, such as bioturbation, will also affect sediment distribution, where microorganisms disturb the sediment because of their activities.

6.2.1.2 Transformation Processes

Photolysis, exposure to ultraviolet (UV) sunlight, may degrade certain EDCs to simpler compounds, rendering them more susceptible to biodegradation. Suspended particulates are responsible for a large amount of turbidity in the watercourse. They decrease the depth and degree of irradiating light to the EDCs, as a high clarity of the water is usually required in order for UV to reach EDCs in the water column. However, some particulates will diffuse light instead of adsorbing it, increasing the area and depth that irradiating light can reach. Turbidity is dependent on the fluvial conditions; fast flowing surface waters will mix suspended particles while slow to zero flowing systems will promote settling of particles. For volatilization to occur, usually only dissolved EDCs will volatilize at the water-air interface. Another important chemical transformation process is hydrolysis, the rate being dependent on the EDCs sorption potential and the temperature and pH of the water system. Adsorption to particles of NOM can catalyze hydrolysis.[23]

Biodegradation is dependent on numerous factors, including:

1. Dissolved oxygen (DO) content
2. Oxidation-reduction potential
3. Temperature
4. pH
5. Availability of other organic compounds
6. Presence of particulate matter
7. Concentration of EDCs
8. Type and concentration of the microorganism[25]

Adsorption to particulates may inhibit biodegradation if strongly bonded as in the case of organochlorine insecticides. However, the whole sediment layer is home to much biodegradation. The mean residence time for a compound transported over 10 km in the water phase is between 4 and 5 hours.[26] This limited residence time allows for little degradation to occur and biological degradation is therefore considered to be only important in low flow conditions.

In the groundwater environment, photolysis and volatilization cannot take place. Degradation processes are also not as evident with the depleted oxygen content, resulting in a decreased prevalence of microorganisms.[23] Percolation through soil strata to the groundwater is considered to be the main method of removing undesirable organic compounds as a result of sorption processes. Hence, groundwater is often considered a purer source for drinking water treatment than surface waters. Prior to using groundwater sources, the retention time for EDCs within the underground water source can be years. The degradative resistance of certain EDCs mean that they are able to persist for a similar length of time.[27]

Estrogenic effects can be reduced through dilution, degradation, and sorption processes. Partitioning to the solid phase is the dominant process for the majority of EDCs (with the exception of polychlorinated biphenyls [PCBs]) as a result of their hydrophobic character and low solubility. In fluvial systems, EDC-associated particulates may be transported downstream, introducing contamination to a previously unpolluted environment. In sublayers of whole sediments, the oxygen deficient or anaerobic conditions impede the rate of degradation. Therefore, sediments may act as a secondary source of EDCs through fluvial resuspension, molecular diffusion, desorption, or bioturbation. Such behavior has ecotoxicological implications, especially for organisms whose lifestyle is sediment dependent.

6.2.2 Effects to Biota in Receiving Surface Waters

During the 1980s, anglers observed that several wild roaches exhibited a range of deformities in certain stretches of some U.K. rivers. When hermaphrodite and intersex fish were discovered near sewage outfalls in the early 1990s, concern arose over the presence of environmental estrogens entering the aquatic environment. A link between the production of VTG in male fish normally detected only in fertile females, and sewage effluent was established.[7,28]

The majority of data on EDCs are related to the aqueous environment from observations of abnormalities in biota, predominantly fish, present in surface waters. Adverse effects have included VTG induction in male fish, which is normally identified only in fertile females,[29] intersex where the organism possesses both male and female characteristics, various physical deformities,[30] and behavioral effects.[31] Any behavioral change while not necessarily affecting the individual may adversely influence a population from the reproductive implications.[32] Many effects have also been demonstrated under laboratory conditions for a wide range of organisms, both aquatic and terrestrial. Selected examples of abnormalities observed in aquatic organisms are shown in Table 6.1.

Though an EDC can be present in the aquatic system, it may not necessarily assert an estrogenic response to any organisms present within. The response observed in biota attributed to EDCs depends on numerous factors. Bioavailability plays an important role (Section 6.2.2.1), as does the EDC, environmental parameters, and the target organism. Examples of these factors and how they may influence the effect of EDCs to a target organism are given in Table 6.2.

6.2.2.1 Bioavailability

In order to have an endocrine disrupting effect to biota in surface waters, EDCs must first become bioavailable to the aquatic organism. Bioavailability is the response that a compound elicits from an organism over a range of concentrations and is synonymous with toxicity.[60] The degree of bioavailability is dependent on:

1. Chemical structure and properties (sorption capability, persistence)
2. Route of exposure (biomagnification, bioconcentration)
3. Aquatic life form of interest (benthic, demersal, pelagic organisms)

The physicochemical properties of the organic contaminant determine whether the EDC will favor the solid or aqueous phase and its persistence in the environment. Many EDCs have high K_{oc} and therefore will sorb to the sediment. This allows lower concentrations of EDCs to be abstracted for drinking water treatment or water reuse practices. However, this has an adverse effect on bottom dwelling biota that have demonstrated increased levels of VTG as a result.[29] Water currents or bioturbation allows for resuspension of the sediment and resultant bioavailability to biota present in water column.

The classic picture of pollutant transfer from sediments to organisms involves an intermediate stage in the water column. However, it is now believed that direct transfer from sediments to organisms occurs to a large extent.[61] The bioavailability of many organic compounds in natural solids decline with increasing time. This process is known as sequestration, and EDCs such as polyaromatic hydrocarbons (PAHs), PCBs, and pesticides undergo this, decreasing the availability of the compound to biota.[62] The importance of lifestyle and feeding habits of biota and their interaction between the dissolved and solid (suspended/whole sediment) phase is illustrated in Figure 6.2.

TABLE 6.1
Selected Examples of Abnormalities Observed in Aquatic Organisms as a Result of EDCs

Organism	Type	Effect	Cause	Observation
Rainbow trout (*Oncorhynchus mykiss*)	Fish	Reproductive	Sewage effluent	Downstream of effluent discharge has induced the production of VTG in the male[14,20,24]
Roach (*Rutilus rutilus*)	Fish	Reproductive	Sewage effluent	Downstream of effluent discharge, intersex has been induced (in some cases 100% of the male fish contained oocytes in their testes)[4]
Flounder (*Platichthys flesus*)	Fish	Reproductive	Sewage effluent	Downstream of effluent discharge has induced the production of VTG in the male[33]
Japanese medaka (*Oryzias latipes*)	Fish	Behavioral	Estradiol	Decreased fecundity in the female resulting in altered sexual activity[34]
	Fish	Reproductive	Nonylphenol	Abnormal gonad and anal fin (female-like) observed in male[35]
White sucker	Fish	Reproductive	Pulp mill effluent	Reduced gonadal size in the female exposed to less than 1% effluent (pulp mill), though at other sites, similar concentrations have not induced endocrine effects[36]
Fathead minnows (*Pimephales promelas*)	Fish	Reproductive	Methylmercury	Reduced gonadal development and spawning success of adult female[37]
Barnacle (*Elminius modestus*)	Invertebrate	Developmental	Nonylphenol Estradiol	The timing of larval development to the cypris stage was disrupted (accelerated)[38]
	Invertebrate	Physiological	Nonylphenol Estradiol	Long-term (12 months) exposure led to significant reduction in adult barnacle size[38]
Dogwhelk (*Nucella lapillus*)	Invertebrate	Developmental Reproductive	Tributyltin Tributyltin	Severely retarded larval development[39] Masculinization of the female (imposex)[39]
Water flea (*Daphnia magna*)	Invertebrate	Reproductive	Nonylphenol and ethoxylate	Induction of metabolic androgenization in the female[40]

(continued)

TABLE 6.2
Factors Affecting Toxic Responses in Organisms from EDCs

Factors EDC	Effects to Organisms from EDC Exposure
Type of EDC/ Metabolite	The different groups of EDCs or metabolites in the same group require different concentrations in order to cause a response in organisms assuming bioavailability.
	For fish (*Oryzias latipes*), the LC50 (48 hours) of the surfactant nonylphenol-16-ethoxylate ($NP_{16}EO$) and metabolites nonylphenol-9-ethoxylate (NP_9EO) and nonylphenol (NP) are 110, 11.2, and 1.4 mg l^{-1}, respectively, illustrating that increasing toxicity occurs with decreasing ethoxylate units.[41]
	The lowest observed effective concentration (LOEC) for VTG synthesis in primary fish cells for estradiol (E2), NP and bisphenol A was 1, 14, and 25 µg l^{-1}, respectively.[42]
Physicochemical properties	Factors such as the K_{ow}, K_{oc}, and water solubility greatly impact on the effects an EDC may induce because of the bioavailability of the EDC. EDCs with low water solubility will favor partitioning to suspended/whole sediment and so will be in contact with benthic organisms more than pelagic.[43]
	In blue mussels (*Mytilus edulis*), PBDEs and PCBs of similar hydrophobicity can have very different bioaccumulation factors (BAF), with preference to PBDE bioaccumulation.[44]
Individual/mixtures	EDCs are able to act together to produce significant effects even when they are present at concentrations below their individual effect threshold.[45,46]
	The effect of a mixture of substances can be additive, synergistic, and even antagonistic.[47]
	In soil-water systems, pesticide behavior in the presence of surfactants is dependent on the degree of hydrophobicity of the pesticide, surfactant type, and concentration.[48]
	VTG induction occurred at lowest mixtures concentrations (E2 and NP, E2 and methoxychlor, NP and methoxychlor) even when concentrations were below individual LOECs.[49]
Environmental concentration	The concentrations required for EDCs to induce endocrine disrupting effects to biota may not always be present in watercourses.
	For NP and bisphenol A, there is a safety margin of 100 and 3000 between concentration in effluent and effects monitored by receptor and indicator assays. However, for E2 there is no safety margin due to a much smaller LOEC.[42]

Environment	Seasonal variability	Concentrations of EDCs can be higher during the winter months compared to the summer period due to several factors including temperature and microorganism content. E2 and E1 are threefold higher during the winter months (7 to 220 ng l^{-1}) compared to the summer (4 to 56 ng l^{-1}) months.[50] However, the estrogenicity can also be less during the winter season because of increased rainfall and thus greater dilution of effluent in receiving waters.
	Temperature	For every 10°C change in temperature, between a two- to fourfold change is exhibited by toxicants. Correlation is usually positive though the toxicity of some organochlorine and pyrethroid pesticides may have an inverse correlation to temperature increase.[51] VTG induction by EE2 using rainbow trout showed VTG induction at 10 and 16.5°C occurred at concentrations of 10 and 0.5 ng l^{-1}, respectively.[7] So in the summer months with increased temperature, lower concentrations of steroids are required to induce VTG induction in target organisms.
	pH	Methylation of mercury (Hg) is increased at low pH forming methyl mercury, which is the highly toxic form of Hg that accumulates in biota such as fish. Bioaccumulation of Hg is greater for fish from aquatic environments that have high pH.[52]
	Dilution factor	Currently, the dilution of effluent entering receiving waters is greatly depended on to decrease any estrogenicity that the effluent contains to no effect levels. However, no effect levels are different for different biota and in periods of low flow (summer), the level of dilution is greatly decreased. For male roach exposed to river water with varying concentrations of effluent, levels of VTG induction increased with effluent concentration.[50] For receiving waters containing greater than 20% effluent, an increase in VTG in fish has been shown to occur.[42]
	Salinity	The partitioning behavior of EDCs to sediments can be influenced by salinity. Increased salinity resulting in an increase of partitioning to the sediment phase (suspended and whole).[53] This will result in EDCs being less bioavailable to pelagic organisms though more bioavailable to demersal and especially benthic organisms and filter feeders.
Organism	Species	VTG induction has been observed in rainbow trout, which have been proven to be a more sensitive than carp and roach. In the flounder, deformed testes have also been observed.[33] Bioaccumulation varies according to the organism. For NP, nonylphenol ethoxylate (NP$_1$EO) and nonylphenol diethoxylate (NP$_2$EO), concentrations in fish and ducks were 0.03 to 7.0 mg kg^{-1} and 0.03 to 2.1 mg kg^{-1}, respectively.[54] According to in vitro studies, variations may be due to differences between fish and mammals regarding binding to the receptor estrogen.[55]

(continued)

TABLE 6.2 (continued)
Factors Affecting Toxic Responses in Organisms from EDCs

Factors		Effects to Organisms from EDC Exposure
Organism	Age/Size	The proportion of male and female fish in length groups for *Acerina cernua* in the lower part of the River Elbe was 100% female for those 20 cm in length; 6:1:3 ratio of female:hermaphrodite:male at 16 cm and 20% female at 13 cm. Different sex ratio relating to length also observed for the fish *Cottus gobio*, with just over 20% female at lengths > 9 cm while 70% female at lengths < 7 cm.[56]
	Sex	The majority of EDCs induce estrogenic responses, primarily affecting male organisms such as VTG induction, which had previously only been observed in fertile female fish.[3,7,28] Other EDCs, such as tributyltin, have caused masculinization or imposex in female marine snails.[57]
	Feeding habits	The majority of EDCs in the aquatic environment have a tendency to sorb to suspended particulates/sediment due to their high K_{ow}. Hence, organisms whose lifestyle revolves around whole and suspended sediment, such as benthic dwellers and filter feeders, are particularly susceptible through direct contact and ingestion of suspended/whole sediment compared to pelagic organisms.
	Timing of exposure	The timing of EDC exposure to organism is very important because at certain points in development, sex is liable and exposure may interfere with the sex determination.[28] Roach spawn in the spring with sexual differentiation of juveniles occurring during the Summer period when effluent concentrations are at their highest.[4]
	Duration of exposure	Tetrabromobisphenol A is rapidly taken up by fish, though if placed in clean water, it is rapidly eliminated from the organism (depurated).[58] For other EDCs depuration may not occur and irreversible effects can be observed after only short contact with the EDC. Depuration in clean water for 150 days after exposure of male roach to full strength effluent still exhibited elevated VTG concentrations and retained female-like reproductive ducts.[59]

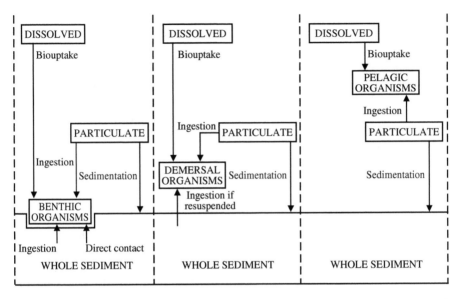

FIGURE 6.2 Interaction of the dissolved phase, particulate phase (suspended sediment), and whole sediment with different aquatic organisms

6.2.2.2 Bioaccumulation

Bioaccumulation is the sum of bioconcentration and biomagnification.[63] Bioconcentration is the uptake of contaminants from the surrounding phase, including direct contact with sediment and aqueous phase (diffusion through the surface membrane) and ingestion of sediment particles. Biomagnification is the uptake via the food chain. The bioconcentration factor (BCF) is calculated under steady state conditions shown in Equation 6.1 and is specific to the compound and test organism.[64] In the presence of dissolved organic matter (DOM), a decrease in bioconcentration is observed due to the greater affiliation between the EDC and the humic acid content of the DOM.

$$BCF = \frac{\text{EDC concentration in the organism}}{\text{EDC concentration in water}} \qquad (6.1)$$

The biomagnification factor (BMF) is calculated according to Equation 6.2, when the system is at steady state. Studies have been carried out for several EDCs such as the PCBs.[65] Hydrophobic character, resistance to degradation, and a high octanol/water partition coefficient (K_{ow}) will enhance the EDCs' ability to bioconcentrate and biomagnify within the food chain. However, bioaccumulation does not necessarily mean that an EDC is bioavailable, as residence in tissue fat does not equate to availability to hormone receptor systems required to initiate an endocrine disrupting effect.[28]

$$BMF = \frac{\text{EDC concentration in the organis}}{\text{EDC concentration in food sour}} \qquad (6.2)$$

The presence of EDCs in receiving waters greatly impacts the organisms present. Effects may be reproductive, developmental, and physiological, favoring an estrogenic rather than androgenic response. Factors affecting the response of organism to EDC are manifold, being compound, environmental, and organism dependent. Bioconcentration and biomagnification of EDCs is able to occur with biomagnification illustrating a preference for lipophilic part of the organism.

6.2.3 ORGANIC OXYGEN COMPOUNDS

6.2.3.1 Phthalates

Since the 1930s, phthalate acid esters (PAEs) have been utilized as plasticizers in a range of plastics, such as polyvinyl chloride (PVC), to cosmetics and medical products. For some, estrogenicity has been demonstrated,[66,67] and several PAEs are considered priority pollutants.[68] Di-(2-ethylhexyl) phthalate (DEHP) is the main PAE identified in the environment, because of its high production (up to 90% volume) and physicochemical properties that favor low water solubility and high sorption to the solid phase.

The fate and behavior of PAEs in the aqueous environment is greatly influenced by the alkyl chain length. Table 6.3 illustrates the potential for sorption, bioconcentration, dissolving, and evaporation against alkyl chain length. Actual physicochemical properties for PAEs are in Chapter 4, Table 4.2. Abiotic degradation by photolysis or hydrolysis are unlikely to be significant in the aquatic environment, since the hydrolysis half-life has been measured in years for many PAEs.[69]

PAEs are detected more frequently in Western European water samples compared to Northern American waters, with concentrations typically between 0.01 and 1 μg l^{-1}.[70] Median concentrations in German surface waters for DEHP and di-n-butylphthalate (DnBP) were identified at 2.3 and 0.5 μg l^{-1} with peaking concentrations of 97.8 and 8.8 μg l^{-1}.[71] The wide variation in concentrations is a result of the geography of PAE inputs into the watercourse.

Biodegradation is considered to be the major transformation process in all aquatic compartments. The shorter chains, such as dihexylphthalate (DHP) and DnBP, degrade at higher rates than the longer chain PAEs (e.g., DEHP). The complete degradation pathway is shown in Figure 6.3. A wide range of microorganisms has been shown to degrade PAEs under both aerobic and anaerobic conditions. Primary degradation half-lives in surface waters has been estimated from less than 1 day to several weeks.[69]

DnBP has an estimated log K_{oc} of 4.5 and is likely to partition significantly to the solid phase.[69] All PAEs have demonstrated partitioning to suspended and whole sediment, where the degree of sorption is dependent on alkyl chain length. Similar to surface waters, PAEs concentrations in sediments range over 3 to 4 orders of magnitude due to the vicinity of sediment to PAE source.[72] Degradation half-lives in sediment is longer compared to surface water half-lives because of the limited

TABLE 6.3
Relating Alkyl Chain Length of PAEs to Several Fate Potentials

PAE	Name	Alkyl Chain Length	Dissolving Potential (mg l⁻¹)	Evaporation Potential (atm m³ mole⁻¹)	Sorption Potential (log K_{oc})	Bioconcentration Potential (log K_{ow})
DnBP	di-n-butylphthalate		INCREASING ↑	INCREASING ↑		
BBP	Butylbenzylphthalate	↓				
DHP	Dihexylphthalate					
DEHP	Di-(2-ethylhexyl) phthalate	INCREASING			INCREASING ↓	INCREASING ↓

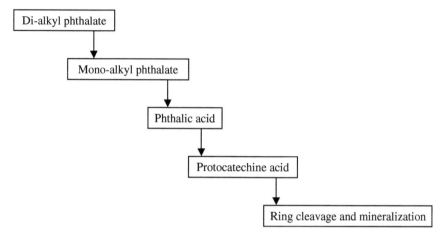

FIGURE 6.3 The complete degradation pathway for PAEs.

oxygen content. Enzymatic degradation by lipase has been demonstrated under laboratory conditions for sediment based DnBP.[73] Under simulated estuarine conditions, DEHP partitioning was dependent on salinity and particle concentration.[74] Greater adsorption was demonstrated when particle concentration was 1 mg l^{-1} in salt-water conditions compared to river water, suggesting a salting out effect.

The toxicity of PAEs to aquatic organisms increases with increasing alkyl chain length because of a corresponding increase in log K_{ow} and decreasing solubility. For three freshwater invertebrates exposed to DEHP at approximately 0.003 mg l^{-1} in water, no significant survival reductions were observed.[75] Similar tests in a sediment environment on freshwater invertebrates were carried out at mg kg^{-1} concentrations. For short alkyl chain PAEs such as DnBP, survival reductions were observed, while DEHP proved to be non-toxic.[76]

For DEHP, no observed effect concentrations (NOECs) for mussel (*Mytilus edulis*) and brook trout (*Salvelinus fontinalis*) were experimentally determined as 42 and 52 μg l^{-1} for test endpoints of mortality and growth respectively.[77] The lowest NOEC for DnBP was determined as 100 mg l^{-1} for rainbow trout (*Oncorhynchus mykiss*) using the test endpoint of growth and survival.[77] Calculated predicted no effect concentrations (PNECs) for PAEs of varying alkyl chain length suggest that there is little concern for aquatic organisms.[70]

6.2.3.2 Bisphenol A

Bisphenol A (BPA) is used in production of polycarbonates and epoxy resins, which are utilized as lacquers to coat metal products such as bottle tops, water supply pipes, and food cans. As a consequence of its prolific use, BPA is ubiquitous in the environment. BPA can also be produced through biodehalogenation of the flame retardants, tetrabromobisphenol A (TBBA) and tetrachlorobisphenol A (see Section 6.2.6.1 for TBBA).

BPA concentrations in domestic effluent have been identified at 490 ng l^{-1},[78] while on-site wastewater disposal in Cape Cod identified BPA at 1 g l^{-1} levels in septage and wastewater.[79] STW effluent from various sites in Germany detected mean concentration in all samples at 16 ng l^{-1}. Analysis of the receiving water that diluted effluent by approximately 100-fold contained a mean value of 4.7 ng l^{-1}.[80] With a log K_{oc} of 3.5, partitioning to the sediment phase is likely and BPA has been identified in bay sediments.[81] BPA has moderate water solubility of 120 mg l^{-1} and low volatility.[82] Hydrolysis and photolysis processes are unlikely to be significant.[83]

BPA has been shown to degrade relatively easily in surface waters receiving a continuous discharge of BPA with greater than 90% degradation achieved within a 4-day period.[84] Abiotic degradation is thought to be negligible since BPA does not contain any hydrolyzable functional groups.[83] In 44 river water microcosms (and 3 activated sludge), 19 BPA-degrading bacteria were isolated.[85] The extent of BPA degradation by the microcosms collected from unpolluted and less polluted sites was generally lower than those collected from heavily polluted sites. This implies that the microbial community has the ability according to its environment. BPA-degrading bacteria have been identified as the bacterial strain MV186 and three others belonging to the genera *Arthrobacter, Enterobacteriaceae*, and *Pseudomonas*.[87] Complete degradation rarely occurred, leaving the metabolites, 2,3-bis (4-hydroxyphenol)-1,2-propanediol and *p*-hydroxyphenacyl alcohol. The microbes capable of degrading these metabolites seemed to be present only in the heavily polluted environments.[87]

Anaerobic biodegradation of BPA was investigated in anoxic estuarine sediments for 162 days.[88] Under methanogenic, sulfate-reducing, iron-reducing, and nitrate-reducing conditions, no degradation of BPA was observed. BPA is readily biodegradable within a short period in the aqueous phase, limiting partitioning to particulates and settling to the sediment phase. However, environmental dehalogenation of TBBA to BPA in anoxic sediments has been demonstrated,[88] and BPA is likely to accumulate within this environmental compartment.

Using the E-screen assay, BPA concentrations of 2 to 5 µg l^{-1} elicited hormonal effects.[89] BPA is moderately toxic to fish with a LC50 and EC50 of 1.1 and 10 mg l^{-1}, respectively.[85] Under the U.S. Toxic Substances Control Act, the environmental concentrations resulting from treated process effluent discharge should not exceed 0.1mg l^{-1}.[84] Reproductive effects on fathead minnow (*Pimephales promelas*) have been induced with BPA concentrations of 640 g l^{-1} administered via the water.[90] In male fathead minnow, the proportion of sex cell types in the testis was altered at BPA dose of 16 g l^{-1} suggesting inhibition of spermatogenesis. BPA levels in surface water are typically one to several orders of magnitude lower than chronic effects observed in test organisms.[82] With rapid biodegradation and experimental and estimated BCFs of <100 and 42 to 196, respectively,[91] the potential for bioaccumulation in biota is low.

6.2.4 ORGANO METALS

6.2.4.1 Methylmercury

Mercury has three important chemical states as given in Table 6.4. In the elemental chemical state, mercury has a minimum residence time of 7 days within the

TABLE 6.4
Chemical States and Properties of Mercury

Chemical State	Form	Symbol	Properties
Elemental	Mercury	Hg^0	Very low water solubility; volatilizes at room temperature
Inorganic	Divalent mercury	Hg^{2+}	Readily associates with particles and water; forms salts which have low water solubility
Organometal	Mono-methylmercury	CH_3Hg^+	Soluble in water; stable due to covalent C-Hg bond
	Di-methylmercury	$(CH_3)_2Hg$	Less water soluble and stable than mono-methylmercury; volatile

atmosphere and can therefore be transported great distances. Entrance to aquatic environments is primarily via wet and dry deposition from the atmosphere.[92] Loss from the water to the atmosphere can also occur for elemental mercury and di-methylmercury.

In the aquatic environment, mercury (Hg^0) undergoes transformations, the most significant being methylation.[93] The addition of methyl groups to divalent mercury results in mono-methylmercury and di-methylmercury. The percentage of methyl-mercury to total mercury averages 5 to 1% in industrial areas.[94] Sediments and to a lesser extent, the water column are the major sites for methylation. Methylmercury concentrations in surficial sediments ranging from <0.7 to 13.2 mg kg^{-1}.[95] Within anoxic sediments, sulfate-reducing bacteria are considered to be the main cause for methylation. Microbial methylation is dominant in the water column with abiotic methylation accounting for approximately 10% of methylmercury production.[52]

Several environmental factors influence the rate of methylation, such as low pH and organic carbon, both of which can promote methylmercury production.[96] The amount of Hg available for transformation also increases with low pH because of preferential binding of Hg to particulates resulting in less volatilization from the water.[52] The distribution coefficient (K_d) gives the relative affinity of Hg for dissolved and particulate phases. Studies of both fresh- and saltwater environments have determined a log K_d of between 5 and 6, indicating a high affinity for the particulate phase.[97–99] For dissolved mercury, association is predominantly strongly bound with organic compounds,[98] limiting availability for methylation.[100]

Mono-methylmercury is the most harmful form of mercury to organisms, demonstrating both bioconcentration and biomagnification. It is readily bioavailable and possesses a long residence time. The vast majority of mercury in fish and their eggs exist as the toxic mono-methylmercury.[101,102] Regardless of fish age and location, around 85% of mercury in eel (*Anguilla anguilla*) was present as organometal.[103] Its ubiquitous presence throughout the water system means that many organisms are likely to come into contact with it. Preference for the particulate phase means that the benthos and particle ingesting organisms are particularly susceptible. Accumulation in tissues results in high concentrations, particularly within fish muscle.[52,104] Fish are very resistant to any adverse effects from methylmercury, more so than the wildlife and humans that consume them.[105]

Biomagnification has been shown as the main source of methylmercury for adult fish.[106] The reproductive performance (gonadal development of female fish and spawning success) of adult fathead minnows (*Pimephales promelas*) was reduced due to the presence of methylmercury in their diet.[37] Methylmercury exposure to fish embryos may occur via environmental interaction or maternal transfer. The success of walleye egg hatching has been shown to decrease with increasing water-borne methylmercury ranging from 0.1 to 7.8 ng l[-1]. Comparatively, maternal transfer did not influence walleye hatching success,[107] and a similar study on fathead minnows formed the same conclusion.[37]

Consequences of EDCs in receiving waters can have adverse implications as evidenced by methylmercury poisoning in humans. Well-known examples include the accidental consumption of treated grain that occurred in Iraq in 1971 and the Japanese Minamata disease in 1953. In Japan, mercury was released from a chemical plant into a nearby watercourse. In this case, environmental methylation of Hg was not the main cause for methylmercury, as the plant was discharging that form of Hg. The presence of methylmercury in the watercourse gave rise to bioaccumulation in biota. The Minamata villagers' staple diet consisted of fish and shellfish, and severe neurological problems were observed with nearly 100 deaths over a several year period.

6.2.4.2 Tributyltin

Tributyltin (TBT) compounds are extremely poisonous — an attribute that has seen them utilized as the active ingredients in marine anti-fouling paint formulations. The potency ensures long lifetimes between repainting. However, TBT's extreme toxicity has resulted in numerous adverse biological effects on nontarget species. TBT is extremely surface reactive and readily adsorbs onto the particulate matter of suspended/whole sediments[108]

Because of its hydrophobic character, photolysis, and fast aqueous related degradation rates, little TBT is observed in the water column. Hence, the fate of TBT is dependent on the adsorption and degradative factors associated with the sediments.[109–111] The physical and chemical properties of TBT indicate the preference for partitioning to sediments and the desire to bioaccumulate (Table 6.5).

Adsorption is important in removing TBT from the water column relatively quickly and is determined by the hydrophobic character and polarity of TBT. Randall and Weber determined that between 72 to 100% of the TBT burden was adsorbed onto suspended and whole sediment.[115] Adsorption has been found to be weak and reversible, though is enhanced with pH or salinity decrease.[116] The particle size of the sediment particle is of particular importance in adsorption processes. The higher the surface area, the more binding capacity TBT has to the particle.

TABLE 6.5
Physicochemical Properties of TBT

Properties	Value	Inference
Water solubility	1 mg l^{-1} [112]	Insoluble associating with particulate matter of suspended/whole sediments
Log K_{ow}	3.7[61]	Likely to partition significantly to sediments
Log K_{oc}	5.3[113]	Primarily adsorbed to sediment
BCF (Mackay equation)	500[114]	May moderately bioaccumulate
BCF (Field data)	1000 to 6000[61] (species dependent)	Likely to bioaccumulate significantly

Degradation can occur via biotic and abiotic processes, but it is primarily microbially mediated and temperature dependent.[117] Transformation involves debutylation to dibutyltin (DBT), monobutyltin (MBT), and inorganic tin and all are less toxic than the parent TBT. The degradation of TBT and its persistence in the aquatic environment varies depending on whether the sediment is aerobic or anaerobic. The degradation half-life of TBT is several days in the water column, months in aerobic sediments and years in anaerobic sediments.[118] The slower degradation rates in anaerobic sediments have been attributed to the limited biotic degradation activity found in the deeper levels.

TBT persistence, desorption from sediments, and high BCF has implications for mobility and partitioning into biota. The biota most affected by TBT are benthic organisms[119] and filter feeders[120] because they are in direct contact with the contaminated sediment. However, resuspension of TBT from currents, dredging, and biotic activity allows bioavailability to aquatic organisms in the water column. Speciation is important; although degradation products DBT and MBT are as persistent as TBT, they are not as toxic while inorganic tin does not accumulate in biota.[121]

TBT is the best example of severe population level effects in the wild induced by an EDC.[122] Both lethal and sublethal effects have been observed in biota, including severely retarded larval development, masculinization of female (imposex),[39] behavioral changes, and shell deformation in oysters. In some areas the severity has resulted in sterility leading to extinction.[123] Imposex has been determined at levels of 3 ng l^{-1} TBT, and a NOEC limit has yet to be identified.[57]

6.2.5 PESTICIDES

From 1992 to 1996, the U.S. National Water-Quality Assessment Programme assessed the status of surface and groundwater resources. Approximately 76 pesticides and 7 pesticide degradates in more than 8000 samples were analyzed, accounting for 75% of the U.S. national agricultural use.[124] More than 95% of river and stream sample contained more than 1 pesticide compared to 50% of groundwater samples. The most frequently detected pesticides in agricultural areas were the herbicides, atrazine (and degradation product deethylatrazine), metolachlor, cyanazine, and alachlor. These herbicides accounted for most detection in larger rivers, major aquifers and many streams. The majority of detections in streams were for simazine, prometon, 2,4-dichlorophenoxyacetic acid (2,4-D), and tebuthiuron. Individual drinking water standards for pesticides were rarely exceeded. In contrast, the aquatic life criteria often were in streams.

Methoxychlor's half-life is up to 1 year and hydrolysis does not play a significant role in abiotic degradation.[91] Photolysis is accelerated in the presence of humic acids with a half-life of several hours. Under anaerobic conditions, the half-life for biodegradation is less than 28 days. Methoxychlor does not accumulate in fish because of rapid metabolism. However, the estrogenic activity of methoxychlor is considered to be due its main metabolite 2,2-*bis*-(*p*-hydroxyphenyl)-1,1,1-trichloroethane.

The physicochemical properties of dieldrin promote partitioning to the sediment phase. For hydrolysis, the half-life of dieldrin is 4 years and biodegradation is minimal.[91] The photodegradative half-life of aqueous phase dieldrin is 2 months, but this transformation process is not important for particulate associated dieldrin. The persistency and high lipophilicity of dieldrin results in bioaccumulation. The behavior of chlordane is similar to dieldrin, though biodegradation in the water phase is thought not to occur, leading to an even more persistent compound. Dechlorination can occur under anaerobic conditions illustrated by hexachlorobenzene by methane-producing bacteria.[125]

Laboratory microcosms with chalk were incubated for 258 days under aerobic conditions. The chalk aquifer microcosm experienced lag phases of a minimum 40 days, after which 51 and 33% of 2,4-D and mecoprop was mineralized.[126] Hydrolysis is a significant process for the transformation of 2,4-D with mineral surfaces acting as catalysts.[127] Photodegradation of 4-chloro-2-methylphenoxyacetic acid occurs with a half-life of approximately 14 days.[127]

Atrazine (along with alachlor) is used throughout the world, resulting in high concentrations in the environment both in surface waters and groundwater. Levels of atrazine have been identified in groundwater at 16.6 and 47 $\mu g\ l^{-1}$ for surface waters.[128] The effects of atrazine to algae in streams showed photosynthesis sensitivity to atrazine contamination.[129] Concentrations in the range of 1 to 5 mg l^{-1} impeded the self-purification processes by proving toxic to algae.

Degradation of organochlorine insecticides can be both biotic and abiotic with a half-life of 4 to 7 days.[91] In acidic environments or under anaerobic conditions, there is a significant decrease in the degradation rate. Competitive effects in the presence of several pesticides may be observed when partitioning to the solid phase. Desorption is limited in older contaminated sediments by 2 to 3 orders of magnitude compared to recent contamination.[130]

There is a wide amount of evidence correlating organochlorine insecticide exposure to endocrine disrupting effects on the wildlife. Exposure results in both physiological abnormalities, such as thinning of eggshells and damage to the male reproductive system, to behavioral changes that are just as potentially dangerous to survival.[131,132]

6.2.6 Polyaromatic Compounds

6.2.6.1 Polybrominated Flame Retardants

Polybrominated diphenyl ethers (PBDEs) and TBBA are polybrominated flame retardants (PBFRs) utilized in plastics, textiles, and other materials for fire prevention.[133] Compositions of PBDE-based flame retardants favor the tetra-, penta-, octa-, and deca-BDE congeners. Their log K_{ow} is 5.9 to 6.2 for tetra-BDEs, 6.5 to 7.0 for penta-BDEs, 8.4 to 8.9 for octa-BDEs, and 10 and 4.5 for deca-BDEs and TBBA.[58] Thus, PBDEs are persistent with a low water solubility and high binding affinity to particles, resulting in a ubiquitous presence in sediment and biota.

PBDEs have been identified in sewage sludge,[134,135] and if applied to crops, may enter watercourses through association with particles in agricultural runoff. Percolation of PBDEs through to groundwater sources is unlikely because of their high hydrophobicity. In sewage effluent, PBDEs have been determined in both the filtrate and residue.[136] Laboratory studies have shown that UV light and sunlight causes debromination to lower PBDE congeners, though has yet to be proven in the environment.[137] UV and sunlight exposure to deca-BDE in sediment yielded similar results with half-lives of 53 hours and 81 hours, respectively.[138]

As predicted from their physicochemical properties, deca-BDE-associated particles have been determined at concentrations up to 4600 ng g^{-1}.[136] Microorganisms from PBDE-contaminated sediment were added to anaerobic sediment containing deca-BDE. Over a 4-month period at regular intervals, no transformation of deca-BDE was observed.[139] Continued incubation over a 2-year period still revealed no degradation of deca-BDE. There is little information on the transformation processes of sediment associated PBDEs.

For TBBA, degradation is microbially metabolized in a two-step process with anaerobic debromination to BPA followed by aerobic mineralization.[140] Under methogenic conditions debromination of TBBA to BPA has been observed with near complete loss of TBBA occurring within 55 days.[88] Under sulfate-reducing conditions, a lag phase of 28 days was observed prior to biotransformation of TBBA to BPA. The dimethyl ether derivative of TBBA has been identified in river and marine sediments, thought to be a result of microbial transformation.[141]

PBDEs have been identified in several aquatic organisms at ng g^{-1} concentrations and there is evidence of bioconcentration.[142] BDEs ranging from tetra- to hexa- have demonstrated similar bioavailability, while deca-BDE did not seem to be bioavailable.[143] Tetra- and hexa-BDEs exhibit the greatest bioaccumulation.[144] Concentrations in Virginia freshwater fish for tetra- and hexa-BDEs ranged from <5 to 47,900 g kg^{-1}, correlating with the PBDE profile in sediments.[145] Within an organism, deca-BDE is metabolized to lower BDE congeners, however the total uptake of deca-BDE is only 0.02 to 0.13%.[146] The low uptake is attributed to the large molecular size of deca-BDE thus bioaccumulation is not as significant as the lower BDE congeners. Pike fed with rainbow trout, which had been injected with a mixture of tetra-BDE, penta-BDE (99), and hexa-BDE (153), had percentage uptakes of 90, 62, and 40%, respectively.[147] TBBA is rapidly taken up from water through the gills of fish and is able to undergo depuration.[58] Consequently, the ecotoxicity of TBBA may not be as pronounced as the lower PBDE congeners.

6.2.6.2 Polychlorinated Biphenyls

PCBs consist of 207 congeners and are utilized in electrical insulating materials as plasticizers and rubbers. Estrogenic potency is dependent on chlorination, with less chlorinated PCBs possessing estrogenic activity and highly chlorinated PCBs having no estrogenic potential. PCBs are highly hydrophobic and persistent leading to a ubiquitous presence in the aquatic environment.[148,149] The half-lives for specific PCB congeners range from 2 to 6 years.[150]

A mass balance of Lake Superior predicted volatilization was the main fate process rather than sedimentation.[151] A 2-year study of a small rural lake in the U.K. found hydraulic transport to be the most important loss mechanism.[152] Similar to Lake Superior, volatilization rates exceeded sedimentation for PCBs. Sediments were also identified as being a secondary source for PCB due to an upward sediment-water flux. The solubility and sorption of 2,2′, 5,5′-tetrachlorobiphenyl was studied along an estuarine salinity gradient. Increasing salinity resulted in salting out of the PCB and an increase in partitioning to particulates.[153] Laboratory and *in situ* sediment-water partitioning of PCBs concluded that 100% is available for desorption.[154]

Concentrations in sediment have been detected ranging from 10.3 to 148 ng g^{-1} (dry weight),[81] with tri-, tetra-, and penta-chlorinated biphenyls accounting for 81% of total PCB concentrations. Total PCB in water and sediment from Daya Bay, China, were 91.1 to 1355 ng l^{-1} and 0.85 to 27.37 ng g^{-1} (dry weight), respectively.[155] Annually, more than 50% of the total PCB burden in the water of Lake Superior is transported by settling particulates to within 35 m of the lake bottom, though only 2 to 5% accumulating in whole sediment.[156] In water samples, highly chlorinated PCBs were dominant compared to lesser chlorinated congeners. This can be explained by the lower volatility, preference to partitioning to solids, and greater resistance to biodegradation compared to lesser chlorinated congeners.[157]

PCBs exhibit a high affinity for total organic carbon (TOC), which influences their bioavailability.[158] Generally, increasing TOC content decreases sediment bioavailability and toxicity.[159] Total PCB content in the benthic annelids, ragworm (*Nereis diversicolor*) and lugworm (*Arenicola marina*) was 6.8 to 6.9 ng g^{-1}. Bioaccumulation was evident with a total PCB concentration in the surrounding whole sediment of only 1.2 ng g^{-1}.[160] With ragworms forming the main diet for many fish, the likelihood of biomagnification occurring is high. Biomagnification has been demonstrated in a marine food chain, increasing total PCB concentrations correlating to increasing trophic level.[161]

6.2.6.3 Polyaromatic Hydrocarbons

Entrance into groundwater from land applied sludge is unlikely to occur for low molecular weight PAHs because of high volatilization rates.[162] For high MW PAHs, rain plays an important role in the weathering process and may result in percolation through to groundwater sources. In groundwater environments, NOM in the aqueous phase has been shown to facilitate PAH transport, while solid phase NOM retards such transport.[163]

In surface waters, concentrations of dissolved PAH in the water column are minimal due to photolysis, volatilization, and their hydrophobic character. The low polarity and high sorption potential of PAHs means partitioning to the sediment phase is significant.[164] For suspended sediment, adsorption favors debris or plant origin.[165] Concentrations in PAHs can be up to several orders of magnitude higher in sediment than the water column.[166] In the U.K.'s Severn Estuary, aqueous PAH levels were 104 to 1152 ng l[-1],[167] while concentrations of sediment associated PAHs were over 5400 mg kg[-1].[168] Total PAH in U.S. sediments have been detected at 145 and 57 mg kg[-1] for rivers and estuaries.[169]

Regardless of PAH sources in the surrounding area of a river, contamination of sediments occurs along the entire length of a river.[62] This suggests that transport of PAH-associated particles occurs with subsequent settling downstream of the initial source. PAH concentrations on settling particles and biodegradation are influenced by seasonal variation. In the Philadelphia Naval Reserve Basin, there was a net deposition of 12.7 g PAH and 0.25% biodegradation in the fall compared to 2.1g PAH in the spring and biodegradation of 50% for settling PAH.[170]

Biodegradation in sediment is an important transformation process and occurs under both aerobic and anaerobic conditions. A wide diversity of PAH-degrading bacteria exist, including bacteria from the *Ralstona*, *Spingomonas*, *Burkholderia*, *Pseudomonas*, *Comamonas*, *Flavobacterium*, and *Bacillus* genera.[171] In anaerobic sediments, the PAHs, phenanthrene and naphthalene degraded under both nitrate- and sulfate-reducing conditions. Degradation rates in anaerobic conditions are 1 to 2 orders of magnitude lower than in aerobic conditions.[172] Partial mineralization of naphthalene occurred under nitrifying conditions in the presence of pure bacterial cultures containing the *Pseudomonas* genera.[173]

The historical contamination of 14 PAHs, using sediment core samples from Tilbury docks in the Thames Estuary, U.K., was determined. Total PAH levels were 9.45 to 33.8 µg g[-1] at an ordnance datum (OD) depth of −5.0 to −8.4 m, corresponding to the mid-1960s to the 1940s. The decrease in total PAH levels over time were related to reduced coal combustion and improvements in nearby STWs through the introduction of activated sludge treatment.[174] Total PAH concentrations from sediment core sampled in 1995 close to Green Bay, Wisconsin, ranged from 0.34 to 19.3 mg g[-1].[175] The shallowest sediment core at 1.1 m had evidence of aerobic biodegradation and photolysis of phenanthrene.

In the aquatic environment, desorption of PAHs from the sediment phase may occur. Laboratory experiments on the interaction between benz[a]anthracene and DOM in sediments with low organic carbon content were shown to be reversible.[176] However, phenanthrene desorption from soils was significantly inhibited by DOM.[177] The presence of other PAHs may also affect desorption phenomena. With increasing phenanthrene concentration, desorption of the more hydrophobic pyrene was significantly enhanced, suggesting competitive inhibition.[177] For *in situ* sediment-associated pyrene and phenanthrene, between 1 and 40% were available for sediment-water partitioning.[154]

PAH associated with sediment is bioavailable to both benthic and demersal aquatic organisms.[178,179] The lifestyle of the organism influences the exposure to sediment-associated PAH and the direct intake of particulates contributing a significant proportion.[168] Single injections of benzo[a]pyrene into Japanese medaka (*Oryzias latipes*) compromised the immune response of the fish to bacterial infection.[180] In the bile of common (*Anguilla anguilla*) and conger eels (*Conger conger*) and European flounder (*Pleuronectes flesus*), 6 metabolites of PAHs were identified. The major metabolite found was 1-hydroxypyrene accounting for 75 to 94% of all metabolites detected.[181] The rate of metabolism was determined for fish by biota-sediment accumulation factors and biota-suspended solids accumulation factors. The relative order was pyrene, benz[a]anthracene, triphenylene, fluranthene, and phenanthrene, with pyrene exhibiting the slowest rate of metabolism.[182]

6.2.7 STEROID ESTROGENS

Natural (E1, E2) and synthetic (EE2) steroid estrogens have been implicated as the major contributors to estrogenic activity in sewage effluent[50,183] and surface waters.[10-12] The importance of estrogens from animal sources has to date been of less concern than their discharge from STWs. However, runoff from manure may contribute to the estrogenic burden entering the aqueous environment.[8,184] In the Netherlands, manure runoff and percolation to waterways used for the abstraction of drinking water has led to the government placing a ban on the number of livestock per farming household. A significant contribution of E2 to aquatic waters can occur from runoff of poultry litter applied to pasture land. E2 runoff was found to be persistent for up to 7 days after an initial application of poultry litter.[8] With a soil-binding capability for E1 and E2 of approximately 60%, thought to be due to the phenolic groups, percolation through to groundwater sources will be limited.[185]

The majority of steroid estrogens have been predicted to be in the dissolved aqueous phase because of the greater surface volume.[26] However, being predominantly hydrophobic organic compounds of low volatility, sorption to the solid phase is likely to be a significant process,[18] especially with regard to the more stable and persistent EE2.[186-188] Steroid estrogen sorption to sediments has been shown to correlate to the TOC content, though the presence of organic carbon is not a prerequisite for sorption.[186] Sorption to whole sediment occurs in the first 24 hours with smaller quantities being adsorbed for a further five days. Though desorption occurs to some extent, adsorption is the favored process.[18] TOC in suspended sediment can be 5 times more effective at sorbing E2 than whole sediment.[26] Particulate size further influences the degree of sorption.[18] In river sediments, E1 and EE2 have been identified at 11.9 and 22.8 ng g^{-1}, respectively.[72,187] With a detection limit of 1ng g,$^{-1}$ E2 was unable to be detected in any samples.

Interconversion between the steroids E1 and E2 occurs, favoring E1. This explains why concentrations of E1 are generally the highest of all steroids in aqueous samples. E1 may also be converted to E3 (maximum 4%) and other metabolites via hydroxylation.[189] Laboratory studies of E2 and EE2 in river water observed degradation under aerobic conditions.[18] Minimal and zero degradation occurred when incubated under anaerobic conditions. At 20°C in aerobic conditions, the half-lives for E2 and EE2 were 27 and 46 days, respectively. E1 was detected as a metabolite of E2 and further degraded during incubation, though the end products could not be identified.[18,19] The persistency of EE2 is partly due to the ethinyl group in position 17β that blocks oxidation to E1. UV radiation from the sun may reduce estrogenicity observed in rivers, since UV treatment has been shown to reduce E1, E2, and EE2 present in sewage effluent.[191]

Natural and synthetic steroid estrogens are present in sewage effluent and receiving waters at far lower concentrations than industrial chemicals.[192] Table 6.6 summarizes steroid concentrations observed in receiving waters worldwide. However, their potency is generally 3 orders of magnitude higher than other EDCs.[32,193] Effluent from Swedish STWs from a predominately domestic origin identified EE2 at levels 45 times greater than that shown to be estrogenic to fish.[78] EE2 has been shown to induce VTG production in male rainbow trout at 0.1 ng l^{-1} levels.[7] The growth and development of testes in maturing male trout has been shown to be retarded by 50%, due to a single dose of EE2 at 2 ng l^{-1}.[194]

Bioaccumulation has been shown to occur, resulting in sex reversal in fish.[195] The BCF for E1, E2, and EE2 for caged juvenile rainbow trout downstream of a STW ranged between 104 and 106.[78] For the algae *Chlorella vulgaris*, biotransformation showed a preference to E1 with an average BCF of 27.[196] For the invertebrate *Daphnia magna*, a BCF of 215 for E1 was established with initial bioaccumulation via the ingestion of *C. vulgaris*, indicating a BMF of 24.[197] In contrast to other EDCs, the sources of steroid estrogens are more difficult to control. For natural steroid estrogens, which are produced and excreted by humans and mammals, regulation by banning or replacement is not an option.[198]

6.2.8 SURFACTANTS

The surfactants, APEOs, are mainly used in industrial applications, with nonylphenol ethoxylate being the most common. Use in domestic detergents has been phased out since the late 1980s, but spermicides and some cleaning agents may still contain APEOs.[205] APs are the degradation products of APEOs, 4-nonylphenol (4-NP) being the most stable and estrogenically potent.

Of the 83% of U.K. nonylphenol ethoxylate production entering the environment, 37% enters the aquatic environment.[206] The point source for APs and APEOs entering the aquatic environment is via industrial discharges to sewer. An estimated 60 to 65% of all nonylphenolic compounds entering the STW are discharged to the aquatic receiving environment with the effluent.[207] APEOs are degraded in STWs to the more potent endocrine disrupting form of APs. Concentrations of octylphenol (OP) in the aqueous phase are generally one order of magnitude lower than NP.[13]

Dilution effects in estuaries resulted in lower concentrations by an order of magnitude. Research carried out in Switzerland has shown that the three main breakdown groups —APEOs (with fewer ethoxylate groups), alkylphenoxy carboxylic acids, and APs — persist in rivers and sediments and groundwater.[207–209] Degradation by microorganisms occurs in two stages: the first is the rapid hydrolysis of the ethoxylate groups forming the more toxic NP, nonylphenol ethoxylate (NP_1EO), and nonylphenol diethoxylate (NP_2EO). The second step is much slower and does not always occur.

TABLE 6.6
Environmental Concentrations for Steroid Estrogens in Receiving Waters

EDC	Receiving Water and Site	Concentration (ng l⁻¹)	Comments	Reference
Estrone (E1)	River water, England	0.27–0.44	EXAMS model	26
	River water, Israel	27–56	—	50
	Sediment, Spain	11.9 µg kg⁻¹ (dw)ᵃ	Sonic extraction and LC-MS	187
	Rivers and streams, Germany	<0.5–1.6	—	199
	Estuarine and freshwater, Netherlands	<0.1–3.4	—	200
17β-Estradiol (E2)	River water, England	0.21–0.37	EXAMS model	26
	Lake water, China	1.64–5.48	HPLC (electrochemical detection)	10
	River water, Israel	4–8.8	—	50
	River water, Israel	48–141	—	185
	Lake water, Japan	0.8	ELISA	201
	Surface water, California, U.S.	0.2–4.1	GC-MS-MS and ELISA	202
	Reservoir water, Church Wilne, England	0.8	GC-MS-MS Detection limit 0.3 ng l⁻¹	203
	Rivers and streams, Germany	0.5	—	199
	Estuarine and freshwater, Netherlands	0.3–5.5	—	200
Estriol (E3)	River water, England	ND	EXAMS model	26
	Lake water, China	ND	HPLC (electrochemical detection)	10
17α-Ethinylestradiol (EE2)	River water, England	0.024–0.038	EXAMS model	26
	Lake water, China	7.3–30.8	HPLC (electrochemical detection)	10
	Sediment, Spain	22.8 µg kg⁻¹ (dw)	Sonic extraction and LC-MS	187

(continued)

TABLE 6.6 (continued)
Environmental Concentrations for Steroid Estrogens in Receiving Waters

EDC	Receiving Water and Site	Concentration (ng l⁻¹)	Comments	Reference
EE2 (*cont.*)	Impounding reservoir water, England	1–3	Immunoassay	204
	Surface water, South East England	2–15	Immunoassay	204
	Surface water, California, U.S.	0.2 – 2.4	GC-MS-MS and ELISA	202
	Reservoir water, Church Wilne, England	ND	GC-MS-MS detection limit 0.3 ng l⁻¹	203
	Rivers and streams, Germany	<0.5	—	199
	Estuarine and freshwater, Netherlands	<0.1–4.3	—	200

Note: ND = not detected

ᵃdw = dry weight

TABLE 6.7
Environmental Concentrations for APs and APEOs in Receiving Waters

EDC	Receiving Water and Site	Concentration (ng l⁻¹)	Comments	Reference
Nonylphenol (NP)	River water, England	20–180,000	Results from 6 rivers	15
	River water, U.S.	<111–640	Results from 30 rivers	216
	River sediment, U.S.	<2.9–2960 µg kg⁻¹ (dw) ᵃ	Results from 30 rivers	216
	Lake Ontario sediment, U.S.	0.09–22.15 mg kg⁻¹ (dw)	Six sites sampled	217
	River sediment, Canada	0.17–72 mg kg⁻¹ (dw)	100% samples contained NP	13
	River water, Canada	<10–920	Higher NP in sediment than water	13

Compound	Matrix/Location	Concentration	Notes	Reference
Octylphenol (OP)	Lake Ontario sediment, U.S.	$0.004–23.7$ mg kg^{-1} (dw)	5 sites sampled	217
	River sediment, Canada	$<0.010–1.8$ mg kg^{-1} (dw)	89% samples contained OP	13
	River water, Canada	$<5–84$	Higher OP in sediment than water	13
Nonylphenol-monoethoxylate (NP$_1$EO)	River water, Switzerland	$<30–69,000$	River Glatt	207
	River water, Eastern U.S.	ND–1700	Results from 10 rivers	218
	River water, U.S.	$<60–600$	Results from 30 rivers	216
	River sediment, U.S.	$<2.3–175$ μg kg^{-1} (dw)	Results from 30 rivers	216
	River sediment, Canada	$<0.015–38$ mg kg^{-1} (dw)	66% samples contained NP$_1$EO	13
	River water, Canada	$<20–7800$	58% samples contained NP$_1$EO	13
Nonylphenol-diethoxylate (NP$_2$EO)	River water, Switzerland	$<30–30,000$	River Glatt	207
	River water, Eastern U.S.	ND–11,800	Results from 10 rivers	218
	River water, U.S.	$<70–1200$	Results from 30 rivers	216
	River sediment, Canada	$<0.015–6$ mg kg^{-1} (dw)	66% samples contained NP$_2$EO	13
	River water, Canada	$<20–10,000$	32% samples contained NP$_2$EO	13
Total APEOs	River and estuary water, England	15,000–76,000	River Aire	219
	River and estuary water, England	6000–11,000	River Mersey	219
	River and estuary water, England	ND–76,000	Tees Estuary	219

Note: ND = not detected

[a]dw = dry weight

A significant proportion of APs are identified in suspended and whole sediment. NP concentrations in river water and sediments were 0.051 to 1.08 µg l^{-1} and 0.5 to 13 µg g^{-1} (dry weight), respectively.[210] Partitioning to particulates is preferential to NP over OP, which is explained by their hydrophobic character. A lack of oxygen in sediment has been shown to reduce the degradation rate of NP by half.[211] There is evidence of anaerobic degradation of NP_1EO to NP in the whole sediment layer,[210] which has implications to benthic organisms. Sediment has been shown to sorb larger quantities of OP when the TOC content of clay and silt particles are high.[212] The same study observed a sorption-desorption hysteresis effect between OP and whole sediment. Several concentrations for APs and APEOs in receiving watercourses are given in Table 6.7.

Surfactants are more toxic to aquatic organisms than mammals, with toxicity for APEOs increasing with decreasing number of ethoxylate units and increasing hydrophobic chain length.[41] With a log K_{ow} of 4.0 and 4.6 for NP_1EO and NP_2EO, there is a tendency for bioconcentration in organisms. The BCF in fish is approximately 300, but can increase to 1300.[41] NP is thought to have the greatest potential for bioconcentration.[213]

Experiments have shown that adult male rainbow trout exposed to 30 µg l^{-1} concentrations of either OP or NP produced the female egg yolk protein VTG. The concentrations are similar to that currently found in U.K. rivers.[214] This was clarified in a similar study examining the effects of exposure to 30 µg l^{-1} of the above chemicals on male rainbow trout. A reduction in testicular growth was observed, and OP was also shown to increase VTG production in the presence of water at levels of 4.8 µg l^{-1}.[215] A recent survey of wild roach in U.K. rivers has found a high percentage of males with eggs in their testes and female egg yolk in their blood.[4]

Linking these effects induced by endocrine disruption to humans is difficult because of the long period (in years) between cause and effect. History has shown that EDCs such as dioxin, lindane, and methylmercury can adversely effect humans through different exposure routes (occupational, environmental, and dietary, respectively). EDCs have been observed in varying concentrations in receiving waters, which are often used as drinking water sources. As a result, attention has focused on human health implications arising from exposure to EDCs via drinking water supplies

REFERENCES

1. Tabak, H.H., Bloomhuff, R.N., and Bunch, R.L., Steroid hormones as water pollutants. II. Studies on the persistency and stability of natural urinary and synthetic ovulation-inhibiting hormones in untreated and treated wastewaters, *Dev. Ind. Microbiol.*, 22, 497, 1981.
2. Rathner, M. and Sonneborn, M., Biologically active oestrogens in potable water and wastewater, *Forum Stadte-Hyg.*, 30, 45, 1979.
3. Tyler, C.R. and Routledge, E.J., Natural and anthropogenic environmental oestrogens: the scientific basis for risk assessment: oestrogenic effects in fish in English rivers with evidence of their causation, *Pure Appl. Chem.*, 70, 1795, 1998.

4. Jobling, S., Nolan, M., Tyler, C.R., Brighty, G., and Sumpter, J. P., Widespread sexual disruption in wild fish, *Environ. Sci. Technol.*, 32, 2498, 1998.

5. Skakkebaek, N.E., Leffers, H., Rajpert-De Meyts, E., Carlsen, E., and Grigor, K.M., Should we watch what we eat and drink? Report on the international workshop on hormones and endocrine disrupters in food and water: possible impact on human health, Copenhagen, Denmark, 27–30 May 2000, *Trends Endocrinol. Metab.*, 11, 291, 2000.

6. Environmental Data Services, Agency urges precautionary action on endocrine disrupters, *ENDS Rep.*, 276, 17, 1998.

7. Purdom, C.E., Hardiman, P.A., Bye, V.J., Eno, N.C., Tyler, C.R., and Sumpter, J. P., Estrogenic effects of effluents from sewage treatment works, *Chem. Ecol.*, 8, 275, 1994.

8. Nichols, D.J., Daniel, T.C., Moore, P.A., Edwards, D.R., and Pote, D.H., Runoff of oestrogen hormone 17β-estradiol from poultry litter applied to pasture, *J. Environ. Qual.*, 26, 1002, 1997.

9. Wilson, S.C., Duarte Davidson, R., and Jones, K.C., Screening the environmental fate of organic contaminants in sewage sludges applied to agricultural soils. 1. The potential for downward movement to groundwaters, *Sci. Total Environ.*, 185, 45, 1996.

10. Shen, J. H., Gutendorf, B., Vahl, H.H., Shen, L., and Westendorf, J., Toxicological profile of pollutants in surface water from an area in Taihu Lake, Yangtze Delta, *Toxicology*, 166, 71, 2001.

11. Snyder, S.A., Villeneuve, D.L., Snyder, E.M., and Giesy, J. P., Identification and quantification of estrogen receptor agonists in wastewater effluents, *Environ. Sci. Technol.*, 5, 3620, 2001.

12. Thomas, K.V., Hurst, M.R., Matthiessen, P., and Waldock, M.J., Characterization of estrogenic compounds in water samples collected from United Kingdom estuaries, *Environ. Toxicol. Chem.*, 20, 2165, 2001.

13. Bennie, D.T., Sullivan, C.A., Lee, H.B., Peart, T.E., and Maguire, R.J., Occurrence of alkylphenols and alkylphenol mono- and diethoxylates in natural waters of the Laurentian Great Lakes basin and the upper St. Lawrence River, *Sci. Total Environ.*, 193, 263, 1997.

14. Harries, J.E., Sheahan, D.A., Jobling, S., Matthiessen, P., Neall, P., Routledge, E., Rycroft, R., Sumpter, J.P., and Tylor, T., A survey of estrogenic activity in United Kingdom inland waters, *Environ. Toxicol. Chem.*, 15, 1993, 1996.

15. Blackburn, M. and Waldock, M., Concentrations of alkylphenols in rivers and estuaries in England and Wales, *Water Res.*, 29, 861, 1995.

16. Lee, H.-B. and Peart, T., Determination of 4-nonylphenol in effluent and sludge from sewage treatment plants, *Anal. Chem.*, 67, 1976, 1995.

17. Ferguson, P.L., Iden, C.R., McElroy, A.E., and Brownawell, B.J., Determination of steroid estrogens in wastewater by immunoaffinity extraction coupled with HPLC-electrospray-MS, *Anal. Chem.*, 73, 3890, 2001.

18. Jurgens, M.D., Williams, R.J., and Johnson, A.C., Fate and behaviour of steroid oestrogens in rivers: a scoping study, Environment Agency, Bristol, London, 1999, p. 80.

19. Sumpter, J.P., Xenoendocrine disrupters — environmental impacts, *Toxicol. Letters*, 102–103, 1998.

20. Harries, J.E., Jobling, S., Mattiessen, P., Sheahan, D.A., and JP, S., Effects of Trace Organics of Fish: Phase 2, Report No. FR/D0022, Foundation for Water Research, Marlow, London, 1995.

21. Harries, J.E., Janbakhsh, A., Jobling, S., Mattiessen, P., Sumpter, J.P., and Tyler, C., Estrogenic potency of effluent from two sewage treatment works in the UK, *Environ. Toxicol. Chem.*, 18, 932, 1999.
22. Ternes, T.A., Occurrence of drugs in German sewage treatment plants and rivers, *Water Res.*, 32, 3245, 1998.
23. Bedding, N.D., McIntyre, A.E., Perry, R., and Lester, J.N., Organic contaminants in the aquatic environment. II. Behaviour and fate in the hydrological cycle, *Sci. Total Environ.*, 26, 255, 1983.
24. Harries, J.E., Sheahan, D.A., Jobling, S., Matthiessen, P., Neall, P., Sumpter, J.P., Tylor, T., and Zaman, N., Estrogenic activity in five United Kingdom rivers detected by measurement of vitellogenesis in caged male trout, *Environ. Toxicol. Chem.*, 16, 534, 1997.
25. Lester, J.N. and Birkett, J.W., *Microbiology and Chemistry for Environmental Scientists and Engineers*, 2nd ed., E & FN Spon, London, 1999.
26. Williams, R.J., Jürgens, M.D., and Johnson, A.C., Initial predictions of the concentrations and distribution of 17-oestradiol, oestrone and ethinyl oestradiol in 3 English rivers, *Water Res.*, 33, 1663, 1999.
27. Veeramachaneni, D.N. R., Deteriorating trends in male reproduction: idiopathic or environmental?, *Animal Reprod. Sci.*, 60–61, 121, 2000.
28. Tyler, C.R., Jobling, S.R., and Sumpter, J.P., Endocrine disruption in wildlife: a critical review of the evidence, *Crit. Rev. Toxicol.*, 28, 319, 1998.
29. Allen, Y., Matthiessen, P., Scott, A.P., Haworth, S., Feist, S., and Thain, J.E., The extent of oestrogenic contamination in the UK estuarine and marine environments further surveys of flounder, *Sci. Total Environ.*, 233, 5, 1999.
30. Meregalli, G. and Ollevier, F., Exposure of *Chironomus riparius* larvae to 17β-ethynylestradiol: effects on survival and mouthpart deformities, *Sci. Total Environ.*, 269, 157, 2001.
31. Bjerselius, R., Lundstedt-Enkel, K., Olsen, H., Mayer, I., and Dimberg, K., Male goldfish reproductive behaviour and physiology are severely affected by exogenous exposure to 17β-estradiol, *Aquatic Toxicol.*, 53, 139, 2001.
32. Miyamoto, J. and Klein, W., Environmental exposure, species differences and risk assessment, *Pure Appl.Chem.*, 70, 1829, 1998.
33. Lye, C.M., Frid, C.L., Gilli, C.E., and McCormick, D., Abnormalities in the reproductive health of flounder *Platichthys flesus* exposed to effluent from sewage treatment works, *Marine Poll. Bull.*, 34, 34, 1997.
34. Nimrod, A.C. and Bensen, W.H., Reproduction and development of Japanese medaka following an early life stage exposure to xenoestrogens, *Aquatic Toxicol.*, 44, 141, 1998.
35. Tabata, A., Kashiwada, S., Ohnishi, Y., Ishikawa, H., Miyamoto, N., Itoh, M., and Magara, Y., Estrogenic influences of estradiol-17β, *p*-nonylphenol and bisphenol-A on Japanese Medaka (*Oryzias latipes*) at detected environmental concentrations, *Water Sci. Technol.*, 43, 109, 2001.
36. Munkittrick, K.R., Vanderkraak, G.J., McMaster, M.E., Portt, C.B., Vandenheucel, M.R., and Servos, M.R., Survey of receiving water environmental impacts associated with discharges from pulp mills. II. Gonad size, liver size, hepatic EROD activity and plasma sex steroid levels in white sucker, *Environ. Toxicol. Chem.*, 13, 1089, 1994.
37. Hammerschmidt, C.B., Sandheinrich, M.B., Wiener, J. G., and Rada, R.G., Effects of dietary methylmercury on reproduction of fathead minnows, *Environ. Sci. Technol.*, 36, 877, 2002.

38. Billinghurst, Z., Clare, A.S., and Depledge, M.H., Effects of 4-*n*-nonylphenol and 17β-estradiol on early development of the barnacle Elminius modestus, *J. Exp. Marine Biol. Ecol.*, 257, 255, 2001.
39. Oehlmann, J., Schulte-Oehlmann, U., Stroben, E., Bauer, B., Bettin, C., and Fiorni, P., Androgenic effects of organotin compounds in molluscs, in *Endocrinically Active Chemicals in the Environment*, Gies, A., Ed., Umwelt Bundes Amt, Berlin, 1995, p. 111.
40. Baldwin, W.S., Graham, S., Shea, D., and LeBlanc, G.A., Metabolic androgenization of female *Daphnia magna* by the xenoestrogen 4-nonylphenol, *Environ. Toxicol. Chem.*, 16, 1905, 1997.
41. Fent, K., Endocrinally active substances in the environment: state of the art, in *Endocrinically Active Chemicals in the Environment*, Gies, A., Ed., Umwelt Bundes Amt, Berlin, 1995, pp. 69.
42. Hansen, P.-D., Dizer, H., Hock, B., Marx, A., Sherry, J., McMaster, M., and Blaise, C., Vitellogenin-a biomarker for endocrine disrupters, *Trac — Trends Anal. Chem.*, 17, 448, 1998.
43. Allchin, C.R., Law, R.J., and Morris, S., Polybrominated diphenylethers in sediments and biota downstream of potential sources in the UK, *Environ. Poll.*, 105, 197, 1999.
44. Gustafsson, K., Bjork, M., Burreau, S., and Gilek, M., Bioaccumulation kinetics of brominated flame retardants (polybrominated diphenyl ethers) in blue mussels (*Mytilus edulis*), *Environ. Toxicol. Chem.*, 18, 1218, 1999.
45. Silva, E., Rajapakse, N., and Kortenkamp, A., Something from "nothing": eight week estrogenic chemicals combined at concentrations below NOECs produce significant mixture effects, *Environ. Sci. Technol.*, ASAP Article, A, 2002.
46. Kortenkamp, A. and Altenburger, R., Approaches to assessing combination effects of estrogenic environmental pollutants, *Sci. Total Environ.*, 233, 131, 1999.
47. Hutchinson, T., European research on endocrine disruptors in the aquatic environment, in *Endocrine Disrupters and Pharmaceutical Active Compounds in Drinking Water Workshop*, Center for Health Effects of Environmental Contamination., 2000.
48. Iglesias-Jiménez, E., Sánchez-Martín, K.M. J., and Sánchez-Camazano, M., Pesticide adsorption in a soil-water system in the presence of surfactants, *Chemosphere*, 32, 1771, 1996.
49. Thorpe, K.L., Hutchinson, T.H., Hetheridge, M.J., Scholze, M., Sumpter, J.P., and Tyler, C.R., Assessing the biological potency of binary mixtures of environmental estrogens using vitellogenin induction in juvenile rainbow trout (*Oncorhynchus mykiss*), *Environ. Sci. Technol.*, 35, 2476, 2001.
50. Rodgers Gray, T.P., Jobling, S., Morris, S., Kelly, C., Kirby, S., Janbakhsh, A., Harries, J.E., Waldock, M.J., Sumpter, J.P., and Tyler, C.R., Long-term temporal changes in the oestrogenic composition of treated sewage effluent and its biological effects on fish, *Environ. Sci. Technol.*, 34, 1521, 2000.
51. Wright, D.A. and Welbourn, P., Factors affecting toxicity, in *Environmental Toxicology*, Wright, D.A. and Welbourn, P., Eds., Cambridge University Press, London, 2002, p. 218.
52. Wright, D.A. and Welbourn, P., Metals and other inorganic chemicals, in *Environmental Toxicology*, Wright, D.A. and Welbourn, P., Eds., Cambridge University Press, London, 2002, p. 249.
53. Karickhoff, S.W., Brown, D.S., and Scott, T.A., Sorption of hydrophobic pollutants on natural sediments, *Water Res.*, 13, 241, 1979.
54. Ahel, M., McEvoy, J., and Giger, W., Bioaccumulation of the lipophilic metabolites of nonionic surfactants in freshwater organisms, *Environ. Poll.*, 79, 243, 1993.

55. Thomas, P. and Smith, J., Binding of xenobiotics to the estrogen receptor of spotted seatrout: a screening assay for potential estrogenic effects, *Marine Environ. Res.*, 35, 147, 1993.

56. Pluta, H.J., Endocrine effects of environmental chemicals on fish: current investigations, in *Endocrinically Active Chemicals in the Environment*, Gies, A., Ed., Umwelt Bundes Amt, Berlin, 1995, p. 81.

57. Burton, G.A. and Scott, K.J., Sediment toxicity evaluation: their niche in ecological assessments, *Environ. Sci. Technol.*, 26, 2068, 1992.

58. de Wit, C.A., An overview of brominated flame retardants in the environment, *Chemosphere*, 46, 583, 2002.

59. Rodgers Gray, T.P., Jobling, S., Kelly, C., Morris, S., Brighty, G., Waldock, M.J., Sumpter, J. P., and Tyler, C.R., Exposure of juvenile roach (*Rutilus rutilus*) to treated sewage effluent induces dose-dependent and persistent disruption in gonadal duct development, *Environ. Sci. Technol.*, 35, 462, 2001.

60. Depledge, M.H., Weeks, J.M., and Bjerregaard, P., Heavy metals, in *Handbook of Ecotoxicology*, Calow, P., Ed., Blackwell Science, London 1998, p. 547.

61. Manahan, S.E., *Environmental Chemistry*, 6th ed. Lewis Publishers, London, 1994.

62. White, J.C. and Triplett, T., Polycyclic aromatic hydrocarbons (PAHs) in the sediments and fish of the Mill River, New Haven, Connecticut, USA, *Bull. Environ. Contam. Toxicol.*, 68, 104, 2002.

63. Jorgenson, S.E., Halling-Sorensen, B., and Mahler, H., *Handbook of Estimation Methods in Ecotoxicology and Environmental Chemistry*, CRC Press, Boca Raton, FL, 1998.

64. Opperhuizen, A., Bioaccumulation kinetics: experimental data and modelling, in *Organic Micropollutants in the Aquatic Environment*, Angeletti, G. and Bjorseth, A., Eds., Academic Publishers, London, 1991, p. 379.

65. Wang, J., Chou, H., Fan, J., and Chen, C., Uptake and transfer of high PCB concentrations from phytoplankton to aquatic biota, *Chemosphere*, 36, 1201, 1998.

66. Jobling, S., Reynolds, T., White, R., Parkett, M.G., and Sumpter, J.P., A variety of environmentally persistent chemicals, including some phthalate plasticisers are weakly oestrogenic, *Environ. Health Persp.*, 103, 582, 1995.

67. Soto, A.M., Sonnenschein, C., Chung, K.L., Fernandez, M.J., Olea, N., and Serrano, F.O., The E-screen assay as a tool to identify oestrogens: an update on oestrogenic environmental pollutants, *Environ. Health Persp.*, 103, 113, 1995.

68. Cicmanec, J.L., Removal of endocrine disrupting compounds using drinking water treatment processes, in *2nd International Conference on Pharmaceuticals and Endocrine Disrupting Chemicals in Water*, Minneapolis, MN, 2001.

69. Staples, C.A., Peterson, D.R., Parkerton, T.F., and Adams, W.J., The environmental fate of phthalate esters: a literature review, *Chemosphere*, 35, 667, 1997.

70. Staples, C.A., Parkerton, T.F., and Peterson, D.L., A risk assessment of selected phthalate esters in North American and Western European surface waters, *Chemosphere*, 40, 885, 2000.

71. Fromme, H., Kuchler, T., Otto, T., Pilz, K., Muller, J., and Wenzel, A., Occurrence of phthalates and bisphenol A and F in the environment, *Water Res.*, 36, 1429, 2002.

72. Petrovic, M., Elijarrat, E., Lopez de Alda, M.J., and Barcelo, D., Analysis and environmental levels of endocrine-disrupting compounds in freshwater sediments, *Trac — Trends Anal. Chem.*, 20, 637, 2001.

73. Tanaka, T., Yamada, K., Tonosaki, T., Goto, G.H., and Taniguchi, M., Enzymatic degradation of alkylphenols, bisphenol A, synthetic estrogen and phthlatic ester, *Water Sci. Technol.*, 42, 89, 2000.

74. Turner, A. and Rawling, M.C., The behaviour of di-(2ethylhexyl) phthalate in estuaries, *Marine Chem.,* 68, 203, 2000.

75. Call, D.J., Markee, T.P., Geiger, D.L., Brooke, L.T., VandeVenter, F.A., Cox, D.A., Genisot, K.I., Robillard, K.A., Gorsuch, J. W., Parkerton, T.F., Reiley, M.C., Ankley, G.T., and Mount, D.R., An assessment of the toxicity of phthalate esters to freshwater benthos. 2. Aqueous exposures, *Environ. Toxicol. Chem.,* 20, 1798, 2001.

76. Call, D.J., Cox, D.A., Geiger, D.L., Genisot, K.I., Markee, T.P., Brooke, L.T., Polkinghorne, C.N., VandeVenter, F.A., Gorsuch, J. W., Robillard, K.A., Parkerton, T.F., Reiley, M.C., Ankley, G.T., and Mount, D.R., An assessment of the toxicity of phthalate esters to freshwater benthos. 2. Sediment exposures, *Environ. Toxicol. Chem.,* 20, 1805, 2001.

77. Staples, C.A., Adams, W.J., Parkerton, T.F., Gorsuch, J. W., Biddinger, G.R., and Reinert, K.H., Aquatic toxicity of eighteen phthalate esters, *Environ. Toxicol. Chem.,* 16, 875, 1997.

78. Larsson, D.G. J., Adolfsson-Erici, M., Parkkonen, J., Pettersson, M., Berg, A.H., Olsson, P.E., and Forlin, L., Ethinyloestradiol: an undesired fish contraceptive?, *Aquatic Toxicol.,* 45, 91, 1999.

79. Rudel, R.A., Melly, S.J., Geno, P.W., Sun, G., and Brody, J. G., Identification of alkylphenols and other estrogenic phenolic compounds in wastewater, septage, and groundwater on Cape Cod, Massachusetts, *Environ. Sci. Technol.,* 32, 861, 1998.

80. Kuch, H.M. and Ballschmiter, K., Determination of endocrine-disrupting phenolic compounds and estrogens in surface and drinking water by HRGC-(NCI)-MS in the picogram per liter range, *Environ. Sci. Technol.,* 35, 3201, 2001.

81. Khim, J.S., Kannan, K.D.L., V., Koh, C.H., and Giesy, J.P., Characterization and distribution of trace organic contaminants in sediment from Masan Bay, Korea.1. instrumental analysis, *Environ. Sci. Technol.,* 33, 4199, 1999.

82. Staples, C.A., Dorn, P.B., Klecka, G.M., O'Block, S.T., and Harris, L.R., A review of the environmental fate, effects, and exposures of bisphenol A, *Chemosphere,* 36, 2149, 1998.

83. Young, W.F., Fawell, J.K., and Davis, R., Oestrogenic Chemicals and Their Behaviour During Sewage Treatment. Report No. 98/TX/01/4, UKWIR, London, 1998, p. 55.

84. Dorn, P.B., Chou, C.S., and Gentempo, J.J., Degradation of Bisphenol A in natural waters, *Chemosphere,* 16, 1501, 1987.

85. Alexander, H.C., Dill, D.C., Smith, L.W., Guiney, P.D., and Dorn, P., Bisphenol A: acute aquatic toxicity, *Environ. Toxicol. Chem.,* 7, 19, 1988.

86. Lobos, J.H., Lieb, T.K., and Su, T.M., Biodegradation of bisphenol A and other bisphenols by a gram negative aerobic bacterium, *Appl. Environ. Microbiol.,* 58, 1823, 1992.

87. Ike, M., Jin, C.S., and Fujita, M., Biodegradation of bisphenol A in the aquatic environment, *Water Sci. Technol.,* 42, 31, 2000.

88. Voordeckers, J.W., Fennell, D.E., Jones, K., and Haggblom, M.M., Anaerobic biotransformation of tetrabromobisphenol A, tetrachlorobisphenol A, and bisphenol A in estuarine sediments, *Environ. Sci. Technol.,* 36, 696, 2002.

89. Krishnan, A.V., Starhis, P., Permuth, S.F., Tokes, L., and Feldman, D., Bisphenol A: an oestrogenic substance is released from polycarbonate flasks during autoclaving, *Endocrinology,* 132, 2279, 1993.

90. Sohoni, P.C.R., T., Hurd, K., Caunter, J., Hetheridge, M., Williams, T., Woods, C., Evans, M., Toy, R., Gargas, M., and Sumpter, J.P., Reproductive effects of long-term exposure to bisphenol A in the fathead minnow (*Pimephales promelas*), *Environ. Sci. Technol.,* 35, 2917, 2001.

91. Gülden, M., Turan, A., and Seibert, H., Endocrinically Active Chemicals and Their Occurrence in Surface Waters, Report No. UBA FB 97–068, Federal Environmental Agency (Umweltbundesamt), Berlin, 1998, p. 246.

92. Downs, S.G., MacLeod, C.L., Jarvis, K., Birkett, J.W., and Lester, J.N., Comparison of mercury bioaccumulation in eel (*Anguilla anguilla*) and roach (*Rutilus rutilus*) from river systems in East Anglia, UK. II. Sources of mercury to fish and the role of atmospheric deposition, *Environ. Technol.*, 20, 1201, 1999.

93. Croston, N.J., Bubb, J.M., and Lester, J.N., Spatial distribution and seasonal changes in methylmercury concentrations in shallow lakes, *Hydrobiologia*, 321, 35, 1996.

94. Downs, S.G., MacLeod, C.L., and Lester, J.N., Mercury in precipitation and its relation to bioaccumulation in fish: a literature review, *Water, Air Soil Pollut.*, 108, 149, 1998.

95. Bubb, J.M., Rudd, T., and Lester, J.N., Distribution of heavy metals in the River Yare and its associated Broads. I. Mercury and methylmercury, *Sci. Total Environ.*, 102, 147, 1991.

96. Braga, M.C.B., Shaw, G., and Lester, J.N., Mercury modeling to predict contamination and bioaccumulation in aquatic ecosystems, *Rev. Environ. Contam. Toxicol.*, 164, 69, 2000.

97. Hurley, J.P., Benoit, J.M., Babiarz, C.L., Shafer, M.M., Andrens, A.W., Sullivan, J.R., Hammond, R., and Webb, D.A., Influence of watershed characteristics on mercury levels in Wisconsin rivers, *Environ. Sci. Technol.*, 29, 1867, 1995.

98. Coquery, M., Cossa, D., and Sanjuan, J., Speciation and sorption of mercury in two macro-tidal estuaries, *Marine Chem.*, 58, 213, 1997.

99. Coquery, M., Cossa, D., and Martin, J.M., The distribution of dissolved and particulate mercury in three Siberian estuaries and adjacent Arctic coastal waters, *Water, Air Soil Pollut.*, 80, 653, 1995.

100. Olson, B.H., Cayless, S.M., Ford, S., and Lester, J.N., Toxic element contamination and the occurrence of mercury-resistant bacteria in Hg-contaminated soil, sediments, and sludges, *Arch. Environ. Contam. Toxicol.*, 20, 226, 1991.

101. Hammerschmidt, C.B., Wiener, J.G., Frazier, B.E., and Rada, R.G., Methylmercury content of eggs in Yellow Perch related to maternal exposure in four Wisconsin lakes, *Environ. Sci. Technol.*, 33, 999, 1999.

102. Johnston, T.A., Bodaly, R.A., Latif, M.A., Fudge, R.J.P., and Strange, N.E., Intra- and interpopulation variability in maternal transfer of mercury to eggs of walleye (*Stizosledion vitreum*), *Aquatic Toxicol.*, 52, 73, 2001.

103. Edwards, S.C., MacLeod, C.L., and Lester, J.N., Mercury contamination of the eel (*Anguilla anguilla*) and roach (*Rutilus rutilus*) in East Anglia, UK, *Environ. Monit. Assess.*, 55, 371, 1999.

104. Edwards, S.C., MacLeod, C.L., and Lester, J.N., Mercury contamination of the eel (*Anguilla anguilla*) and roach (*Rutilus rutilus*) in East Anglia, UK, *Environ. Monit. Assess.*, 55, 371, 1999.

105. Clarkson, T.E., Mercury toxicity: an overview, in *Proceedings, National Forum on Mercury in Fish*, EPA 823/R-95/002, U.S. Environmental Protection Agency, Washington, D.C., 1995, p. 91.

106. Hall, B.D., Bodaly, R.A., Fudge, R.J.P., Rudd, J.W.M., and Rosenberg, D.M., Food as the dominant pathway of methylmercury uptake by fish, *Water, Air Soil Pollut.*, 100, 13, 1997.

107. Latif, M.A., Bodaly, R.A., Johnston, T.A., and Fudge, R.J.P., Effects of environmental and maternally derived methylmercury on the embryonic and larval stages of walleye (*Stizosledion vitreum*), *Environ. Poll.*, 111, 139, 2001.

108. Harrison, R.M., *Understanding Our Environment*, 3rd ed., Royal Society of Chemistry, London, 1999.

109. Dowson, P.H., Pershke, D., Bubb, J.M., and Lester, J.N., Spatial distribution of organotin in sediments of lowland river catchments, *Environ. Poll.*, 76, 259, 1992.

110. Dowson, P.H., Bubb, J.M., and Lester, J.N., A study if the partitioning and sorptive behaviour of butyltins in the aquatic environment, *Appl. Organomet. Chem.*, 7, 623, 1993.

111. Dowson, P.H., Bubb, J.M., and Lester, J.N., Organotin distribution in sediments and waters of selected east coast estuaries in the U.K., *Marine Poll. Bull.*, 24, 492, 1992.

112. Laughlin, R.B., French, W., and Guard, H.E., Accumulation of *bis*(tributyltin) oxide by the marine mussel *Mytilus edulis*, *Environ. Sci. Technol.*, 20, 884, 1986.

113. Unger, M.A., MacIntyre, W.G., and Huggett, R.J., Sorption behaviour of tributyltin on estuarine and freshwater sediments, *Environ. Toxicol. Chem.*, 7, 907, 1988.

114. Mackey, D., Correlation of bioconcentration factors, *Environ. Sci. Technol.*, 16, 274, 1982.

115. Randell, L. and Weber, J.H., Adsorptive behaviour of butyltin compounds under simulated estuarine condition, *Sci. Total Environ.*, 57, 191, 1986.

116. Dowson, P.H., Bubb, J.M., and Lester, J.H., Temporal distribution of organotins in the aquatic environment five years after the 1987 UK retail ban on TBT based antifouling paints, *Marine Poll. Bull.*, 26, 1993.

117. Stewart, C. and de Mora, S.J., A review of the degradation of tri (*n*-butyl) tin in the marine environment, *Environ. Technol.*, 11, 565, 1990.

118. Environment Canada, Initiate Study of the Toxicity of Tributyltin in Contaminated Sediment to Benthic Invertebrates, Notable Freshwater Snails, http://www.cciw.ca/nwri/aepb/source_fate.html, 1989 (accessed Feb. 8, 2002).

119. Macguire, R.J. and Tkacz, R.J., Degradation of the tri-*n*-butyltin species in water and sediment from Toronto Harbor, *J. Agric. Food Chem.*, 33, 947, 1985.

120. Langston, W.J. and Burt, G.R., Bioavailability of butyltin compounds, *Analyst*, 120, 667, 1992.

121. Alzieu, C., Biological effects of tributyltin on marine organisms, in *Tributyltin: Case Study of an Environmental Contaminant*, de Mora, S.J., Ed., Cambridge University Press, London, 1996, p. 172.

122. Van der Kraak, G., Observations of endocrine effects in wildlife with evidence of their causation, *Pure Appl. Chem.,* 70, 1785, 1998.

123. Minchin, D., Stroben, E., Oehlmann, J., Bauer, B., Duggan, C.B., and Keatinge, M., Biological indicators used to map organotin contamination in Cork Harbour, Ireland, *Marine Poll. Bull.*, 32, 188, 1996.

124. Gilliom, R.J., Alley, W.M., and Gurtz, M.E., Design of the National Water-Quality Assessment program: Occurrence and Distribution of Water Quality Conditions, U.S. Geological Survey, http://water.wr.usgs.gov/pnsp.austrat, 1995 (accessed Feb. 26, 2002).

125. Chang, B.-V., Chen, Y.-M., Yuan, S.-Y., and Wang, Y.-S., Reductive dechlorination of hexachlorobenzene by an anaerobic mixed culture, *Water, Air Soil Pollut.*, 100, 25, 1997.

126. Kristensen, G.B., Sorensen, S.R., and Aamand, J., Mineralization of 2,4-D, mecoprop, isoproturon and terbuthylazine in a chalk aquifer, *Pest Manage. Sci.*, 57, 2001.

127. Stangroom, S.J., Collins, C.D., and Lester, J.N., Abiotic behaviour of organic, micro pollutants in soils and the aquatic environment. A review. II. Transformations, *Environ. Technol.*, 21, 865, 2000.

128. Paterson, K.G. and Schnoor, J.L., Fate of alachlor and atrazine in a riparian zone field site, *Water Environ. Res.*, 64, 274, 1992.

129. Zagorc-Koncan, J., Effects of atrazine and alachlor on self-purification processes in receiving streams, *Water Sci. Technol.*, 33, 181, 1996.

130. Gao, J.P., Maguhn, J., Spitzauer, P., and Kettrup, A., Sorption of pesticides in the sediment of the Teufelsweiher pond (Southern Germany). II. Competitive adsorption, desorption of aged residues and effect of dissolved organic carbon, *Water Res.*, 32, 2089, 1998.

131. Colburn, T., Pesticides: how research has succeeded and failed to translate science in to policy: endocrinological effects on wildlife, *Environ. Health Persp.*, 103 (Suppl. 6), 81, 1995.

132. LeBlanc, G.A., Are environmental sentinels signalling?, *Environ. Health Persp.*, 103, 888, 1995.

133. Rahman, F., Langford, K.H., Scrimshaw, M.D., and Lester, J. N., Polybrominated diphenyl ether (PBDE) flame retardants, *Sci. Total Environ.*, 275, 1, 2001.

134. Nylund, K., Asplund, L., Jansson, B., Jonssen, P., Litzén, K., and Sellström, U., Analysis of some polyhalogenated organic pollutants in sediment and sewage sludge, *Chemosphere*, 24, 1721, 1992.

135. Sellström, U., Kierkegaard, A., Alsberg, T., Jonsson, P., Wahlberg, C., and de Wit, C., Brominated flame retardants in sediments from European estuaries, the Baltic Sea and in sewage sludge, *Organohal. Comp.*, 40, 383, 1999.

136. de Boer, J., van der Horst, A., and Wester, P.G., PBDEs and PBBs in suspended particulate matter, sediments, sewage treatment plant in- and effluents and biota from the Netherlands, *Organohal. Compd.*, 47, 85, 2000.

137. Watanabe, I. and Tatsukawa, R., Formation of brominated dibenzofurans from the photolysis of flame retardant decabromobiphenyl ether in hexane solution by UV and sunlight, *Bull. Environ. Contam. Toxicol.*, 39, 953, 1987.

138. Sellström, U., Soderstrom, G., de Wit, C.A., and Tykslind, M., Photolytic debromination of decabromodiphenyl ether (DeBDE), *Organohal. Comp.*, 35, 447, 1998.

139. de Wit, C.A., Brominated Flame Retardants, Report 5065, Swedish Environmental Protection Agency, Stockholm, Sweden, 2000, p. 94.

140. Ronen, Z. and Abeliovich, A., Anaerobic-aerobic process for microbial degradation of tetrabromobisphenol A, *Appl. Environ. Microbiol.*, 66, 2372, 2000.

141. Watanabe, I. and Kashimoto, T., The flame retardant tetrabromobisphenol-A and its metabolite found in river and marine sediments in Japan, *Chemosphere*, 12, 1533, 1983.

142. Haglund, P.S., Zook, D.R., Buser, H.R., and Hu, J., Identification and quantification of polybrominated diphenyl ethers and methoxy-polybrominated diphenyl ethers in baltic biota, *Environ. Sci. Technol.*, 31, 3281, 1997.

143. Dodder, N.G., Stranberg, B., and Hites, R.A., Concentrations and spatial variations of polybrominated diphenyl ethers and several organochlorine compounds in fishes from the Northeastern United States, *Environ. Sci. Technol.*, 36, 146, 2002.

144. Watanabe, I., Kashimoto, T., and Tatsukawa, R., Polybrominated biphenyl ethers in marine fish, shellfish and river and marine sediments in Japan, *Chemosphere*, 16, 2389, 1987.

145. Hale, R.C., Guardia, M.J., Harvey, E.P., Mainor, T.M., Duff, W.H., and Gaylor, M.O., Polybrominated diphenyl ether flame retardants in Virginia freshwater fishes (USA), *Environ. Sci. Technol.*, 2001.

146. Kierkegaard, A., Balk, L., Tjärnlund, U., de Wit, C.A., and Jansson, B., Dietary uptake and biological effects of decabromodiphenyl ether in rainbow trout (*Oncorhynchus mykiss*), *Environ. Sci. Technol.*, 33, 1612, 1999.

147. Burreau, S., Axelman, J., Broman, D., and Jakibsson, E., Dietary uptake in pike (Esox lucius) of some polychlorinated bipiphenyls, polychlorinated naphthalenes and poly-brominated diphenyl ethers administered in natural diet, *Environ. Toxicol. Chem.*, 16, 2508, 1997.

148. Scrimshaw, M.D., Bubb, J. M., and Lester, J. N., Magnitude and distribution of contaminants in salt marsh sediments of the Essex coast. IV. organochlorine insecti-cides and polychlorinated biphenyls, *Sci. Total Environ.*, 155, 73, 1994.

149. Scrimshaw, M.D., Bubb, J.M., and Lester, J. N., Organochlorine contamination of UK Essex Coast salt marsh sediments, *J. Coastal Res.*, 12, 246, 1996.

150. Blanchard, M., Rteil, M.-J., Ollivon, D., Garban, B., Chesterikoff, C., and Chevreuil, M., Origin and distribution of polyaromatic hydrocarbons and polychlorobiphenyls in urban effluents to wastewater treatment plants of the Paris area (France), *Water Res.*, 35, 3679, 2001.

151. Jeremiason, J.D., Hornbuckle, K.C., and Eisenreich, S.J., PCBs in Lake Superior, 1978–1992 — Decreases in water concentrations reflect loss by volatilisation, *Environ. Sci. Technol.*, 28, 903, 1994.

152. Gevao, B., Hamilton-Taylor, J., and Jones, K.C., Towards a complete mass balance and model for PCBs and PAHs in a small rural lake, Cumbria UK, *Limnol. Oceanogr.*, 45, 881, 2000.

153. Turner, A. and Rawling, M.C., The influence of salting out of the sorption of neutral organic compounds in estuaries, *Water Res.*, 35, 4379, 2001.

154. McGroddy, S.E., Farrington, J. W., and Gschwend, P.M., Comparison of the *in situ* and desorption sediment-water partitioning of polycyclic aromatic hydrocarbons and polychlorinated biphenyls, *Environ. Sci. Technol.*, 30, 172, 1995.

155. Zhou, J.L., Maskaoui, K., Qiu, Y.W., Hong, H.S., and Wang, Z.D., Polychorinated biphenyl congeners and organochlorine insecticides in the water column and sedi-ments of Daya Bay, China, *Environ. Poll.*, 113, 373, 2001.

156. Jeremiason, J.D., Eisenreich, S.J., Baker, J.E., and Eadie, B.J., PCB decline in settling particles and benthic recycling of PCBs and PAHs in Lake Superior., *Environ. Sci. Technol.*, 32, 3249, 1998.

157. Tyler, A.O. and Millward, G.E., Distribution and partitioning of polychlorinated dibanzo-p-dioxins, polychlorinated dibenzofurans and polychlorinated biphenyls in the Humber Estuary, *Marine Poll. Bull.*, 32, 397, 1996.

158. Ferraro, S.P., Lee, H., Ozretich, R.J., and Specht, D.T., Predicting bioaccumulation potential: a test of a fugacity-based model, *Arch. Environ. Contam. Toxicol.*, 19, 386, 1990.

159. Boese, B.L., Winsor, M., Lee, H., Echols, S., Pelletier, J., and Randall, R., PCB congeners and hexachlorobenzene biota sediment accumulation factors for *Macoma nasuta* exposed to sediments with different total organic carbon contents, *Environ. Toxicol. Chem.*, 14, 303, 1995.

160. Scrimshaw, M.D. and Lester, J.N., Organochlorine contamination in sediments of the Inner Thames Estuary, *J. Chart. Inst. Water Environ. Manage.*, 9, 519, 1995.

161. Borga, K., Gabrielson, G.W., and Skaare, J.U., Biomagnification of organochlorines along a Barents Sea food chain, *Environ. Poll.*, 113, 187, 2001.

162. Smith, K.E.C., Green, M., Thomas, G.O., and Jones, K.C., Behavior of sewage sludge-derived PAHs on pasture, *Environ. Sci. Technol.*, 35, 2141, 2001.

163. Liu, H.M. and Amy, G., Modeling partitioning and transport interactions between natural organic-matter and polynuclear aromatic-hydrocarbons in groundwater, *Environ. Sci. Technol.*, 27, 1553, 1993.

164. Law, R.J. and Biscaya, J. L., Polycyclic aromatic hydrocarbons (PAH) - Problems and progress in sampling, analysis and interpretation, *Marine Poll. Bull.*, 29, 235, 1994.

165. Ariese, F., Ernst, W.H. O., and Sijm, D.T.H.M., Natural and synthetic organic compounds in the environment — a symposium report, *Environ. Toxicol. Pharm.*, 10, 65, 2001.

166. James, M.O. and Kleinow, K.M., Trophic transfer of chemical in the aquatic environment, in *Aquatic Toxicology: Molecular, Biochemical, and Cellular Perspectives*, Malins, D.C. and Ostrander, G.K., Eds., CRC Press, Boca Raton, FL, 1993, p.1.

167. Law, R.J., Dawes, V.J., Woodhead, R.J., and Matthiessen, P., Polycyclic aromatic hydrocarbons (PAH) in seawater around England and Wales, *Marine Poll. Bull.*, 34, 306, 1997.

168. Woodhead, R.J., Law, R.J., and Matthiessen, P., Polycyclic aromatic hydrocarbons in surface sediments around England and Wales, and their possible biological significance, *Marine Poll. Bull.*, 38, 773, 1999.

169. Huntley, S.L., Bonnevie, N.L., and Wenning, R.J., Polycyclic aromatic hydrocarbon and petroleum hydrocarbon contamination in sediment from Newark Bay Estuary, New Jersey, *Arch. Environ. Contam. Toxicol.*, 28, 93, 1995.

170. Pohlman, J.W., Coffin, R.B., Mitchell, C.S., Montgomery, M.T., Spargo, B.J., Steele, J.K., and Boyd, T.J., Transport, deposition and biodegradation of particle bound polycyclic aromatic hydrocarbons in a tidal basin of an industrial watershed, *Environ. Monit. Assess.*, 75, 155, 2002.

171. Widada, J., Nojiri, H., Kasuga, K., Yoshida, T., Habe, H., and Omori, T., Molecular detection and diversity of polycyclic aromatic hydrocarbon-degrading bacteria isolated from geographically diverse sites, *Appl. Microbiol. Biotechnol.*, 58, 202, 2002.

172. Rockne, K.J. and Strand, S.E., Biodegradation of bicyclic and polycyclic aromatic hydrocarbons in anaerobic enrichments, *Environ. Sci. Technol.*, 32, 3962 1998.

173. Rockne, K.J., Chee-Sanford, J. C., Hedlund, B.P., Staley, J.T., and Strand, S.E., Anaerobic naphthalene degradation by microbial pure cultures under nitrate-reducing conditions, *Appl. Environ. Microbiol.*, 65, 1595, 2000.

174. Taylor, P.N. and Lester, J.N., Polynuclear aromatic hydrocarbons in a River Thames sediment core, *Environ. Technol.*, 16, 1155, 1995.

175. Su, M.C., Christensen, E.R., Karls, J. F., Kosuru, S., and Imamoglu, I., Apportionment of polycyclic aromatic hydrocarbon sources in lower Fox River, USA, sediments by a chemical mass balance model, *Environ. Toxicol. Chem.*, 19, 1481, 2000.

176. Johnson, W.P., Sediment control of facilitated transport and enhanced desorption, *J. Environ. Eng.–ASCE*, 126, 47, 2000.

177. Hwang, S.C. and Cutright, T.J., Impact of clay minerals and DOM on the competitive sorption/desorption of PAHs, *Soil Sed. Contam.*, 11, 269, 2002.

178. Munkittrick, K.R., Blunt, B.R., Leggett, M., Huestis, S., and McCarthy, L.H., Development of a sediment bioassay to determine bioavailability of PAHs to fish, *J. Aquatic Ecosystem Health*, 4, 169, 1995.

179. Landrum, P.F., Eadie, B.J., and Faust, W.R., Toxicokinetics and toxicity of a mixture of sediment-associated polycyclic aromatic hydrocarbons to the amphipod *Diporeia* sp., *Environ. Toxicol. Chem.*, 10, 35, 1991.

180. Carlson, E.A., Li, Y., and Zelikoff, J.T., Exposure of Japanese medaka (Oryzias latipes) to benzo[a]pyrene suppresses immune function and host resistance against bacterial challenge, *Aquatic Toxicol.*, 56, 289, 2002.

181. Ruddock, R.J., Bird, D.J., and McCalley, D.V., Bile metabolites of polycyclic aromatic hydrocarbons in three species of fish from the Severn Estuary, *Ecotoxicol. Environ. Safety*, 51, 97, 2002.

182. Burkhard, L.P. and Lukasewycz, M.T., Some bioaccumulation factors and biota-sediment accumulation factors for polycyclic aromatic hydrocarbons in lake trout, *Environ. Toxicol. Chem.*, 19, 1427, 2000.

183. Desbrow, C., Routledge, E.J., Brighty, G.C., Sumpter, J.P., and Waldock, M., Identification of estrogenic chemicals in STW effluent. 1. Chemical fractionation and *in vitro* biological screening, *Environ. Sci. Technol.*, 32, 1549, 1998.

184. Finlay-Moore, O., Hartel, P.G., and Cabrera, M.L., 17β–estradiol and testosterone in soil and runoff from grasslands amended with broiler litter, *J. Environ. Qual.*, 29, 1604, 2000.

185. Shore, L.S., Gurevitz, M., and Shemesh, M., Estrogen as an environmental pollutant, *Bull. Environ. Contam. Toxicol.*, 51, 361, 1993.

186. Lai, K.M., Johnson, K.L., Scrimshaw, M.D., and Lester, J.N., Binding of waterborne steroid estrogens to solid phases in river and estuarine systems, *Environ. Sci. Technol.*, 34, 3890, 2000.

187. Lopez de Alda, M. and Barcelo, D., Use of solid-phase extraction in various of its modalities for sample preparation in the determination of estrogens and progestogens in sediment and water, *J. Chromatogr. A.*, 938, 145, 2001.

188. Bowman, J. C., Zhou, J.L., and Readman, J.W., Sediment-water interactions of natural oestrogens under estuarine conditions, *Marine Chem.*, 77, 263, 2002.

189. Panter, G.H., Thompson, R.S., Beresford, N., and Sumpter, J.P., Transformation of a non-oestrogenic steroid metabolite to an oestrogenically active substance by minimal bacterial activity, *Chemosphere*, 38, 3579, 1999.

190. Layton, A.C., Gregory, B.W., Seward, J.R., Schultz, T.W., and Sayler, G.S., Mineralization of steroidal hormones by biosolids in wastewater treatment systems in Tennessee U.S.A., *Environ. Sci. Technol.*, 34, 3925, 2000.

191. Walker, D., Oestrogenicity and wastewater recycling: experience from Essex and Suffolk water, *J. Chart. Inst. Water Environ. Manage.*, 14, 427, 2000.

192. Garcia-Reyero, N., Grau, E., Castillo, M., Lopez de Alda, M.J., Barcelo, D., and Pina, B., Monitoring of endocrine disrupters in surface waters by the yeast recombinant assay, *Environ. Toxicol. Chem.*, 20, 1152, 2001.

193. Christiansen, T. and Korsgaard, B., Effects of nonylphenol and 17β-oestradiol on vitellogenin synthesis, testicular structure and cytology in male eelpout *Zoarces viviparus*, *J. Exp. Biol.*, 201, 179, 1998.

194. Jobling, S., Sheahan, D.A., Osborne, J.A., Matthiessen, P., and Sumpter, J.P., Inhibition of testicular growth in rainbow trout (*Oncorhynchus mykiss*) exposed to environmental estrogens, *Environ. Toxicol. Chem.*, 15, 194, 1995.

195. Martin-Robichaud, D.J., Peterson, R.H., Benfey, T.J., and Crim, L.W., Direct feminization of lumpfish (*Cyclopterus-lumpus*) using 17β-estradiol-enriched artemia as food, *Aquaculture*, 123, 137, 1994.

196. Lai, K.M., Scrimshaw, M.D., and Lester, J.N., Biotransformation and bioconcentration of steroid estrogens by *Chlorella vulgaris*, *Appl. Environ. Microbiol.*, 68, 859, 2002.

197. Gomes, R.L., Deacon, H.E., Lai, K.M., Birkett, J. W., Scrimshaw, M.D., and Lester, J. N., An assessment of the bioaccumulation of estrone in *Daphnia magna*, *Environ. Chem. Toxicol.*, submitted.

198. Lai, K.M., Scrimshaw, M.D., and Lester, J.N., The effects of natural and synthetic steroid estrogens in relation to their environmental occurrence, *Crit. Rev. Toxicol.*, 32, 113, 2002.

199. Ternes, T.A., Kreckal, P., and Mueller, J., Behaviour and occurrence of estrogens in municipal sewage treatment plants. II. Aerobic batch experiments with activated sludge, *Sci. Total Environ.*, 225, 91, 1999.

200. Belfroid, A.C., Van der Horst, A., Vethaak, A.D., Schäfer, A.J., Rijs, G.B.J., Wegener, J., and Cofino, W.P., Analysis and occurrence of estrogenic hormones and their glucuronides in surface water and waste water in the Netherlands, *Sci. Total Environ.*, 225, 101, 1999.

201. Matsui, S., Takigami, H., Matsuda, N., Taniguchi, N., Adachi, J., Kawami, H., and Shimizu, Y., Estrogen and estrogen mimics contamination in water and the role of sewage treatment, *Water Sci. Technol.*, 42, 173, 2000.

202. Huang, C.-H. and Sedlak, D., Analysis of estrogenic hormones in wastewater and surface water using ELISA and GC/MS/MS, *Environ. Toxicol. Chem.*, 20, 133, 2000.

203. Fawell, J. K., Sheahan, D., James, H.A., Hurst, M., and Scott, S., Oestrogens and oestrogenic activity in raw and treated water in Severn Trent water, *Water Res.*, 35, 1240, 2001.

204. Aherne, G.W. and Briggs, R., The relevance of the presence of certain synthetic steroids in the aquatic environment, *J. Pharm. Pharmacol.*, 41, 735, 1989.

205. Environment Agency, Endocrine-Disrupting Substances in the Environment: What Should Be Done?, London, 1998, p. 13.

206. Consultants in Environmental Sciences Ltd., Uses, Fate and Entry to the Environment of Nonylphenol Ethoxylates, Beckenham, Kent, England, 1993.

207. Ahel, M., Giger, W., and Schaffner, C., Behaviour of alkylphenol polyethoxylate surfactants in the aquatic environment. II. Occurrence and transformation in rivers, *Water Res.*, 28, 1143, 1994.

208. Ahel, M., Giger, W., and Koch, M., Behaviour of alkylphenol polyethoxylate surfactants in the aquatic environment. I. Occurrence and transformations in sewage treatment, *Water Res.*, 28, 1131, 1994.

209. Ahel, M., Schaffner, C., and Giger, W., Behaviour of alkylphenol polyethoxylate surfactants in the aquatic environment. III. Occurrence and elimination of their persistent metabolites during infiltration of river water to groundwater, *Water Res.*, 30, 37, 1996.

210. Isobe, T., Nishiyama, H., Nakashima, A., and Takada, H., Distribution and behavior of nonylphenol, octylphenol and nonylphenol monoethoxylate in Tokyo metropolitan area: their association with aquatic particles and sedimentary distributions, *Environ. Sci. Technol.*, 35, 1041, 2001.

211. Ekelund, R., Granmo, A., Magnusson, K., Berggren, M., and Bergman, A., Biodegradation of 4-nonylphenol in seawater and sediment, *The Alkylphenols and Alkylphenol Ethoxylates Rev.*, 1, 13, 1998.

212. Johnson, A.C., White, C., Besien, T.J., and Jurgens, M.D., The sorption potential of octylphenol, a xenobiotic oestrogen, to suspended and bed-sediments collected from industrial and rural reaches of three English rivers, *Sci. Total Environ.*, 210–211, 271, 1998.

213. Weeks, J.A., Adams, W.J., Guiney, P.D., Hall, J.F., and Naylor, C.G., Risk assessment of nonylphenol and its ethoxylates in U.S. river water and sediment, *The Alkylphenols and Alkylphenol Ethoxylates Rev.*, 1, 64, 1998.

214. Ashfield, L.A., Pottinger, T.G., and Sumpter, J.P., Exposure of rainbow trout to alkylphenolic compounds: Effects on growth and reproductive status, in *Environmental Endocrine Disrupters and Oestrogen Mimics*, SETAC, The Society of Environmental Toxicology and Chemistry, Liverpool, England, 1995.

215. Jobling, S., Sheahan, D., Osborne, J.A., Matthiessen, P., and Sumpter, J.P., Inhibition of testicular growth in rainbow trout (*Oncorhynchus mykiss*) exposed to estrogenic alkylphenolic chemicals, *Environ. Toxicol. Chem.*, 15, 194, 1996.

216. Naylor, C.G., Mieure, J.P., Adams, W.J., Weeks, J.A., Castaldi, F.J., Ogle, L.D., and Romano, R.R., Alkylphenol ethoxylates in the environment, *The Alkylphenols and Alkylphenol Ethoxylates Rev.*, 1, 32, 1998.

217. Bennett, E.R. and Metcalfe, C.D., Distribution of alkylphenol compounds in Great Lakes sediments, United States and Canada, *Environ. Toxicol. Chem.*, 17, 1230, 1998.

218. Field, J.A. and Reed, R.L., Nonylphenol polyethoxy carboxylate metabolites of nonionic surfactants in U.S. paper mill effluents, municipal sewage treatment plant effluents, and river waters, *Environ. Sci. Technol.*, 30, 3544 1996.

219. Blackburn, M., Kirby, S.J., and Waldock, M.J., Concentrations of alkyphenol polyethoxylates entering UK estuaries, *Chemosphere*, 38, 109, 1999.

7 Endocrine Disrupters in Drinking Water and Water Reuse

R.L. Gomes and J.N. Lester

CONTENTS

1-56670-601-7/03/$0.00+$1.50
© 2003 by CRC Press LLC

219

7.1 INTRODUCTION

When surface waters are used as sources for drinking water, abstraction for water treatment may often be downstream of effluent discharge points.[1] Drinking water abstraction from groundwater can also contain endocrine disrupting chemicals (EDCs) from diffuse (percolation/infiltration) or point (recharge) sources.[2] Materials in contact with drinking water supplies can be an additional source for EDC contamination. Drinking water is a direct route into the human body and for any EDCs that may be present. Other pathways via ingestion (eating crops irrigated with effluent) or bodily interaction (bathing in waters containing effluent) can place the human body in contact with EDCs.

Where once treated wastewater effluent was deemed a nuisance, it is increasingly being recognized as a valuable resource. With increasing human populations and unevenly matched water supply and demand, water reuse is a sustainable and economical option. The reuse of wastewater occurs by either indirect (environmental intervention) or direct means and can have potable (drinking) or nonpotable applications. Indirect water reuse accounts for up to one third of water taken from U.K. lowland rivers containing effluent.[3]

In receiving waters, sewage effluent is subjected to dilution and any EDCs present are available to transformation processes including microbial activity, sorption, and photocatalysis.[4] The concept of direct water reuse ("pipe to pipe") means that without environmental intervention, transformation processes and dilution is limited or unable to take place. Currently, the dilution and self-purification processes of receiving waters are greatly depended upon for the removal of EDCs from the hydrological cycle, which sewage treatment works (STWs) have failed to remove.

7.2 ENDOCRINE DISRUPTERS AND DRINKING WATER

Water treatment is the process of cleaning water that will be fit for public use and consumption. Contaminants in water may be of biological or chemical origin, with microorganisms having traditionally been of greatest concern and the driver for standards and regulations. Regulations do exist for several EDCs, mostly pesticides because of their persistent nature and ubiquitous presence. The concept of EDCs and possible related human health implications arising through drinking contaminated water is still being investigated. According to the Centers for Disease Control and Prevention, potential health effects include adverse reproductive outcomes, developmental disabilities, endometriosis, and breast and testicular cancer.[5]

7.2.1 Sources of Water for Drinking Water Supply

In certain regions, surface water can constitute up to 98% of the water used for drinking water supply.[6] However, groundwater is a preferred source for drinking water, because it generally has a higher quality than surface waters and requires less treatment. Groundwater contamination is mainly from diffuse sources, such as waste disposal practices, irrigation, spills, and leaks from pipelines. By the end of the 1980s, more than 50% of the U.S. population depended on groundwater as the main source of drinking water. In the 1990s, estimates of more than 50% of the U.S. total water quality problems were being attributed to agricultural, industrial, and residential nonpoint sources.[7] In Europe, groundwater for potable supply is more than 70% in Austria, Denmark, Iceland, and Portugal and between 50 to 75% for Belgium, Finland, France, Germany, Luxembourg, and the Netherlands.[8,9]

Reservoirs often serve as sources of water for drinking water treatment. The water present in the reservoir has originated from surface water and groundwater sources. The retention time of reservoirs, often of longer duration than in surface waters, allows for self-purification processes to occur. Pollution from nonpoint sources, such as agricultural runoff, is unlikely compared to surface waters. A further advantage to reservoir storage is that the quality of the water can be monitored more closely. Any pollution events in the water sources supplying the reservoir can be closed off until the water quality is improved.

7.2.2 Drinking Water Treatment

Similar to STWs, drinking water treatment (DWT) varies greatly according to the water source, possible contaminants present, and regulations imposed in each country. Depending on the source and level of pollution (real or assumed), pretreatment (e.g., disinfection) may also be required. The basic water treatment procedure is summarized in Figure 7.1.

DWT has turned to ozone (O_3) as an alternative to chlorine due to lower trihalomethane (THM) formation. O_3 is generated by ultraviolet (UV) irradiation of air or oxygen and corona discharge. It removes taste, odor, and color-forming compounds; improves flocculation; and enhances biodegradability. However, the effects of O_3 are short-lived and chlorine is still required for the residual effect. O_3 is an

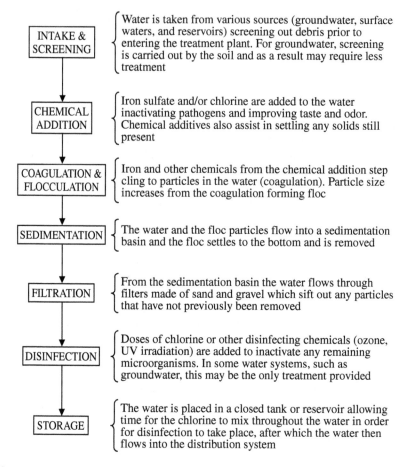

INTAKE & SCREENING — Water is taken from various sources (groundwater, surface waters, and reservoirs) screening out debris prior to entering the treatment plant. For groundwater, screening is carried out by the soil and as a result may require less treatment

CHEMICAL ADDITION — Iron sulfate and/or chlorine are added to the water inactivating pathogens and improving taste and odor. Chemical additives also assist in settling any solids still present

COAGULATION & FLOCCULATION — Iron and other chemicals from the chemical addition step cling to particles in the water (coagulation). Particle size increases from the coagulation forming floc

SEDIMENTATION — The water and the floc particles flow into a sedimentation basin and the floc settles to the bottom and is removed

FILTRATION — From the sedimentation basin the water flows through filters made of sand and gravel which sift out any particles that have not previously been removed

DISINFECTION — Doses of chlorine or other disinfecting chemicals (ozone, UV irradiation) are added to inactivate any remaining microorganisms. In some water systems, such as groundwater, this may be the only treatment provided

STORAGE — The water is placed in a closed tank or reservoir allowing time for the chlorine to mix throughout the water in order for disinfection to take place, after which the water then flows into the distribution system

FIGURE 7.1 Summary of basic drinking water treatment process.

effective method for the removal of organic pollutants from both wastewater and drinking water due to the following characteristics:

1. O_3 is a powerful oxidizing agent.
2. O_3 reacts rapidly with most organic pollutants.
3. O_3 contains important bactericidal and viricidal effects.

Increasingly, a variety of other processes, such as advanced oxidation, adsorption and membrane filtration are being utilized to aid in the removal of both biological and chemical contaminants between the filtration and disinfection stages. These water treatment technologies may also act as tertiary treatment for wastewater in STWs and in water reuse. Advanced oxidation utilizes a combination of O_3, hydrogen peroxide (H_2O_2) or UV irradiation. The free radical formation is enhanced, improving the removal efficiency of some organic contaminants.[10] Similar to O_3, UV irradiation can

TABLE 7.1
Types of Membrane Filtration and Applications

Membrane Pressure	Low		High	
Filtration Type	Ultra (UF)	Micro (MF)	Reverse osmosis (RO)	Nano (NF)
Operating Pressure (psi)	10–30	75–250	>250	75–150
Nominal Pore Size (nm)	10	200	<1	<RO–>UF
Removes	Particulate matter; microorganisms		>99 total dissolved solids; trace contaminants; desalting	>90% natural organic matter; color-causing compounds disinfection by-products (DBPs)
Does Not Remove	Dissolved organic matter (DOM); taste, odor, color-causing compounds; anthropogenic chemicals			
Applications	Limited to surface water sources; disinfection for UF		Long history of desalination	Groundwater softening
Comments	Exception: UF introduced PAC which is injected into influent		Recent innovations do not require high psi (TFC)	Thin film composite (TFC) of RO membranes covering pore size between RO and UF

be used as a disinfectant though both require a secondary disinfectant, such as chlorine, to maintain a residual effect in the distribution system.[11] Photodegradation can be enhanced with the use of titanium dioxide (TiO_2).[12] Adsorption processes usually favor granular activated carbon (GAC) with the use of powdered activated carbon (PAC) to a lesser extent. Membrane technology can be classified into two classes: low pressure and high pressure. Table 7.1 summarizes the main points within each class.[13] Development and implementation of water treatment technologies are driven by three primary factors: the discovery of new water contaminants, promulgation of new water quality standards and economic restraints.[13]

7.2.3 ENDOCRINE DISRUPTERS IDENTIFIED IN DRINKING WATER

Environmental fate parameters that determine whether a chemical will be present in the aqueous phase of reservoirs or surface waters used for drinking water are:

1. The sorption potential to suspended or whole sediment (log K_{oc})
2. Dissolving potential or water solubility (mg l^{-1})
3. Evaporation potential from water (Henry's Law constant)

Many EDCs have a high sorption (and bioconcentration) potential and low water solubility and volatility. Preference will be to suspended/whole sediment with low quantities in the dissolved aqueous phase. EDCs entering DWT are likely to be associated with particulates. During DWT, the pH is altered, which may cause EDCs' desorption from particulates.

Contaminants in drinking water may originate from three points:

1. Prior to drinking water treatment, EDCs are present in the source water entering the DWT facility.
2. During drinking water treatment, chlorinated and oxygenated compounds are produced as a result of disinfection (disinfection by-products [DBPs]).
3. Upon post-drinking water treatment, EDCs are introduced via materials in contact with the drinking water supply.

The vast majority of data on EDCs in drinking water are due to their presence in the water source. The inefficient removal during DWT results in these compounds being present in the finished product. DBPs produced during treatment include THMs and other chlorinated products. The precursors of THMs are thought to be humic and fulvic acids, which once chlorinated form THMs (e.g., carcinogenic chloroform). The presence of EDCs in drinking water may also arise through leaching from materials in contact with the water supply.

7.2.3.1 Water Sources

In 1990, it was found that 28% of drinking water samples taken exceeded the European Commission's (EC) safe limit of 0.1 mg l^{-1} for pesticides.[14] In San Francisco delta water, analysis failed to show the presence of synthetic organic chemicals (SOCs), several of which are EDCs. Adverse effects in fish revealed their presence and raised the concerns about the feasibility of analyzing for trace concentrations (see Chapter 4). The presence of SOCs were from the substantial quantities of pesticides applied annually to land that had drained into the delta.[15] In the Rivers Rhine and Meuse there have been 72 different pesticides in excess of the EEC limit. Most widely identified pesticides were triazines, phenylurea herbicides, and chlorophenoxy carbonic acids.[16] U.K. monitoring of ground water between the Chalk and Bunter Sandstone aquifers established that the predominant pesticides were the triazines.[17]

Analysis of the drinking water in the U.S. has shown the total concentration of alkylphenolic compounds to be nearly 1 mg l^{-1}.[18] In groundwater down-gradient of an infiltration bed for secondary treated sewage effluent, nonylphenol (NP) and octylphenol (OP) and their ethoxylates were detected at 30 g l^{-1} concentrations. Bisphenol A (BPA), nonylphenol monoethoxycarboxylate, and nonyl- and octylphenol tetraethoxylate were detected in some drinking water wells at concentrations ranging from below the quantification limit to 32.9 g l^{-1}.[19]

There is concern over the presence of steroid estrogens in drinking water. If 17α-ethinylestradiol (EE2) is present in drinking water at only nanogram concentrations, this may be enough to present a health risk.[20] In the U.K., analysis was undertaken at drinking water intakes and storage reservoirs that abstract from rivers and received significant amounts of sewage effluent. The vitellogenin (VTG) assay detected no estrogenicity in any samples.[21] Final drinking water from six DWT works, which employed a range of treatment processes including only conventional treatment, O_3, granular activated carbon (GAC) and a combination of O_3 and GAC, were monitored for estrone (E1), 17β-estradiol (E2), and EE2. No steroid estrogens could be identified using detection limits of 0.2 ng l^{-1} for E1 and E2 and 0.4 ng l^{-1} for EE2.[22] Male roach were immersed in tap water for 4 months with sampling at day 0 and 1, 2, and 4 months. VTG levels were quantified using a validated enzyme-linked immunosorbent assay,[23] which showed zero elevation of VTG levels in the tap water.[24]

Studies investigated the link between surface water, which are the receptacles for sewage effluent, and the sources for drinking water treatment.[25] The estrogenic activity in river and drinking water of the U.K.'s Severn Trent, was biologically (both *in vivo* and *in vitro*) and chemically analyzed. Water samples were taken along the River Trent, in the Church Wilne reservoir, at the inlet and outlet point, and in the raw and final Strensham DWT. With chemical detection limits of 0.3 ng l^{-1} for all steroid estrogens, EE2 was not detected in any samples. At the raw water intake for DWT, 0.3 ng l^{-1} of E2 was determined. In the reservoir 25 ng l^{-1} of E2 was observed, but was believed to be contaminated because there were no other detections. E1 was at concentrations up to 7 ng l^{-1} in the river and 1.8 ng l^{-1} at the raw water intake with no detection in all other samples. This study observed no steroid estrogens in final drinking water, a fact that is substantiated by another U.K. study.[26]

In addition to monitoring steroid estrogens at Severn Trent, other EDCs were investigated. No BPA and NP and its ethoxylates were observed using chemical detection limits of 5.1 µg l^{-1}, 0.2 µg l^{-1}, and 4 µg l^{-1}, respectively.[25] However, di-(2-ethylhexyl) phthalate (DEHP) was determined at levels between 1.3 and 2.6 mg l^{-1} in river and reservoir water.[25] At the raw water intake for DWT, concentrations were 4.9 µg l^{-1} with a final drinking water content of 2.5 µg l^{-1}. The U.S. Environmental Protection Agency (EPA) classifies DEHP as a water priority pollutant with a maximum contaminant level (MCL) at 6 µg l^{-1} and MCL goal of zero.[27]

7.2.3.2 Production during Drinking Water Treatment

Though designed to remove contaminants, treatment processes may also produce by-products that may concern human health. Polychlorinated dibenzo-*p*-dioxins (PCDDs) and polychlorinated dibenzofurans (PCDFs) were formed during photolysis due to the presence of pentachlorophenol (PCP).[28] The Orange and Los Angeles counties conducted a 5-year health effects study on indirect potable reuse using infiltration groundwater recharge. Toxicological studies attempted to detect, characterize, and trace previously unidentified carcinogens in groundwater recharge sources. Ames and *Salmonella lester* strains toxicity tests determined that half the mutagenic contaminants were derived from the chlorination process.[29]

Chlorination acts as both an oxidizing and chlorinating agent. Due to chlorination, oxygenated and chlorinated compounds are produced, the most notable being THMs. Chlorine reacts electrophilically with certain aromatic and hetercyclic ring systems and is especially apparent with compounds possessing the hydroxyl moiety.[30] Oxygenated and chlorinated polyaromatic hydrocarbons (PAHs) can be formed by chlorination, resulting in compounds that are more toxic than the parent EDC.[31] Chlorination may therefore generate more persistent (and toxic?) compounds from parent EDCs such as steroid estrogens. However, for polychlorinated biphenyls (PCBs), PCDDs, and PCDFs, chlorination was shown to have no influence on these EDCs.[32]

The Endocrine Disrupter Screening and Testing Advisory Committee (EDSTAC) recommends that DBPs in drinking water require evaluation for endocrine disruption.[33] The *in vitro* MVLN assay was used as recommended by EDSTAC to evaluate the estrogenic effect of DBPs by comparing chlorinated and unchlorinated lake water. Filtered lake water was chlorinated at an initial concentration of 1 mg l[-1]. After 24 hours, residual chlorination was around 0.1mg l[-1]. The estrogenic effect of chlorinated water was approximately 2.3 times as strong as the unchlorinated water. Possible reasons for the increase in estrogenicity were:

1. Chlorination produced estrogenic by-products such as organochlorine substances.
2. The low molecular fraction that binds to the estrogen receptor increases due to chlorination through oxidation and hydrolysis.
3. Chlorine releases estrogenic compounds which interact with estrogenic humic substances in the aqueous environment.

The study by Itoh et al. concluded that investigation into endocrine disrupting effects of chlorination by-products is required in addition to identification of EDCs in drinking water samples.[33]

7.2.3.3 Materials in Contact with Drinking Water

EDCs may also be introduced into drinking water supplies via materials in contact with the drinking water. Substances that are used in plastic and other materials for lining drinking water supplies have been implicated as EDCs. Leaching from these materials may therefore occur when in contact with water. Possible health concerns over EDCs leaching from plastic pipes were raised at the first readings of proposals for the new European Union law on drinking water standards.[34] EDCs with the potential to leach from materials in contact with water supplies are phthalates, BPA, alkylphenols (APs), and PAHs.[35]

The limited use of phthalates result in a small surface area in contact with water and are less likely to cause leaching into water supplies.[35] A study by Brotons et al. highlighted the potential problem of BPA leaching from lacquer in food cans into water supplies.[36] BPA is also used to reline water mains and as coatings for many fittings in contact with drinking water. BPA may occur in water supplies at concentrations up to 1 mg l[-1].[37] Alkylphenolic compounds leach from certain types of plastics. NP is added to polystyrene and polyvinyl chloride (PVC), increasing the stability of the plastic. These compounds have been detected in water, which have passed through flexible PVC tubing.[38] APs may be found in PVC and other polymer tubes and may leach into drinking water supplies.[39]

The majority of ductile iron distribution pipes are internally coated with coal-tar pitch. Installation of new coal lined pipes ceased in Germany and the U.K. during the 1970s.[40] Coal tar pitch can contain up to 50% PAHs and can leach into the drinking water supply.[3] PAH concentrations from 190 to 302 ng l^{-1} have been detected in areas where chlorinated water is distributed.[41] Biofilms on the coal tar surface exhibit protective effects. Hostile environments, such as disinfection, stagnation, and anaerobic conditions are factors that promote the release of PAHs into drinking water.[40] Such factors include:

1. A decrease in pH from neutral to acidic resulted in remobilization of PAHs.
2. The concentration of free residual chlorine is related to the PAH level in drinking water. Chlorine dioxide disinfection led to the highest PAH levels.
3. There are disturbances in the hydraulic regime, such as a rapid increase in flow velocity or change in the direction of the flow.
4. The residence time of drinking water must be kept to a minimum. Stagnation periods of more than 7 hours lead to elevated PAH levels.

PAH-free drinking water occurs when disinfection is stopped, or UV is used which leaves no residual.[40] However, chlorination and its residual effect is important for protection against pathogens in drinking water supplies.

7.2.4 REMOVAL OF ENDOCRINE DISRUPTERS DURING DRINKING WATER TREATMENT

The majority of EDCs are highly lipophilic and unlikely to be present in significant quantities in water, tending to sorb to solids present in watercourses. Such qualities increase the probability of removal during DWT. Once in the DWT, the strongest influences on EDCs are considered to be their biodegradability and physicochemical properties.[42]

The conventional process involves flocculation, sedimentation, and filtration. Lipophilic pesticides (Cl-pesticides) can be partially removed during this process illustrated by alachlor, decreasing by 24% after flocculation, sedimentation, filtration, and chlorination. Conversely, more polar pesticides are not significantly reduced by these practices.[43] EE2 has been shown to bind rapidly to sludge flocculent, remaining stable provided the floc does not break down. E2 disappears rapidly in approximately 10 minutes from sludge forming E1. The inactive conjugate, estrone-3-sulfate (E1–3S,) is removed within 12 hours with only limited formation of estrogenic E1.[44]

Microbiological treatment processes using conditioned or adapted bacteria have been investigated at the laboratory or pilot scale. However, there is no evidence as yet for atrazine removal across slow sand filters.[45] The sensitivity to toxic matter and the increased skill required for handling this treatment method makes this an unattractive option in spite of lower investment and running costs.

Measuring toxicity with a Microtox toxicity analyzer, O_3 was shown to reduce acute toxicity while increasing pH resulted in greater toxicity reduction.[46] O_3 is used primarily for biological contaminant removal, but has been shown to reduce steroid estrogens.[47] Comparison of O_3 and O_3 with UV irradiation led to greater decomposition of 2,4-dichlorophenoxyacetic acid (2,4-D) using the combined applications than the sum of use individually.[48] The combination of O_3 combined with H_2O_2 in comparison to O_3 alone can lead to a greater reduction in pesticide concentrations. However, this is dependent on factors such as the mass ratio of H_2O_2/O_3 and pH.[49,50]

Using O_3 on pesticides identified in the Onga River, Japan, a dose of 1 mg l^{-1} achieved less than 10% reductions in the concentrations of simazine and atrazine. Increasing the dose to 5 mg l^{-1} resulted in 85% removal for each pesticide.[51] Water treatment pilot plant studies investigated selected pesticides, APs, and ethoxylates. The reaction time with O_3 for atrazine is less than other pesticides, requiring higher doses of O_3 to increase elimination. The percentage removed by ozonation was shown to depend on the O_3 dose, contact time, and pH.[52]

GAC is the treatment usually selected for organic contaminant removal.[53] For contaminants covered in the National Primary Drinking Water Standards, the recommended water treatment is activated carbon. In 1962, the U.S. Public Health Service measured a variety of organic compounds (natural, synthetic, toxic, and nontoxic) in water by passing them through GAC.[15] Developed after WWII, GAC is considered the main process for the removal of many organic compounds. Activated carbon is able to adsorb PCBs, PAHs, and pesticides.[54] Atrazine is an apolar pesticide and can be well removed by GAC filtration.[55] A variation, PAC, is considered not to be as effective as GAC for the removal of pesticides.[56] The performance of different carbon types revealed that bituminous coal based GAC was superior for pesticide removal. The bed length and contact time are important factors in organic pollutant removal. The longer the empty bed contact time (EBCT) or bed length, the greater the percentage removal.[57,58]

Reverse osmosis (RO) has been used since the late 1950s for desalination. The performance of four different membranes in removing a feed concentration of 1 µg l^{-1} each of simazine, atrazine, diuron, bentazon, and dinosed were tested. The membrane material in the retention mechanism was identified as the component most responsible for removal. The most effective membrane proved to be PVD1, which rejected the above pesticides by more than 90%.[59] The different removal methods used for a range of EDCs are summarized in Table 7.2. Advanced oxidation is favored, especially in the presence of a catalyst and adsorption processes in the form of GAC.

Removal processes working in combination generally offer the greatest removal efficiency. Ozonation and GAC are sufficient for the removal of high concentrations of APs and alkylphenol ethoxylates (APEOs) from water.[52] A report by Walker[60] (see Section 7.5, Case Study) concluded that minute traces of estrogens could be removed by the culmination of:

1. Dilution and self purification in rivers and reservoirs
2. Activated carbon and UV radiation in the treated wastewater recycling plant
3. O_3 and activated carbon in the water treatment plant

Research on removal methods of organic contaminants has primarily focused on pesticides, with APEOs and steroids to a lesser extent. The best removal processes for EDCs are oxidation and adsorption. The EPA and National Primary Drinking Water Standards advocate GAC. EDC removal by O_3 is dependent on dose, contact time, and pH. Though the EDC may be gone, daughter products are often formed that require identification to determine their toxicity. Advanced oxidation using UV irradiation with TiO_2 catalyst is able to mineralize many EDCs.

Polybrominated flame retardants (PBFRs) have not been investigated in drinking water. However, their low solubility, especially with respect to higher congeners, means any presence would very likely associate with the solids and be removed during treatment. VTG induction in fish is one of the main indictors of estrogenicity. Several U.K. water companies maintain caged fish in the vicinity of sewage effluent outflow for testing. Such an approach would be advantageous in reservoirs or by abstraction points for drinking water treatment.

Are levels in drinking water of concern to humans? Assuming low concentrations of EDCs may be present in drinking water, effects would have to be cumulative. More research on the cumulative effects of mixtures in real life situations are required before the risks to humans through drinking water contaminated with EDCs can be defined.

The chlorination process has been implicated as increasing mutagenicity and certain EDCs in drinking water. Disinfection, which leaves a residual, is paramount for protection of water supplies from pathogens. Perhaps, processes such as RO, which have been shown to decrease mutagenic response, are required after disinfection.

Certain EDCs, albeit in trace concentrations, have been identified in drinking water. Relating EDCs and endocrine disruption effects to human health is marred by difficulty. There are numerous pathways for pollutants to interact with humans, and controls cannot be utilized. One study identified a significant decrease in sperm density for men residing within the Thames Water supply area compared to those outside this locale.[81] This study assumed that the level of drinking water was uniform irrespective of likely pollutant loading, drinking water source, or the percentage of water reuse.

According to the Royal Society of Chemistry, while there is currently no direct evidence to support an association between exposure to EDCs and any reported effects on humans, there have been few, if any, studies that have attempted to look for evidence. However, it has been uniformly agreed that the processes involved in water treatment remove estrogenic chemicals. It is concluded that drinking water is not a significant source of exposure to estrogenic chemicals to man.

TABLE 7.2
Removal Methods for EDCs during Drinking Water Treatment

EDC	Removal Method	Treatment Process	Comments
Alkylphenol polyethoxy carboxylates (APnECs)	Reverse osmosis	MF	Highly hydrophilic and persist through lime addition, coagulation, rapid sand filtration, activated carbon adsorption, and chlorination; efficiently removed by reverse osmosis.[61]
Alkylphenols (APs)	GAC	AP	Advocated by the EPA, with PAC being applicable for systems that include mixing basins, precipitation, or sedimentation and filtration[27]
	UV and catalyst		For UV only no change in concentration was observed; with the titanium dioxide (TiO_2) catalyst, 90% decomposition occurred within 60 minutes; octylphenol was the least stable with 90% degradation in <20 minutes; after 5 hours, 80% of initial APs was completely mineralized[62]
Alkylphenol ethoxylates (APEOs)	GAC	AP	Advocated by the EPA, with PAC being applicable for systems that include mixing basins, precipitation or sedimentation and filtration.[27]; nonylphenol ethoxylates not always reduced due to saturation with competitive adsorption favoring other contaminants in the water sample[63]
Bisphenol A (BPA)	UV with catalyst	AO	UV and TiO_2 catalyst resulted in complete mineralization within 20 hours; estrogenic activity decreasing to <1% of initial BPA activity within 4 hours[64]. 90% decomposition occurred within 50 minutes; after 3 hours, 90% mineralization was achieved[62]
17β-estradiol (E2)	UV with catalyst	AO	No change observed for only UV irradiation; 90% reduction after 2 hours of UV and TiO_2 catalyst[62]
	Filtration (sand)	C	Readily transported through sand with 85% in the effluent; some degradation occurred forming a metabolite[65]
Estrone (E1)	UV	AO	20% decomposition was observed.[62]
17α-ethinylestradiol (EE2)	UV with catalyst	AO	With UV and TiO_2 catalyst, photodegradation was faster than for E2; 90% decrease in original concentration for the two steroids occurred within 30 minutes[62]

Compound	Process	Type	Notes
Tetrabromobisphenol A	Filtration (sand)	C	Extensive sorption with only 4.5% identified in the effluent[66]
Pentachlorophenol	GAC	AP	GAC fluidized bed reactor at EBTCs 2.3 hours; Anaerobic degradation to chlorophenol (>99%) with second stage aerobic for complete removal of chlorophenol[67]; adsorption decreases with increasing temperature (10 to 60°C) and decreasing pH (6–11). desorption required for regeneration increases with increasing temperature[68,69]
	Reverse osmosis	MF	Ultra-low pressure RO membrane rejects PCP by over 90%[70]
Atrazine	Ozone	AO	No hydroxy derivatives were observed; 30 minutes required to reach 60% degradation[71]
	Ozone/Hydrogen peroxide	AO	2 minutes were needed to reach the same level of degradation as ozone only; with raw water levels of 0.1 μg l^{-1}, the new EU regulation cannot be met by ozone and ozone/hydrogen peroxide[71]
Lindane	UV with catalyst	AO	Mineralization by TiO_2-UV photocatalytic degradation[72]
2,4-D	UV	AO	Under laboratory conditions, 70 μg ml^{-1} 2,4-D is reduced to 20 μg ml^{-1} in 10 hours; in the same time, 30 μg ml^{-1} degrades to 5 μg ml^{-1} due to photolysis[73]
Methoxychlor	GAC	AP	Wide range of water treatment processes tested and GAC determined the best removal meeting the maximum contaminant level (MCL) of 0.1mg l^{-1}; GAC is the best available technology (BAT) advocated by the EPA[27]
Endosulfan	GAC	AP	GAC is the BAT advocated by the EPA; for small water systems, PAC may be used[27]
DDT	GAC	AP	GAC is the BAT advocated by the EPA[27]
Isoproturon	Hypochlorite	AO	Forms 4 chlorinated and/or hydroxylated ring substituted derivatives; reaction was faster than observed for chlorine dioxide[74]
	Chlorine dioxide	AO	Forms two hydroxylated aromatic ring substituted derivatives[74]
Benzo[e]pyrene (BeP) Benzo[k]fluoranthene (BkF)	Ozone	AO	Two-stage O_3 system (retention time 10 minutes) formed oxidation products though no mutagenicity detected; aerobic biodegradation eliminated ozonation products within one hour[75]

(continued)

TABLE 7.2 (continued)
Removal Methods for EDCs during Drinking Water Treatment

EDC	Removal Method	Treatment Process	Comments
Benzo[a]anthracene (BaA)	Ozone	AO	Varying ozone dosages used; 15 oxidation products resulted.[76]
BaA, BbF, BkF, BaP	UV/Ozone	AO	Destroyed by more than 90% for concentrations between 200 ng l⁻¹ and 12 µg l⁻¹; superior to UV or ozone only treatment[77]
Dibenzo[a,h]anthracene	UV	AO	No significant effect on the PAHs[77]
Phenanthrene	Ozone with catalyst	AO	Baked sand acted as a catalyst removing 90% of which 60% degraded in the first minute[78]
Polychlorinated dibenzo-p-dioxins (PCDDs)	Ozone then powder sorbent	AO and AP	O₃ followed by filtration through powder sorbents removed majority of di-, tetra- and penta-CDDs; 30% to 60% of hexa- and hepta-CDDs remaining[79]
Polychlorinated dibenzofurans (PCDFs)	Filtration through granular sorbents	AP	Filtration through granular sorbents removed 90 to 95% of PCDDs and PCDFs[79]
	Coagulation	C	In coagulation sludge, PCDDs present at higher concentrations than PCDFs; PCDD is congener dependent favoring larger congeners[32]
	GAC	AP	Similar pattern to coagulation sludge, majority of PCDDs an PCDFs having been removed during coagulation.[32]
	UV	AO	No significant degradation observed[77]
	Ozone	AO	No significant degradation observed[77]
	UV/Ozone	AO	No significant degradation observed[77]
2,3,7,8 TCDD	Coagulation, sedimentation and filtration	C	Coagulation, sedimentation, and filtration would be very effective as advocated by the EPA due to low solubility and so preference to the solid phase when entering water treatment.[27]
Diethyl phthalate	GAC	AP	GAC is the BAT advocated by the EPA[27]; 6 GAC evaluated and bituminous coal was the most efficient GAC for removal.[63]

Di-(2ethyl hexyl) phthalate (DEHP)	GAC	AP	GAC is the BAT advocated by the EPA; the Freundlich coefficient K gives GAC removal for a chemical; values >200 are economically feasible; at 8308 $\mu g\,g^{-1}$ (l μg^{-1})$^{1/n}$, this was the highest value for 130 chemicals tested [27]
PCBs	GAC	AP	GAC is the BAT advocated by the EPA[27]
	UV	AO	Some highly chlorinated PCB congeners were resistant to short duration of UV, requiring 300 minutes of photolysis to be completely destroyed; dechlorination is the major photolytic mechanism[80]

Note: C = conventional treatment — coagulation, flocculation, sedimentation, filtration; AO = advanced oxidation; AP = adsorption process; MF = membrane filtration.

7.3 ENDOCRINE DISRUPTERS AND DRINKING WATER QUALITY LEGISLATION

Several EDCs, especially pesticides, have legislation concerning their production, use, or concentration in water resources. However, this has not been on account of their endocrine disrupting properties. Drinking water standards provide limits for unacceptable risks from selected contaminants based on current health information, which do not include endocrine disruption.[82] Recently, legislation is starting to recognize and subsequently regulate water quality contaminants based on their endocrine disrupting properties. Most attention has been toward synthetic chemicals when considering an impact on human health or the environment. However, natural chemicals can be significant sources of EDCs, such as estrogens and phytoestrogens. Legislation can only be implemented when risk has been determined, and EDCs is an area that still requires further research in order to quantify these risks.

There are three approaches to legislation pertaining to drinking water:

1. Legislation directly regulating drinking water such as the Safe Drinking Water Act 1974 (SDWA) and the European Directive on Drinking Water.
2. Legislation and regulations regarding the water sources used for drinking water (surface water, groundwater) such as Surface Water Abstraction and 1986 and 1996 amendments to the SDWA.
3. Safeguarding water sources from pollution can also be from the viewpoint of the compound. Substances found to have adverse effects to humans or the environment may be banned or production greatly reduced such as the case for APs, tributyltin (TBT), and numerous pesticides.

7.3.1 EUROPEAN LEGISLATION AND POLICIES

The European Parliament has called for a precautionary approach to the endocrine disrupting issue. There is a wide legislative framework that can be amended to control EDCs (Table 7.3). Some recent directives include direct references to EDCs. Integrated Pollution Prevention and Control (96/61/EEC) includes a special reference to substances that "may effect reproduction in or via the aquatic environment." EC Directives are not legally enforceable within member states and require implementation into national legislation, no less stringent than agreed in the directive. Several programs (COMPREHEND) on a national and international basis are investigating EDCs using bioassays, chemical analysis and quantitative structural activity relationships.

7.3.1.1 EC Dangerous Substances Directive

The Dangerous Substances Directive (76/464/EEC) and amendments categorize substances into two lists. List I substances are particularly dangerous because of their persistency, toxicity, and bioaccumulation potential.[83] List II substances are less dangerous, but may have a deleterious effect on the aquatic environment. Pollution of List I substances need to be eliminated, while List II substances must be reduced. For control of these substances, most of Europe uses Uniform Emission Standards

TABLE 7.3
Legislation and Regulations that Can Be Amended to Control EDCs

Type of Control	Legislation and Regulations
Water Quality	Habitats Directive (92/43/EEC)
	Surface Water Abstraction Directive (75/440/EEC)
	Groundwater Directive (80/68/EEC)
	Drinking Water Directive (80/778/EEC)
Industrial Sector	Sludge Directive (86/278/EEC)
	Urban Wastewater Treatment Directive (91/271/EEC)
	Integrated Pollution Prevention and Control (96/61/EC)
	Sludge (Use in Agriculture) Regulations 1989 (SI 1263)
	Marketing and Use Directive (76/769/EEC)
Substances	Detergent Directive (76/769/EEC)
	Environmental Protection (Prescribed Processes and Substances) Regulations 1991 (SI 472)
	Dangerous Substances Directive (76/464/EEC)
	Biocidal Products Directive (98/8/EC)
	Control of Pesticides Regulations 1986 (SI 1510)

(UES), while the U.K. uses Environmental Quality Standards (EQS).[84] UES values are expressed as the final concentration in effluent and as the load per production unit that may be discharged. EQS values vary depending on the use to be made of that water body (e.g., use for drinking water abstraction). The EQS is given as the concentration, taking into account the dilution factor or the presence of other inputs into the same water body.

The EQS and Environmental Quality Objective (EQO) contribute to the implementation of the EC Framework Directive for Water Policy (2000/60/EC). EQSs may also specify the highest or lowest occurrence in surface water and groundwater of organisms that can serve as indicators of the state of the environment. The levels specified in EQSs must be complied with after a specified date. Agreed List I substances that are also EDCs include lindane, PCP, endrin, and dieldrin. Endosulfan is a List II substance and has a statutory EQS in surface waters since 1997.[85] The Environment Agency in the U.K. has several operational EQSs for substances, including dioxins, and EQSs for other EDCs are in development.[85,86] Currently, TBT is the only substance to have an EQS derived on the basis of endocrine disruption.

7.3.1.2 EC Water Framework Directive

The EC Water Framework Directive (WFD) (2000/60/EC) complements several recent directives including the Integrated Pollution Prevention and Control, Urban Wastewater Treatment Directive, and the new Drinking Water Directive.[87] The directive covers surface waters, groundwater, and coastal waters. Implementation into national law is required by December 2003.[88] Water quality objectives for these water bodies of good status are to be achieved by 2015. Two components of the WFD that may be amended for EDCs in waters used for drinking water abstraction are:

1. Dealing with the water quality pressures from point and diffuse sources in order to achieve good status
2. Addressing pollution control by controlling the sources by both emission limit values and quality objectives to suit the receiving waters

A list of priority substances has been established to replace the list of dangerous substances, which may be included in List I of the Dangerous Substances Directive. Identification of these substances include EDCs identified under the Oslo and Paris Commission strategy.[89] The list of priority substances includes brominated diphenyl ethers, mercury and its compounds, and APs.

7.3.1.3 EC Drinking Water Directive

The EC Drinking Water Directive (80/778/EEC) and new Drinking Water Directive (98/83/EC) set a maximum admissible concentration of 0.1 µg l^{-1} for any individual pesticide in drinking water and 0.5 µg l^{-1} for the total presence of pesticides including toxic metabolites. The two concentrations have no toxicological significance, but are useful for trend analysis.[3] The concentrations 0.1 µg l^{-1} is the standard for any pesticide specified in the EC Drinking Water Directive and therefore is relevant to treated supplied drinking water rather than environmental waters. Generally over 90% of determinations for pesticides are below the limit of detection (LOD), but some pesticides can be toxic at very low levels, possibly at or below their LOD. The LOD may vary between pesticides and also by sample media and analysis method. Annex 15 states that while there is increasing concern regarding the potential impact of EDCs on humans and wildlife, evidence is insufficient to base parametric values for ECDs at the community level.[90]

7.3.1.4 EC Endocrine Disruption Strategy

The EC's paper proposed a priority list of potential endocrine disrupters as an aid to environmental and product monitoring, followed by amendments to existing legislation.[91] A community strategy for endocrine disruption included a draft priority list of chemical substances.[92] Inclusion was based on criteria of persistence in the environment, production volume, scientific evidence of endocrine disruption and human/wildlife exposure. A final list was published in 2001[93] and is currently being used to identify:[94]

1. High priority substances for further testing when methods become available
2. What can be addressed under present legislation
3. Gaps in knowledge
4. Vulnerable consumer groups

7.3.2 U.S. Legislation and Policies

The influence of European experiences with water supply systems motivated the change in U.S. attitude toward water supplies.[7] Similar to the European approach, water purity was traditionally driven by concerns relating to microorganisms. During the 1930s, with the creation of thousands of SOCs and large quantities entering watercourses, the U.S. began considering chemical contamination. The EPA currently estimates that 40% of U.S. watercourses fail to meet federal clean water standards. The U.S. approach to EDCs has been more proactive than Europe's. As far back as 1986, California passed the Safe Drinking water and Toxic Enforcement Act (Proposition 65), requiring publication of chemicals known to cause cancer and reproductive toxicity.[95]

7.3.2.1 Clean Water Act of 1972 and Amendments

The 1972 Federal Water Pollution Control Act (later renamed the Clean Water Act [CWA]) was the turning point in U.S. water quality legislation. The goal was the "restoration of the chemical, physical and biological integrity" of U.S. water resources.[96] Objectives set in the CWA aimed to decrease the quantity of pollutants discharged to surface waters with water quality in navigable waters suitable for swimming and fishing by 1983, secondary treatment for sewage treatment works by 1988, and zero pollutants discharge goal by 1995. These objectives were not met in the time frame and the mechanisms did not account for diffuse sources of pollution, which greatly impacts on groundwater quality. Hence, the 1987 amendments to the CWA attempted to deal with nonpoint sources of pollution. The total maximum daily load (TMDL) rule set limits on pollution in individual watercourses based on its use, whether it is fishing, boating, swimming, or a source for drinking water. The U.S. Government has postponed implementation of a new TMDL rule devised by the EPA until 2003.[97]

7.3.2.2 Safe Drinking Water Act of 1974 and Amendments

The purpose of the SDWA, first passed in 1974, is to protect the U.S. drinking water from biological and chemical contamination. Responsibility was given to the EPA for establishing quality standards (National Primary Drinking Water Standards) and treatment requirements for drinking water.[98,99] Amendments in 1986 established maximum contaminant level goals (MCLGs) for substances, primarily SOCs, present in public water systems. MCLGs are set to a level expected to cause no known or anticipated adverse effect on public health throughout a lifetime of exposure. Depending on the expected effect of a substance, the driver being the carcinogenic capability of the compound, MCLGs can be assessed in three ways:

1. For non-carcinogens, MCLG is determined on the reference dose (based on no observed effects level).
2. For carcinogens, MCLG is set to zero, based on the zero threshold assumption.
3. For suspected carcinogens, MCLG is calculated using a reference dose and an added safety margin.

Future maximum contaminant limits (MCLs) may be set at lower concentrations than at present if adverse endocrine disrupting effects are observed because of their presence.[27] Once a MCLG is set, the EPA determines a MCL, which unlike the MCLG is legally enforceable. MCLs are as close as feasible to the MCLG taking into account cost and best available technology (BAT). Although the EPA can specify the BAT, its use is not mandated. MCLGs and MCLs are available for several EDCs because of their carcinogenic properties rather than EDC capabilities of which carcinogenicity is only one aspect. The SDWA contaminant regulations are established for 38 (Phase II) SOCs and 23 (Phase V) inorganic synthetic chemicals (IOCs).[100] However, MCLs do not account for more than one compound present in a drinking water sample or possible synergistic effects.[101] Amendments to the SWDA in 1996 and the Food Quality Protection Act (FQPA) included specific language pertaining to endocrine disruption. Section 136 cites screening for drinking water contaminants possessing estrogenic and other endocrine-disrupting effects. If such substances pose a threat to human health, then regulatory action under current legal authorities would be required.

The EPA is required by the 1996 amendments to publish lists of unregulated chemical and microbial contaminants and contaminant groups every 5 years. These chemicals are known or anticipated to occur in public water systems and may pose a risk to drinking water. Published in March 1998, the first drinking water Contaminant Candidate List (CCL) was devised from 8 existing chemical lists within the EPA and narrowed down from 391 to 50 chemical and chemical groups. In developing the draft CCL, the EPA initially prepared a list of contaminants suspected of having adverse effects on endocrine functions of humans and wildlife.[102] Section 136 led to 206 chemicals on the draft CCL being included in the Endocrine Disrupter Priority Setting Database.[103] The EPA withdrew 21 contaminants from the draft CCL based solely on the possibility of their being EDCs. Several contaminants implicated as EDCs, such as dieldrin, are included on the CCL, though for reasons other than being an EDC. The 1998 CCL contained 20 industrial organic chemicals, 22 pesticides, and 6 inorganic chemicals.

7.3.2.3 Endocrine Disrupter Screening Program

In 1996, through the FQPA and SDWA Amendments, the EPA implemented a screening program to determine whether certain substances may have an effect (behavioral, reproductive, or developmental) on humans, similar to that produced by the natural steroid estrogen.[104] EDSTAC made several recommendations and in 1998 the Endocrine Disrupter Screening Program (EDSP) was established.[105] The recommendations were to:

1. Address the effects to both humans and the wildlife.
2. Examine the effects on biological processes.
3. Include pesticides, commercial chemicals, and environmental contaminants.

The SDWA defers to the FQPA as the principal mechanism for developing the screening program. However, the SDWA states that the EDSP should evaluate drinking water source contaminants to which a significant number of people could be exposed.[106] The EDSP is two-tiered (screening and testing) and will also investigate the low-dose issue of EDCs occurring at levels substantially below the no observed effect levels followed in traditional studies.

7.3.2.4   Toxic Substances Control Act of 1976 and Amendments

Implemented by the EPA, companies can be required to assess chemicals for toxic effects. From inception in 1979, more than 30,000 new chemicals have been reviewed to determine whether the manufacture, processing, distribution, use, or disposal may present an unreasonable risk to human health or the environment. Under the new chemical review program, the EPA identifies potential drinking water contaminants from the chemical's physicochemical properties, estimates on the amount released, removal during the sewage treatment process, fate and transport in the environment, and the extent of dilution by the receiving environmental medium.[107]

7.4   ENDOCRINE DISRUPTERS AND WATER REUSE

7.4.1   Water Reuse: An Overview

Water reuse often completes the cycle between receiving waters for effluent and water abstraction for potable or nonpotable use, illustrated in Figure 7.2. Water reuse accounts for up to one third of water taken from lowland rivers containing domestic and industrial effluent. The concept of water reuse originated in Greece more than 2000 years ago when crops were irrigated with sewage effluent.[108] In Europe, reused water is primarily for irrigation (agricultural and recreational) followed by industrial

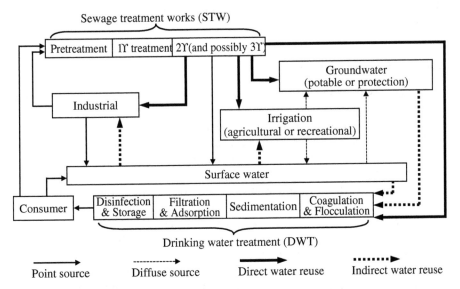

FIGURE 7.2   Relationship between point and diffuse sources, water and wastewater treatment, potable water reuse, and several applications of nonpotable water reuse.

practices. Water resources are finite, and contamination in one source without adequate wastewater or water treatment may lead to pollution of other sources. Reuse is thought to amplify levels of contaminants by keeping them within the cycle.[109]

In the U.K., the unplanned practice of indirect water reuse during the 19th century led to epidemics in waterborne diseases such as typhoid and cholera. In 1854, Dr. John Snow identified a drinking water pump on Broad Street, London, as the source of a cholera outbreak.[110] When the link between water supply and disease was discovered, treatment processes for wastewater were improved and chemical disinfection of drinking water supplies became standard practice. The principle of taking the highest quality water source and protecting it from contamination was also implemented.[15] Similar to drinking water, water reuse standards and regulations have been driven by health implications arising from the presence of microorganisms. However, organic contaminants with the potential to harm public health are becoming increasingly important.[111] Both conventional and reused water may hold significant quantities of organic contaminants that possess toxicological properties.[112] Several water reuse projects have been monitored for contaminants beyond those currently required by regulations. For some projects, the issues of EDCs have been raised during public discussions.[113,114]

7.4.1.1 The Need for Water Reuse

Water resources are finite and unevenly distributed in relation to demand and supply. As the human population expands and their demand for water increases, the need for reusing water becomes necessary. In certain areas of the world, water resources are able to meet the demand.[115] However, in parts of the U.S. where demand has outstripped supply, water has had to be imported from other regions or salt water desalinated. In comparison, from an economical and sustainable viewpoint, water reuse is an attractive option.

7.4.1.2 Types of Water Reuse

Water reuse is the utilization of treated wastewater by either an indirect or direct approach, for potable (drinking) or nonpotable use. The type of water reuse, be it direct or indirect, may influence the concentration and behavior of EDCs present in that water system. Water reuse can also be *planned* or *unplanned*. Planned water reuse would be the purposeful augmentation of a water supply source with reclaimed water derived from treated wastewater. Unplanned reuse is the unintentional addition of wastewater, whether treated or not, to a water supply that is subsequently used as a water source. The cities of Philadelphia, Cincinnati, and New Orleans draw water from the Delaware, Ohio, and Mississippi Rivers, respectively. These rivers are also the recipients of wastewater effluent, thus unplanned indirect potable reuse occurs.[116]

Indirect potable water reuse is the return of highly treated wastewater into an impoundment, such as a reservoir or the natural environment.[117] This allows for dilution and mixing with other waters for a period of time and also gives the opportunity for natural processes to provide additional treatment such as sunlight (surface waters) and filtration through soil (to groundwater). The blended water is then diverted to a water treatment plant prior to distribution. Environmental intervention also increases public confidence through the perception of clean and safe water source. Indirect potable reuse manages EDCs and other contaminants by a culmination of:

1. Secondary and tertiary sewage treatment
2. Dilution and transformation processes in receiving water
3. Removal during drinking water treatment

Indirect nonpotable water reuse has several planned intermediate stages between use and nonpotable reuse. The level of treatment for nonpotable water reuse need not be to the same standard required for potable applications. Dilution of effluent in the watercourse is an important factor that can only occur with indirect water reuse, allowing concentrations of EDCs present in the effluent to be diluted. However, if high amounts of effluent are discharged into surface waters, then little dilution can occur. In the U.K., the Rivers Thames, Lea and Ouse can contain up to 50 to 60% sewage effluent. During dry periods the proportion of total flow to sewage effluent content can be greater.[30]

Direct potable water reuse introduces treated wastewater (after extensive processing beyond the usual wastewater treatment) directly into the water supply system with no environmental intervention.[113] Reclaimed wastewater may be blended with other sources of water to form the final drinking water product.[117] Direct reuse both augments the water supply and protects the receiving waters from potential pollution. Currently, because of health concerns and public perception, this is not viewed as a viable option for potable supply. However, in Windhoek, Nambia, a direct potable water reuse project has been supplementing the potable water supply for the past 25 years.

Direct nonpotable water reuse is treated wastewater piped directly to use without the intervention of the environment or injected in ground water and then extracted further down direct to the user. These methods help industries in recycling water reuse on site. The advantages are from an economic, public perception, and environmental perspective. Other applications of the reused water include the protection of saltwater intrusion and augmentation for agricultural, urban, or recreational use.

7.4.1.3 Applications for Water Reuse

The majority of water reuse projects are for nonpotable applications. Table 7.4 illustrates several water reuse applications and potential concerns related to each application. Irrigation, groundwater recharge, and potable reuse are discussed in further detail. Table 7.5 summarizes several planned water reuse projects for different applications. Several of the projects utilized toxicity testing to monitor for adverse effects. However, current toxicity tests do not reliably detect EDC adverse effects of sexual differentiation, behavior, or reproductive development as the cause and consequence may be separated by a large period of time.[118]

TABLE 7.4
Water Reuse Applications for Wastewater and Potential Concerns Related to Each Application

Water Reuse Application			Potential Concerns for Reusing Wastewater					
			Negative Public Perception	Toxicity to Aquatic Biota	Eutrophication in Receiving Water	Presence of Pathogens	Presence of Heavy Metals	Presence of Toxic Organic Compounds
Potable	Direct	Pipe to pipe	✓			✓	✓	✓
	Indirect	Groundwater recharge (storage)	✓			✓	✓	✓
		Surface water discharge	✓	✓	✓	✓	✓	✓
Nonpotable	Protection	Groundwater recharge				✓	✓	✓
	Agricultural	Fodder crop irrigation				✓	✓	✓
		Edible crop irrigation	✓			✓	✓	✓
		Nurseries	✓	✓	✓	✓	✓	✓
	Industrial	Cooling						
		Washing						
		Construction						
		Stack scrubbing						

Recreational
- Landscape irrigation ✓ ✓ ✓ ✓
- Boating and fishing ✓ ✓ ✓ ✓
- Bathing ✓ ✓ ✓

Urban
- Fire protection ✓ ✓
- Toilet flushing ✓ ✓
- Air conditioning ✓ ✓
- Street/car washing ✓ ✓ ✓ ✓

TABLE 7.5
Selection of Planned Wastewater Reuse Projects that Have Been Implemented Worldwide

Water Reuse Project	Site	Operation	Planned Reuse Type		Planned Application	Comments
			Direct/Indirect	Potable/Nonpotable		
Cedar Creek	New York, U.S.		Indirect	Potable	Groundwater recharge	Groundwater sole source for drinking water in that area[119]
Denver Water Department Research Project	Denver, U.S.	1979–1990	Direct	Potable	Pilot plant: Augment drinking water supply	Demonstration plant to investigate viability of direct potable reuse[120]
Essex and Suffolk Water	Chelmsford, U.K.	1997–1998	Indirect	Potable	Temporary: Augment drinking water supply	Unprecedented estrogenic investigation including observing VTG induction in male trout and chemical/bio analysis (see Section 7.5, Case Study)
Lake Arrowhead Wastewater Reuse Pilot Plant	California, U.S.	3 years	Indirect	Potable	Pilot plant: Augment drinking water supply	Membrane technology used[121]; removal of base neutral organic contaminants such as pyrene investigated[122]
San Diego Aqua II	California, U.S.	Late 1980s to early 90s	Direct	Potable	Pilot plant: Augment drinking water supply	Health effects investigated; chemical risk assessment, epidemiological study, and genetic toxicity testing

Water Factory 21	California, U.S.	1976–present	Direct	Nonpotable	Groundwater recharge: barrier to saltwater intrusion	Also unplanned indirect potable reuse as ≤ 50% augments the water supply used for potable abstraction
West Basin Water Recycling Plant	California, U.S.	1995–present	Indirect	Nonpotable	Industrial/irrigation and groundwater recharge	Recharge used for saltwater protection; 2 processes: Title 22 and Barrier system; investigated base neutral organics
Whittier Narrows	California, U.S.	1962–present	Direct	Nonpotable	Groundwater replenishment	Plans for expansion to indirect potable reuse are constrained by health effects from trace organics in effluent
Windhoek Reclamation Project	Nambia, Africa	1975–present	Direct	Potable	Augment drinking water supply	Biomonitoring, toxicity testing, and epidemiological study on human health

7.4.2 IRRIGATION

Nonpotable reuse for agricultural irrigation has been practiced for centuries. The concept is gaining acceptance because of the water resource imbalance between supply and demand. If irrigation is carried out with wastewater, any contaminants present in the effluent may accumulate in sufficient concentrations in crops, entering the food chain.[123] The migration of chemicals through the soil to groundwater sources is one of the main diffuse sources for contamination. EDCs applied to land also have the potential to contaminate underground water sources that may be used as sources for drinking water treatment. The efficacy of irrigation with treated wastewater is dependent on the site specific crop, soil, and climatic conditions.[124,125]

In Israel, the storage of wastewater in deep seasonal reservoirs prior to irrigation has been practiced since the 1970s. At the reservoir inlet, NP and OP concentrations of 94 and 82 μg l^{-1} have been observed. With a residence time of 48 hours, NP and OP levels at the outlet pipe were decreased to 66 and 51 μg l^{-1}.[126] Transformation and dilution processes are responsible for the decrease, but significant levels of these APs will be present in the water applied to land. There is the potential for uptake of EDCs in plants arising from sewage effluent or sludge application. Harms demonstrated that 4-nonylphenol (4-NP) accumulated to a small degree in plants. When taken up by the root system, the majority of 4-NP remained in that section of the plant.[127] Limited microbial degradation of 4-NP occurred in the soil, with concentrations in excess of 5mg ml^{-1}, proving toxic to the microorganisms.

The potential for endocrine disruption in water reuse has been raised by Anglian Water, U.K. A project was commissioned on APs, specifically NP and its ethoxylates (NPEO) in the Kings Lynn area. Endocrine disruption induced by NP and NPEOs was taken from the perspective of water reuse for agricultural and industrial purposes.[128] Water reuse for irrigation has identified estrogen contamination between 1.3 and 0.1 nmol l^{-1}.[129]

California law states that reused water for irrigation must be to tertiary treatment standard because it is in direct contact with food crops. Up to 80% of the wastewater reused in California is used for irrigation.[130] Possible mechanisms of food crop contamination are:

1. Contaminant build-up on crop through repeated application and evaporation
2. Uptake via the roots from the soil receiving reused wastewater
3. Uptake from the foliage

7.4.2.1 West Basin Water Recycling Plant

Initiated in 1995, West Basin treats water in two ways depending on application: Title 22 and Barrier systems. Title 22 processes are tertiary treatment used for indirect nonpotable applications and require high quality water, such as irrigation and certain industrial processes. Barrier water is for groundwater protection against saltwater intrusion, the quality of blended imported and reused water exceeding EPA drinking standards.[131] Both treatment processes are shown in Figure 7.3. The fate of base neutral compounds was monitored in the Barrier treatment, which utilized lime clarification.

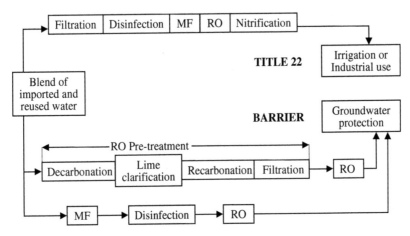

FIGURE 7.3 Treatment process for the West Basin Water Recycling Plant (MF = microfiltration; RO = reverse osmosis).

For DEHP in Title 22 water, levels were 0.8 µg l⁻¹ with no detection in Barrier water. After RO pretreatment, there was a 25% increase in base neutral compounds, followed by 70% removal efficiency following RO. After lime clarification (RO pretreatment), pesticides and phenols increased in solution due to increasing pH. However, this effect was reversible when the pH decreased during recarbonation (pH 7).[132]

7.4.3 GROUNDWATER RECHARGE

The method of recharge with treated wastewater is either direct injection or via infiltration through to the groundwater. Infiltration requires large areas of land with tertiary treatment carried out through the soil strata; this is known as soil aquifer treatment (SAT). SAT can be an excellent method of purifying effluent depending on the geological conditions.[133] For SAT, the majority of contaminant removal occurs within the top 6 feet of the vadose zone.[134] Major purification processes occurring in SAT are slow sand filtration, chemical precipitation, adsorption, ion exchange, biodegradation, and disinfection.[135] For direct injection, a higher quality of treated wastewater is needed to protect both the groundwater source and prevent matter obstructing the injection wells. The degree of treatment is dependant on application, soil geology, dilution available and retention time until extraction.[136] The Rio Hondo spreading basin in Los Angeles and the Livermore Valley Project in San Francisco are examples of groundwater recharge by infiltration and direct injection, respectively.[113] The applications for artificial groundwater recharge can be replenished or protected and include: [137,138]

1. Saltwater intrusion barriers in coastal aquifers
2. SAT for future reuse
3. Storage of sewage effluent
4. Controlling ground subsidence
5. Replenishment for potable or nonpotable supply

Once recharged, there is a retention time in the ground to allow for dilution and transformation processes. Many EDCs are persistent and resist degradation, so these contaminants may be present when the water source is reused. Chlorinated hydrocarbons are not readily removed during infiltration.[139] Groundwater recharge projects in the U.K. at chalk and gravel sites concluded that further investigations on the removal of trace organic chemicals are required to fully assess any implications to human health.[140] Recharge of secondary effluent (activated sludge) to an aquifer in Israel by SAT removed 97% of nonionic surfactants; however, concentrations of 22 to 25 mg l^{-1} were still identified in reclaimed water.[141]

7.4.3.1 Californian Water Factory 21 Project

During the 1960s, the Orange County Water District (OCWD) began a pilot plant scale reclamation project that has developed into Water Factory 21. The source of injection water is a blend of RO-treated reused water, carbon adsorption-treated reused water and deep well water. Direct nonpotable water reuse is utilized in the form of injection of treated wastewater into an aquifer to form a barrier against saltwater intrusion. However, up to 50% of the injected effluent ends up augmenting the water supply used for indirect potable drinking water abstraction. Advanced treatment processes are shown in Figure 7.4. Three RO systems were studied at Water Factory 21.[142]

Water Factory 21 was the first groundwater recharge project to protect from the saltwater intrusion barrier in California. The original requirement was 50:50 wastewater and non-wastewater with the implementation of RO increasing the ratio to 67:33 in favor of wastewater.[143] Since 1991, greater than 67% and up to 100% wastewater may be injected under specific conditions. Research by the OCWD is required on the effects of injecting 100% wastewater and to continue investigations on the fate of organics of wastewater origin during groundwater recharge.[144]

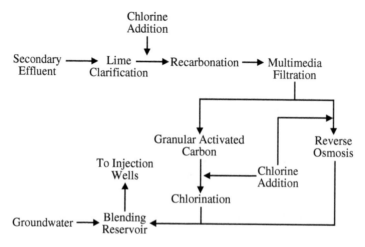

FIGURE 7.4   Water Factory 21 treatment process.

7.4.3.2 Whittier Narrows Groundwater Replenishment Project

Replenishment occurs through surface percolation using a mix of storm water, imported water, and treated wastewater in unlined river channels or specially constructed spreading basins. The reused water is to secondary treatment, chlorinated and dual-media filtered, contributing 16% of the total inflow into the groundwater.[145] Plans for expansion have been constrained by health concerns from the effects of exposure to trace organic contaminants with indirect potable reuse. Research tasks have included toxicological and chemical studies of the groundwater with identification of trace organic and any significance to health. The limitations of only monitoring pesticides were recognized and a range of organic contaminants selected from lists by the EPA and California Department of Health Services were investigated (Table 7.6).[145]

Mutagenic response occurred in the most hydrophobic fraction and there was no evidence of synergistic effects. PCP, 2,3′5 trichlorobiphenyl, and aldrin gave the highest response in reused water compared to other water sources. Though DEHP was observed up to 13 μg l^{-1} in reused water, values were far greater in the other

TABLE 7.6
Trace Organics Monitored in the Waters Used for Groundwater Replenishment at Whittier Narrows

EDCs (μg l^{-1})	LOD	Reused Water	Storm Water	Imported Water	Well Water Cl$_2$	Well Water No Cl$_2$
Pentachlorophenol (PCP)	1.3	BD–16	BD–6.8	BD	BD	BD
Lindane	0.2	BD–1	BD–1.4	BD	BD	BD
2,3′5 trichlorobiphenyl	0.1	BD–0.7	BD	BD	BD	BD
2,2′4,4′ tetrachlorobiphenyl	0.2	BD	BD	BD	BD	BD
Aldrin	0.2	BD–0.9	BD	BD	BD	BD
DDT	0.2	BD	BD	BD	BD	BD
Dieldrin	0.4	BD	BD	BD–0.4	BD	BD
Atrazine	0.5	BD	BD-3.9	BD	BD–0.9	BD–0.9
Simazine	1.1	BD	BD–6.6	BD	NQ	BD–1.7
Di-(2 ethyl hexyl) phthalate	0.5	0.7–13	22–82	BD–170	BD–59	BD–5.7

Note: LOD = limit of detection; NQ = detected but not quantifiable; BD = below detection limit
Source: From Nellor, M.H., Baird, R.B., and Smyth, J.R., Health effects of indirect potable water reuse, *J. Am. Water Works Assoc.*, 77, 88, 1985. With permission.

water samples (with the exception of unchlorinated well water). The data show that EDCs are present not only in reused water, but also in other water sources. Therefore, comparatively, reused water does not seem to be a greater risk to health by the presence of EDCs.

Infiltration studies were also carried out on the efficacy of soil for attenuating inorganic and organic contaminants in the reused wastewater. Percolation allows a final opportunity to clean reused water prior to entrance into the groundwater, which was located at 8 feet.[145] The primary removal mechanism was identified as biodegradation. Research showed no acute health effects, though chronic effects require evaluation. The research resulted in the percentage of reused water for groundwater replenishment to be increased during dry seasons.

7.4.4 POTABLE SUPPLY

Of the water treated to drinking water standard, the amount required to be of potable quality for domestic consumption is below 5%.[146] Direct potable reuse is the release of water into a municipal distribution system immediately after treatment with no environmental intervention. This method is not practiced to the same extent as indirect reuse, since it is deemed to be drinking "dirty" water. Several pilot plants have investigated the feasibility of direct water reuse. However, to the author's knowledge, only one direct potable wastewater reuse project is currently in operation at Windhoek in Nambia.

Indirect water reuse is practiced widely, more from an unplanned than planned perspective. Retention of effluent in receiving waters allows for dilution and transformation processes to decrease EDC concentrations present from the effluent. The Santa Ana River in Orange County is almost 100% effluent with hormonal concentrations in the river similar to that of the effluent.[113] Biotransformation processes are not thought to occur greatly due to the river environment. Removal of these compounds is unlikely to be great before abstraction for drinking water treatment, raising the possibility of EDCs in the drinking water source. Monitoring of organic contaminants was undertaken in the U.K.'s River Trent, which receives both domestic and industrial effluent. The possibility of abstraction to a reservoir prior to water treatment was assessed. Contaminant analysis included the EDCs, PCP, triazines, urons, and acid herbicides. No target or unknown compounds were detected at levels of concern.[147] Essex and Suffolk Water thoroughly assessed the impact of estrogenic substances on an indirect potable reuse scheme that operated during the July 1997 and December 1998 U.K. drought. Steroid analysis and VTG induction in male trout detected insignificant estrogenicity (see Section 7.5, Case Study).

7.4.4.1 Windhoek Reclamation Project

With the full utilization of nearby groundwater and surface water sources, 1968 saw the beginning of direct potable reuse of domestic sewage in Windhoek, Nambia.[148] Domestic sewage underwent conventional biological STW with the effluent discharged into a series of maturation ponds. The effluent would then gravitate to the water reclamation plant downstream. Four different process configurations have been used since 1968 involving a multibarrier treatment sequence. Configuration four was implemented in 1986 with adaptations in 1995. Sewage treatment includes activated sludge and 14 days in maturation ponds. Treatment at the water reclamation plant includes dissolved air flotation, sand filtration, and carbon filtration in addition to chlorination. The final product is blended with surface water from a dam on a minimum 1:1 ratio. A second blending step ensures that a maximum 25% of total drinking water contains reused water. Further extensions including O_3 and a two-stage GAC have been proposed.

In addition to monitoring the reused water by three independent laboratories, an epidemiological study of patients from surrounding hospitals, clinics, and doctors rooms was carried out from 1968 until 1982. Monthly toxicity testing included waterflea lethality, urease enzyme activity and bacterial growth inhibition. Ames salmonella mutagenicity was also determined monthly. From 1981 to 1992, the breathing rhythm of male guppies and several other species were monitored. Results from these studies have revealed no adverse effects. Investigation into EDCs, especially steroid estrogens, which are dominant in domestic effluent, has not been studied.

7.4.4.2 San Diego Aqua II Pilot Plant

The main objective of the San Diego pilot plant was to determine whether the advanced wastewater treatment process could reduce contaminant related health concerns of direct potable reuse to levels no greater than the present water supply. Secondary treatment used water hyacinths followed by advanced wastewater treatment. This treatment included coagulation, filtration, RO, air-stripping, and GAC followed by disinfection, producing water suitable for potable reuse.

Screening for mutagenicity, potentially toxic chemicals, and bioaccumulation of chemical mixtures were monitored in both the reused water and San Diego's untreated potable supply. The project analyzed a wide range of organic contaminants including the EDCs, PCBs, pesticides, PAHs and dioxins. The concentrations of all regulated contaminants were below the U.S. and state drinking water standards and MCLs.[149] Lindane spiked at 70 mg l^{-1} was removed by over 80% after the hyacinth pond with an overall removal of >99.96% after advanced wastewater treatment.[150] DEHP, diethyl phthalate, and di-n-octyl phthalate were identified, with DEHP concentration of 7.6 mg l^{-1} for the reused water, greater than that observed in the untreated water supply.[150]

Mutagenic and genotoxic activity was observed in both the reused water and San Diego's untreated water supply, with a lower activity for the reused water. Fish biomonitoring using juvenile fathead minnows were exposed for 28 days, resulting in no differences between the two waters. Risk assessments resulted in reused water being approximately 40 times less than the cancer risk associated with the untreated San Diego water supply.[150]

7.4.4.3 Denver Water Department Research Project

The semiarid condition of Denver has an annual precipitation of 12 inches. Though there is a large aquifer, pumping costs are prohibitive and the water source is

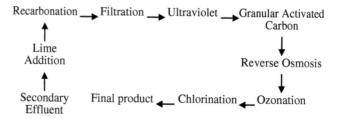

FIGURE 7.5 Multibarrier treatment process used for the Denver Water Department Research Project.

contaminated with chemical waste from dumps.[151] From 1979 to 1990, the viability of direct potable reuse from unchlorinated secondary effluent was undertaken. Water exceeded the Denver drinking water quality for all chemical, physical, and microbial parameters with the exception of nitrogen.[152] Multiple barriers for trace organic contaminants were utilized with the treatment process outlined in Figure 7.5.

The EDCs, 2,4-D, lindane, and methoxychlor, were all below the limit of detection (0.04 to 2.0 μg l^{-1}), meeting the established MCLGs under the SDWA. Whole animal toxicity testing was carried out on product water as well as long-term acute and chronic health effects testing.[109] Fifteen organic contaminants were also dosed 100 times the normal levels. The cumulative percentage removal was 74% for the EDC, methoxychlor.[153]

7.4.5 REUSE REGULATIONS AND CRITERIA

Public health considerations have centered on infections from pathogens as one of the main water quality parameters for water reuse (Table 7.7). Since 1828, when the first organic compound was synthesized, more than 6 million organic compounds have been produced.[154] The safety data for the majority of chemicals is typically limited to whether the chemical causes cancer or gross birth defects. The more subtle effects of endocrine disruption are not taken into account. As a result, health implications arising from chemical contaminants in the water supply are of increasing concern.

There are concerns as to whether drinking water standards are adequate to ensure the safety of all waters, regardless of the source. The concerns are valid since these standards were developed assuming use of the best available water source, not a reclaimed water source.[155] The National Interim Primary Drinking Water Regulations in 1975 stated that water used for drinking water purposes requires continuous protection with priority given to the selection of the purest source. Polluted sources should not be utilized unless other sources are economically unavailable.[98]

The exploitation of planned water reuse varies country to country.[156] Reuse standards vary around the world; an observation has been made that high quality effluent is required for high-income countries rather than lower-income countries.[124] No European regulations or guidelines for water reuse exist with the exception of the Urban Wastewater Treatment Directive (91/271/EEC): "treated wastewater shall be reused wherever appropriate."[157]

TABLE 7.7
Principal Water Quality Parameters for Water Reuse

Category		Water Quality Parameter
General	Suspended solids	Total suspended solids (TSS)
	Nutrients	N_2, P, K
	H^+	pH
Pathogens	Protozoa	*Giardia lambia, Entamoeba histolytica*
	Helminths	*Acaris lumbricodes*
	Bacteria	*Escherichia coli, Shigella, Salmonella*
	Viruses	Hepatitis
Inorganic	Dissolved organics	Total dissolved solids (TDS), Ca, Na, B, Cl
	Heavy metals	Cd, Zn, Hg, Ni
Organic	Biodegradable	Biochemical oxygen demand (BOD), Chemical oxygen demand (COD), Total organic carbon (TOC)
	Stable	Pesticides, PAH, Chlorinated hydrocarbons

California and Florida have been at the forefront of developing criteria specific to planned indirect water reuse.[158] California has had a basic regulatory structure for water recycling and reuse projects since 1969. However, projects for indirect potable reuse were evaluated on a case-by-case basis. In January 1996, the committee comprised of California's Department of Health Services and Department of Water Resources adopted a regulatory framework for potable reuse. Six criteria were established that must be met before a project may proceed. In contrast to planned indirect potable reuse projects, which are subject to intense scrutiny, unplanned indirect potable reuse occurs whenever STWs discharge to watercourses that downstream serve as drinking water sources.

For public water systems using indirect water reuse, 1986 amendments to the safe SDWA required surface water to be disinfected, and in some cases, additionally filtered. The SDWA requires groundwater used for drinking water to be surrounded by wellhead protection areas, which protect the water source from diffuse pollution, and guidelines for injecting wastewater into groundwater are also covered. Direct potable reuse is not advocated by the EPA guidelines on water reuse.[134] Minimum effluent treatment requirements for reused wastewater according to the EPA guidelines are given in Table 7.8. Monitoring should include inorganic and organic compounds that are known or suspected to be toxic, carcinogenic, teratogenic, or mutagenic and are not included in drinking water standards. However, the scope for the monitoring of organic contaminants in water is only limited by the analytical detection methods.[159]

Water reuse for potable application offers a direct route into the body. Rivers are relied on for dilution of effluent and self-purification processes for the removal

TABLE 7.8
Effluent Treatment Processes for Different Water Reuse Applications

Water Reuse Application	Effluent Treatment Processes			
	Secondary	Filtration	Disinfection	Comments
Urban uses	✔	✔	✔	
Industrial uses	✔		✔	
Groundwater recharge nonpotable by infiltration	Minimum primary			Site specific
Groundwater recharge nonpotable by injection	✔			Site specific
Groundwater recharge potable infiltration	✔		✔	Site specific
Groundwater recharge potable injection	✔	✔	✔	Advanced treatment required
Augmentation of surface water supplies	✔	✔	✔	Advanced treatment required

of EDCs. Evidence of endocrine disruption in rivers illustrates that STWs do not adequately remove these compounds. The health risk from contaminants in drinking water is based on extrapolations of animal toxicity tests and human exposure estimates to that contaminant. Drinking water standards are not intended for water reuse due to wastewater introducing new and unknown quantities of contaminants. Water reuse and criteria are far more proactive in the U.S. than Europe, with U.S. projects investigating contaminants beyond the requirements set in drinking water standards.

At the present time, based on the only in-depth water reuse and EDC study (see Section 7.5, Case Study), a dilution of 3:1 was required, in addition to river transformation processes which were needed to decrease EDCs to no-effect levels. Therefore, direct potable reuse, such as Windhoek, may require tests specific for EDCs, such as VTG induction and steroid analysis, which are the most prevalent EDCs in domestic sewage.

7.5 DRINKING WATER TREATMENT AND INDIRECT POTABLE REUSE: A CASE STUDY

7.5.1 OVERVIEW

The U.K.'s Essex and Suffolk Water investigated the presence of EDCs in water systems from both a reuse and drinking water treatment perspective. Wastewater from Chelmsford STW is normally discharged into the Blackwater estuary through an outfall pipe. Installed in 1920, the pipe served to protect the abstraction point at Langford on the River Chelmer from contamination of Chelmsford effluent. From the 23rd July 1997 to 31st December 1998, wastewater underwent UV disinfection at Langford Pumping Station prior to discharging directly into Hanningfield Reservoir, which serves the Hanningfield water treatment works. During this period, water from the River Chelmer and the Blackwater Estuary was also pumped to the Hanningfield Reservoir, along with the UV disinfected wastewater (30% total).[47] This temporary recycling scheme was consented by the Environment Agency (EA) as an emergency measure in response to the 1995–1998 drought.

Male rainbow trout were placed in Chelmsford sewage effluent, Hanningfield Reservoir, pre- and post-UV disinfection of wastewater at Langford, and the River Chelmer. Effluent was faintly estrogenic to male rainbow trout because of the presence of steroid estrogens. The effect was removed when diluted 3:1 with river water. At the Langford abstraction point, the River Chelmer was not found to be estrogenic, nor was the Hanningfield Reservoir. Direct toxicity assessments on effluent, River Chelmer and Hanningfield Reservoir gave no toxicity to rainbow trout, daphnia, and an alga.

7.5.2 Ultraviolet Disinfection Plant

Steroid hormones were monitored before and after an UV disinfection plant receiving sewage effluent. E1, E2, and EE2 underwent UV treatment at doses 32 m Wscm^{-2} for 19 seconds. Concentrations of E1, E2 and EE2 in the sewage effluent entering the UV disinfection plant ranged from 8 to 33 ng l^{-1}, 1.3 to 48 ng l^{-1}, and 1 to 3.4 ng l^{-1}, respectively. Concentrations for E1, E2, and EE2 post-UV varied from undetectable to 20, 26, and 11.1ng l^{-1}, respectively. Percentage reductions of the steroids receiving UV treatment ranged from approximately 30 to 100%. At times UV treatment did not change the steroid concentration and several times increased the presence of steroids by 66 and 91% (E1 from 9 to 15 ng l^{-1} and EE2 from 1 to 11.1 ng l^{-1}).[60]

Further work on UV treatment involved laboratory experiments of E1, E2, and EE2 at treatment doses of 145 mWscm^{-2} for 20 seconds. The mean percentage reduction of these three steroids was 24%, 4%, and 20% respectively. In 2 of the 5 experiments, post-UV treatment resulted in increased E2 concentration by 5 and 10%.[60] Reasons for the varying results may be attributed to small sample size, analytical error deviation, or interconversion between the steroids.

7.5.3 Potable Water Pilot Plant

A potable water pilot plant investigated the removal of spiked steroid estrogens after individual treatment parameters. In raw water, concentrations for E1, E2, and EE2 were <1 <3 and <3 ng l^{-1}, respectively. Table 7.9 shows the percentage removal of the spiked steroid estrogens after each treatment.[60]

The dose for pre- and postozonation was 1.0 mg l^{-1} and a contact time of 4 minutes and 10 minutes, respectively. The increased contact time led to an increased steroid estrogen removal of 17%. GAC proved the best removal treatment, while

TABLE 7.9
Percentage Removal of Spiked Steroid Estrogens after Individual Water Treatment

Treatment Parameter	Estrone (E1)	17β-estradiol (E2)	17α-ethinylestradiol (EE2)
Spiked amount (ng l^{-1})	1580	810	1100
		Treatment Removal (%)	
After preozonation	28.5	71.6	34.5
After clarification	22.1	46.9	28.2
After sand filter	62.0	129.6	90.9
After postozonation	11.4	24.7	17.3
After GAC	0.25	1.2	0.91

sand filtering was the poorest. After sand filter treatment, E2 concentrations increased from an initial spiked concentration of 810 to 1050 ng l^{-1}. Although no explanation is given,[60] as disinfection and adsorption processes usually follow filtration, significant steroid removal would still occur.

The potable water pilot plant also investigated the use of four GAC and O$_3$ treatment on selected pesticide and AP removal.[52] Hanningfield Reservoir water influent was spiked with a mixture of pesticides consisting of atrazine, diuron, isoproturon, chlorotoluron and mecoprop at a dose of 0.5 to 1.0 mg l^{-1} for each pesticide. Influent then passed onto a preozone contactor, O$_3$ dose 1 mg l^{-1} for a contact time of 4.8 minutes, operating at counter current. After the preozone contactor, water was coagulated with ferric sulfate, clarified and softened, and passed through a rapid gravity filter consisting of sand and anthracite and then two O$_3$ contactors. Four GAC were evaluated for the final phase of water treatment. The characteristics are summarized in Table 7.10.

Diuron, isoproturon, and chlorotoluron were preferentially removed with O$_3$ by more than 80%. After GAC, the concentrations were below the detection limit of 0.025 µg g^{-1}. Ozonation removed mecopop by 69% and atrazine by 42%. After ozonation, Cecarbon GAC 830 and CPL Miller gave a cumulative removal of 81% and 93% for mecopop. The comparison of GAC was based on atrazine, the most difficult pesticide to remove. Cecarbon GAC 830 removed more atrazine than other GAC tested. However, the bed life was superior for CPL Miller.

AP and APEO concentrations in raw water measured up to 0.1 and 2.1 µg l^{-1}, respectively. The raw water was spiked to concentrations of 1.1 µg l^{-1} for APs and 25.5 to 40 µg l^{-1} for APEOs. Filtration and ozonation removed 70% of APs and 95.9% for APEs, with GAC further removing them by more than 96%. The maximum AP and APEO levels after spiking followed by ozonation and GAC were 0.2 µg l^{-1} and 1.1 µg l^{-1}. Ozonation and GAC are sufficient for the removal of high concentrations of APs and APEOs from water.

7.5.4 The Future

The EA has granted a permanent wastewater-recycling scheme that discharges 35 Ml day^{-1} of effluent into the River Chelmer, from which Essex and Suffolk Water

TABLE 7.10
Characteristics of the Four Granular Activated Carbon Used in Final Water Treatment Step

Activated Carbon	Carbon Type	Surface Area ($m^2 \, g^{-1}$)	Mesh Size	Activity ($mg \, g^{-1}$)[a]
Cecarbon GAC 830	Bituminous coal based	Between 950 and 1100	8 × 30	920
Hydronorit GCW	Bituminous coal based		12 × 40	1050
CPL Miller	Bituminous coal based		12 × 40	NS
Chemviron F300	Bituminous coal based		8 × 30	NS

[a] Expressed as the iodine number

Note: NS = not stated

will abstract for water treatment. The reasoning for the new proposal was because of decreasing water resources and advances in sewage treatment. The augmentation of Langford abstraction through the diversion of effluent to River Chelmer is thought to be the first example of planned indirect reuse in the U.K.[160] One of the conditions of the consent is the comparison of fish upstream and downstream from the discharge location at periods of low flow for possible endocrine disruption.

REFERENCES

1. Leinster, P., McIntyre, A.E., Lester, J.N., and Perry, R., Analysis of volatile organic compounds in water, wastewater and an industrial effluent, *Chemosphere*, 10, 291, 1981.
2. Erickson, B.E., Analyzing the ignored environmental contaminants, *Environ. Sci. Technol.*, 36, 141A, 2002.
3. Packham, R.F., Drinking water quality and health, in *Pollution: Causes, Effects and Control*, 3rd ed., Harrison, R.H., Ed., The Royal Society of Chemistry, Cambridge, 1996, p. 53.
4. Ternes, T.A., Occurrence of drugs in German sewage treatment plants and rivers, *Water Res.*, 32, 3245, 1998.
5. McGeechin, M.A. and Moll, D.M., CDC perspective on emerging chemical contaminants in drinking water, in *Identifying Future Drinking Water Contaminants*, National Research Council, Ed., National Academy Press, Washington, DC, 1999, p. 43.
6. Wishart, S.J., Mills, S.W., and Elliott, J.C., Considerations for recycling sewage effluent in the UK, *J. Chart. Inst. Water Environ. Manage.*, 14, 284, 2000.
7. Melosi, M.V., Pure and plentiful: the development of modern waterworks in the United States, 1801–2000, *Water Policy*, 2, 243, 2000.

8. Piotrowski, R. and Kraemer, R.A., Policy framework: Germany, in *Drinking Water Supply and Agricultural Pollution: Preventative Action by the Water Supply Sector in the European Union*, Schrama, G.J.I., Ed., Kluwer, Dordrecht, 1998, p. 93.

9. Hoff, M., Water Reuse in the European Union, unpublished MSc. mini project, Brunel University, Uxbridge, London, 2000.

10. Westerhoff, G.P. and Chowdhary, Z.K., Water treatment systems, in *Water Resources Handbook*, Mays, L.W., Ed., McGraw-Hill, New York, 1996, p. 17.1.

11. U.S. Environmental Protection Agency, Ultraviolet Light Disinfection Technologies in Drinking Water Application: An Overview, EPA 811/R-96/002, Washington, DC, 1996.

12. Nishida, K. and Ohgaki, S., Photolysis of aromatic chemical compounds in aqueous TiO_2 suspensions, *Water Sci. Technol.*, 30, 39, 1994.

13. Najm, I. and Trussell, R.R., New and emerging drinking water treatment technologies, in *Identifying Future Drinking Water Contaminants*, National Research Council, Ed., National Academy Press, Washington, DC, 1999, p. 220.

14. Environmental Data Services, Pesticides in drinking water linked to breast cancer, *ENDS Rep.*, 241, 8, 1995.

15. Okun, D.A., Historical overview of drinking water contaminants and public water utilities, in *Identifying Future Drinking Water Contaminants*, National Research Council, Ed., National Academy Press, Washington, DC, 1999, p. 22.

16. Oskam, G., Van Genderen, J., Hopman, R., Noji, T.M., Noordsji, A., and Puijker, L.M., A general view of the problem with special reference to the Dutch situation, *Water Supply*, 11, 1, 1993.

17. Clark, L., Gomme, J., and Oakes, D.B., Pesticide transport investigations in the major aquifers of the UK, *Water Supply*, 11, 19, 1993.

18. Clark, L.B., Rosen, R.T., Hartman, T.G., Louis, J.B., Suffet, I.H., Lippincott, R.L., and Rosen, J. D., Determination of alkylphenol ethoxylates and their acetic acid derivatives in drinking water by particle beam liquid chromatography/mass spectrometry, *Int. J. Environ. Anal. Chem.*, 147, 167, 1992.

19. Rudel, R.A., Melly, S.J., Geno, P.W., Sun, G., and Brody, J.G., Identification of alkylphenols and other estrogenic phenolic compounds in wastewater, septage, and groundwater on Cape Cod, Massachusetts, *Environ. Sci. Technol.*, 32, 861, 1998.

20. Stahlschmidt-Allner, P., Allner, B., Rombke, J., and Knacker, T., Endocrine disrupters in the aquatic environment, *Environ. Sci. Poll. Res.*, 4, 155, 1997.

21. Harries, J. E., Jobling, S., Mattiessen, P., Sheahan, D.A., and JP, S., Effects of Trace Organics of Fish: Phase 2, Report No. FR/D0022, Foundation for Water Research, Marlow, London, 1995.

22. James, H.A., Fielding, M., Franklin, O., Williams, D., and Lunt, D., Steroid Concentration in treated sewage Effluents and Water Courses: Implication for Water Supplies, Report No.98/TX/01/1, U.K. Water Industry Research Limited, London, 1998, p. 32.

23. Tyler, C.R., Van der Eeden, B., Jobling, S., Panter, G., and Sumpter, J.P., Measurement of vitellogenin, a biomarker for exposure to oestrogenic chemical in a wide variety of cyprinid fish, *J. Comp. Physio. B: Biochem. Systemic Environ. Physio.*, 166, 418, 1996.

24. Rodgers Gray, T.P., Jobling, S., Morris, S., Kelly, C., Kirby, S., Janbakhsh, A., Harries, J.E., Waldock, M.J., Sumpter, J.P., and Tyler, C.R., Long-term temporal changes in the oestrogenic composition of treated sewage effluent and its biological effects on fish, *Environ. Sci. Technol.*, 34, 1521, 2000.

25. Fawell, J.K., Sheahan, D., James, H.A., Hurst, M., and Scott, S., Oestrogens and oestrogenic activity in raw and treated water in Severn Trent water, *Water Res.*, 35, 1240, 2001.

26. Hutchinson, J., Harding, L., and Carlile, P., Effect of Water Treatment Processes on Oestrogenic Chemicals, Report No. 95/DW/05/10, United Kingdom Water Industry Research, London, 1996.

27. U.S. Environmental Protection Agency, Removal of Endocrine Disrupting Compounds Using Drinking Water Treatment Processes, EPA/625/R-00/015, Washington, DC, 2001, p. 20.

28. Vollmuth, S., Zajc, A., and Niessner, R., Formation of polychlorinated dibenzo-p-dioxins and polychlorinated dibenzofurans during the photolysis of pentachlorophenol-containing water, *Environ. Sci. Technol.*, 28, 1145, 1994.

29. National Research Council, Ed., Health effect studies of reuse systems, in *Issues in Potable Reuse: The Viability of Augmenting Drinking Water Supplies with Reclaimed Water*, Water Science and Technology, Environment and Resources, National Academy Press, Washington, DC, 1998, p. 164.

30. Bedding, N.D., McIntyre, A.E., Perry, R., and Lester, J.N., Organic contaminants in the aquatic environment II. Behaviour and fate in the hydrological cycle, *Sci. Total Environ.*, 26, 255, 1983.

31. Manoli, E. and Samara, C., Polycyclic aromatic hydrocarbons in natural waters: sources, occurrence and analysis, *Trac — Trends Anal. Chem.*, 18, 417, 1999.

32. Rivera, J., Eljarrat, E., Espadaler, I., Martrat, M.G., and Caixach, J., Determination of PCDF/PCDD in sludges from a drinking water treatment plant — influence of chlorination treatment, *Chemosphere*, 34, 989, 1997.

33. Itoh, S., Ueda, H., Naasaka, T., Nakanishi, G., and Sumitomo, H., Evaluating variation of estrogenic effects by drinking water chlorination with the MVLN assay, *Water Sci. Technol.*, 42, 61, 2000.

34. Rogers, A., Europe to set tougher targets for bathing and drinking water, *Lancet*, 348, 1728, 1996.

35. Fawell, J.K. and Chipman, J.K., Potential endocrine disrupting substances from materials in contact with drinking water, *J. Chart. Inst. Water Environ. Manage.*, 15, 92, 2001.

36. Brotons, J.A., Olea-Serano, M.F., Villalobos, M., Pedraza, V., and Olea, N., Xenoestrogens released from lacquer coatings in food cans, *Environ. Health Persp.*, 103, 608, 1995.

37. Environmental Data Services, Public exposed to oestrogen risks from food cans, *ENDS Rep.*, 246, 3, 1995.

38. Junk, G.A., Svec, H.J., Vick, R.D., and Avery, M.J., Contamination of water by synthetic polymer tubes, *Environ. Sci. Technol.*, 8, 1100, 1974.

39. Soto, A.M., Justicia, H., Wray, J.W., and Sonnenschein, C., *p*-nonyl-phenol: an estrogenic xenobiotic released from "modified" polysterene, *Environ. Health Persp.*, 92, 167, 1991.

40. Maier, M., Maier, D., and Lloyd, B.J., Factors influencing the mobilisation of polycyclic aromatic hydrocarbons (PAHs) from the coal-tar lining of water mains, *Water Res.*, 34, 773, 2000.

41. Maier, D., Maier, M., Lloyd, B.J., and Toms, I.P., Remobilisation of polycyclic aromatic hydrocarbons from the coal-tar lining of water mains, *Ozone Sci. Eng.*, 18, 517, 1997.

42. Giger, W., Emerging chemical drinking water contaminants., in *Identifying Future Drinking Water Contaminants*, National Research Council, Ed., National Academy Press, Washington, DC, 1999, p. 112.

43. Miltner, R.J., Baker, D.B., Speth, T.F., and Frouk, C.A., Treatment of seasonal pesticides in surface waters, *J. Am. Water Works Assoc.*, 1, 43, 1989.

44. Bates, A., Overview of UK research and treatment technologies, in *Endocrine Disrupters and Pharmaceutical Active Compounds in Drinking Water Workshop*, Center for Health Effects of Environmental Contamination, Chicago, IL, 2000.

45. Foster, D., Mice, M., Rachwal, A.J., and White, S.J., New treatment processes for pesticides and chlorinated organics control in drinking water, *J. Chart. Inst. Water Environ. Manage.*, 5, 466, 1991.

46. Di Laconi, C., Lopez, A., Ramadon, R., Ricco, G., and Towei, M.C., Ozonation of secondary effluent of tannery industry: kinetic and effect on biodegradability and toxicity, in *Wastewater Treatment: Standards and Technologies to Meet the Challenges of the 21st Century*, Horan, H., and Haigh, M. Eds., Terence Dalton Publishers, Leeds, England, 2000, p. 625.

47. Walker, D., The impact of disinfected treated wastewater in a raw-water reservoir, *J. Chart. Inst. Water Environ. Manage.*, 15, 9, 2001.

48. Prado, J., Arantegui, J., Chamaro, E., and Esplugas, S., Degradation of 2,4-D by ozone and light, *Ozone Sci. Eng.*, 16, 235, 1994.

49. Glaze, W.H. and Kang, J.W., Advanced oxidation processes for treating groundwater contaminated with TCE and PCE: laboratory studies, *J. Am. Water Works Assoc.*, 5, 57, 1988.

50. Beltran, F.J., Gonzalez, M., Rivas, J., and Marin, M., Oxidation of mecroprop in water with ozone and ozone combined with hydrogen peroxide, *Ind. Eng. Chem. Res.*, 33, 125, 1994.

51. Koga, M., Kadokami, K., and Shinohara, R., Laboratory scale ozonation of water contaminated with trace pesticides, *Water Sci. Technol.*, 26, 145, 1992.

52. Roditaki, K., Removal of Pesticides and Oestrogens Using Ozone and Granular Activated Carbon, unpublished MSc thesis, Cranfield University, London, 1996.

53. De Zuane, J., Water treatment, in *Drinking Water Handbook*, 2nd ed., De Zuane, J., Ed., John Wiley & Sons, New York, 1997, p. 419.

54. Spellman, F.R. and Drinan, J., Water treatment, in *The Drinking Water Handbook*, 1st ed., Spellman, F.R. and Drinan, J., Eds., Technomic Publishing Company, Lancaster, PA, 2000, p. 227.

55. Hopman, R., Meerkerk, M.A., Siegers, W.G., and Kuithof, J.C., The prediction and optimization of pesticide removal by GAC-filtration, *Water Supply*, 12, 197, 1994.

56. Thiem, L.T., Badorek, D.L., Johari, A., and Alkhatib, E., Adsorption of synthetic organic shock loadings, *J. Environ. Eng. — ASCE*, 113, 1302, 1987.

57. Taylor, G.S., Hillis, P., and Walker, I., Pilot-plant trials on River Dee water at Huntingdon, *J. Chart. Inst. Water Environ. Manage.*, 7, 333, 1993.

58. Baldouf, G., Haist-Gulde, B., and Brauch, H.J., Removal of pesticides from raw waters, *Water Supply*, 11, 187, 1993.

59. Hofman, J.A.M. H., Noji, T.H.M., and Schipper, J.C., Removal of pesticides and other organic micropollutants with membrane filtration, *Water Supply*, 11, 129, 1993.

60. Walker, D., Oestrogenicity and wastewater recycling: experience from Essex and Suffolk water, *J. Chart. Inst. Water Environ. Manage.*, 14, 427, 2000.

61. Fujita, Y., Ding, W., and Reinhard, M., Identification of wastewater dissolved organic carbon characteristics in reclaimed wastewater and recharged groundwater, *Water Environ. Res.*, 68, 867, 1996.

62. Tanizaki, T., Kadokami, K., and Shinohara, R., Catalytic photodegradation of endocrine-disrupting chemicals using titanium dioxide photosemiconductor thin films, *Bull. Environ. Contam. Toxicol.*, 68, 732, 2002.

63. Paune, F., Caixach, J., Espadaler, I., Om, J., and Rivera, J., Assessment on the removal of organic chemicals from raw and drinking water at a Llobregat river water works plant using GAC, *Water Res.*, 32, 3313, 1998.

64. Ohko, Y., Ando, I., Niwa, C., Tatsuma, T., Yamamura, T., Nakashima, T., Kubota, Y., and Fujishima, A., Degradation of bisphenol A in water by TiO_2 photocatalyst, *Environ. Sci. Technol.*, 35, 2365, 2001.

65. Larsen, G., Casey, F., Magelky, B., Pfaff, C., and Hakk, H., Sorption, mobility and fate of 17beta-estradiol and testosterone in loam soil and sand, in *2nd International Conference on Pharmaceuticals and Endocrine Disrupting Chemicals in Water*, Minneapolis, 2001, p. 138.

66. Larson, G., Casey, F., Bergman, A., Hakk, H., and Garber, E., Sorption, mobility and fate of tetrabromobisphenol A (TBBPA) in loam soil and sand, in *2nd International Conference on Pharmaceuticals and Endocrine Disrupting Chemicals in Water*, Minneapolis, 2001, p. 144.

67. Wilson, G.J., Khodadoust, A.P., Suidan, M.T., Brenner, R.C., and Acheson, C.M., Anaerobic/aerobic biodegradation of pentachlorophenol using GAC fluidized bed reactors: optimisation of the empty bed contact time, *Water Sci. Technol.*, 38, 9, 1998.

68. Mollah, A.H. and Robinson, C.W., Pentachlorophenol adsorption and desorption characteristics of granular activated carbon. I. Isotherms, *Water Res.*, 12, 2901, 1996.

69. Mollah, A.H. and Robinson, C.W., Pentachlorophenol adsorption and desorption characteristics of granular activated carbon. II. Kinetics, *Water Res.*, 12, 2907, 1996.

70. Ozaki, H. and Li, H., Rejection of organic compounds by ultra-low pressure reverse osmosis membrane, *Water Res.*, 36, 123, 2002.

71. Acero, J.L., Stemmler, K., and Gunten, U.V., Degradation kinetics of atrazine and its degradation products with ozone and OH radials: a predictive tool for drinking water treatment, *Environ. Sci. Technol.*, 34, 591, 2000.

72. Pichat, P., Photocatalytic degradation of aromatic and alicyclic pollutants in water: by-products, pathways and mechanisms, *Water Sci. Technol.*, 35, 73, 1997.

73. Martin, C.A., Cabrera, M.I., Alfano, O.M., and Cassano, A.E., Photochemical decomposition of 2,4-dichlorophenoxyacetic acid (2,4-D) in aqueous solution. II. Reactor modeling and verification, *Water Sci. Technol.*, 35, 197, 1997.

74. Lopez, A., Mascolo, G., Tiravanti, G., and Passino, R., Degradation of herbicides (ametryn and isoproturon) during water disinfection by means of two oxidants (hypochlorite and chlorine dioxide), *Water Sci. Technol.*, 35, 129, 1997.

75. Kornmuller, A. and Wiesmann, U., Continuous ozonation of polycyclic aromatic hydrocarbons in oil/water-emulsions and biodegradation of oxidation products, *Water Sci. Technol.*, 40, 107, 1999.

76. Yao, J.-J., Huang, Z.-H., and Masten, S.J., The ozonation of benz(a)anthracene: pathway and product identification, *Water Res.*, 32, 3235, 1998.

77. Vollmuth, S. and Niessner, R., Degradation of PCDD, PCDF, PAH, PCB and chlorinated phenols during the destruction treatment of landfill seepage water in laboratory model reactor (UV, ozone, and UV/ozone), *Chemosphere*, 30, 2317, 1995.

78. Choi, W., Hong, S.J., Chang, Y.S., and Cho, Y., Photocatalytic degradation of polychlorinated dibenzo-p-dioxins on TiO2 film under UV or solar light irradiation, *Environ. Sci. Technol.*, 34, 4810, 2000.

79. Smirnov, A.D., Schecter, A., Papke, O., and Beijak, A.A., Conclusions from Ufa, Russia, draining water dioxin cleanup experiments involving different treatment technologies, *Chemosphere*, 32, 479, 1996.

80. Shi, Z., LaTorre, K.A., Ghosh, M.M., Layton, A.C., Luna, S.H., Bowles, L., and Sayler, G.S., Biodegradation of UV-irradiated polychlorinated biphenyls in surfactant micelles, *Water Sci. Technol.*, 38, 25, 1998.

81. Ginsburg, J., Okolo, S., Prelevic, G., and Hardiman, P., Residence in the London area and sperm density, *Lancet*, 343, 230, 1994.

82. Ding, W., Fujita, Y., Aeshimann, E., and Reinhard, M., Identification of organic residues in tertiary effluents by GC/EI-MS, GC/CI-MS and GC/TSQ-MS, *Fres. J. Anal. Chem.*, 354, 48, 1996.

83. European Commission, Council Directive 76/464/EEC of 4 May 1976 on pollution caused by certain dangerous substances discharged into the aquatic environment of the Community, *Off. J. Eur. Comm.*, L 12923, 1976.

84. Zabel, T.F., Legislation for the control of discharges to the aquatic environment, in *Use of Adsorbents for the Removal of Pollutants from Wastewaters*, McKay, G., Ed., CRC Press, Boca Raton, FL, 1996, p. 7.

85. Environment Agency, Endocrine-Disrupting Substances in the Environment: What Should Be Done?, London, 1998, p. 13.

86. Young, W.F., Whitehouse, P., Johnson, I., and Sorokin, N., Proposed predicted-no-effect-concentration (PNECs) for natural and synthetic steroid oestrogens in surface waters. Report No. P2-T04/TS1, Environment Agency, London, 2002, p. 172.

87. European Commission, Directive 2000/60/EC of the European Parliament and of the Council of 23 October 2000 establishing a framework for Community action in the field of water policy, *Off. J. Eur. Comm.*, L327/1, 2000.

88. Cleverly, G., The water framework directive: an overview, *Water Environ. Manager*, 6, 8, 2001.

89. European Commission, Decision No. 2455/2001/EC of the European Parliament and of the Council of 20 November 2001 establishing the list of priority substances in the field of water policy and amending Directive 200/60/EC, *Off. J. Eur. Comm.*, L 331/1, 2001.

90. European Commission, Council Directive 98/83/EC of 3 November 1998 on the quality of water intended for human consumption, *Off. J. Eur. Comm.*, L 330/32, 1998.

91. European Commission, Working paper on hormone disrupting chemicals, European Commission, DG XI, Brussels, 1998.

92. Commission of European Communities, Community Strategy for Endocrine Disrupters: A Range of Substances Suspected of Interfering with the Hormone Systems of Humans and Wildlife COM(1999) 706 final, Brussels, 1999.

93. Commission of European Communities, Communication from the Commission to the Council and the European Parliament on the Implementation of the Community Strategy for Endocrine Disrupters: A Range of Substances Suspected of Interfering with the Hormone Systems of Humans and Wildlife (COM(1999) 706) COM(2001) 262 final, Brussels, 2001, p. 31.

94. Friends of the Earth, Endocrine Disrupting Pesticides: European Priority List, London, 2001, p. 7.

95. National Research Council, Setting Priorities for Drinking Water Contaminants, National Academy Press, Washington, DC, 1999, p. 128.

96. Deason, J.P., Schad, T.M., and Cherk, G.W., Water policy in the United States: a perspective, *Water Policy*, 3, 175, 2001.

97. Chasis, S., Kennedy, K., Olson, E., Quintero, A., Rosenberg, D., and Stoner, N., Clean Water, Clear Choice: Upcoming Bush Administration Decisions on Water Pollution, National Resources Defense Council, New York, 2001, p. 1.
98. U.S. Environmental Protection Agency, National interim primary drinking water regulations, *Fed. Regist.*, 40, 59566, 1975.
99. U.S. Environmental Protection Agency, National Primary Drinking Water Standards, EPA 816/F-01/007, Washington, DC, 2001, pp.4.
100. U.S. Environmental Protection Agency, Contaminants regulated under the Safe Drinking Water Act, http://www.epa.gov/safewater/consumer/contam_timeline.pdf, 2001 (accessed Feb. 26, 2002).
101. Miller, T.L. and Wilber, W.G., Emerging drinking water contaminants: overview and role of National Water-Quality Assessment Program, in *Identifying Future Drinking Water Contaminants*, National Research Council, Ed., National Academy Press, Washington, DC, 1999, p. 33.
102. U.S. Environmental Protection Agency, Announcement of draft drinking water contaminant candidate list, *Fed. Regist.*, 62, 52194, 1997.
103. Walker, J.D., Gray, D.A., and Pepling, M.K., Past and future strategies for sorting and ranking chemicals: applications to the 1998 Drinking Water Contaminant Candidate List Chemicals, in *Identifying Future Drinking Water Contaminants*, National Research Council, Ed., National Academy Press, Washington, DC, 1999, p. 51.
104. de Fur, P., U.S. policies and perspectives on endocrine disrupting chemicals, in *Endocrine Disrupters and Pharmaceutical Active Compounds in Drinking Water Workshop*, Center for Health Effects of Environmental Contamination, 2000.
105. U.S. Environmental Protection Agency, Endocrine Disrupter Screening Program: Report to Congress, 2000, p.1.
106. Fenner-Crisp, P.A., Maciorwski, A.F., and Timm, G.E., The endocrine disrupter screening program developed by the U.S. Environmental Protection Agency, *Ecotoxicology*, 9, 85, 2000.
107. Auer, C.M., Blunck, C., Chow, F., and Williams, D.R., Sorting and screening of potential drinking water contaminants: new and existing chemicals under the Toxic Substances Control Act, in *Identifying Future Drinking Water Contaminants*, National Research Council, Ed., National Academy Press, Washington, DC, 1999, p. 103.
108. Biswas, A.K., Role of wastewater reuse in water planning and management, in *Treatment and Reuse of Wastewater*, Biswas, A.K., and Arar, A., Eds., Butterworths, London, 1988, p. 3.
109. Rogers, S.E., Peterson, D.L., and Lauer, W.C., Organic contaminants removal for potable reuse, *J. Water Poll. Cont. Fed.*, 59, 722, 1987.
110. Lilienfeld, D.E., John Snow: the first hired gun?, *Am. J. Epidemiol.*, 152, 4, 2000.
111. Rowe, D.R. and Abdel-Magid, I.M., Health aspects of using reclaimed water in engineering, in *Handbook of Wastewater Reclamation and Reuse*, 1st ed., Rowe, D.R. and Abdel-Magid, I.M., Eds., CRC Press, Boca Raton, FL, 1995, p. 107.
112. National Research Council, Ed., Reliability and quality assurance issues for reuse systems, in *Issues in Potable Reuse: The Viability of Augmenting Drinking Water Supplies with Reclaimed Water*, Water Science and Technology, B.C. O. G., Environment and Resources, National Academy Press, Washington, DC, 1998, p. 208.
113. Sedlak, D., Challenges associated with quantification of trace concentrations of pharmaceuticals in a complex matrix, in *Endocrine Disrupters and Pharmaceutical Active Compounds in Drinking Water Workshop*, Center for Health Effects of Environmental Contamination, Chicago, IL, 2000, p. 6.

114. Davis, M., Endocrine disruptors in wastewater, in *Endocrine Disrupters and Pharmaceutical Active Compounds in Drinking Water Workshop*, Center for Health Effects of Environmental Contamination, 2000, p. 4.

115. Faby, J. A., Brissaud, F., and Bontoux, J., Wastewater reuse in France: water quality standards and wastewater treatment technologies, *Water Sci. Technol.*, 40, 37, 1999.

116. Swayne, M., Boone, G., Bauer, D., and Lee, J., Wastewaters in Receiving Waters at Water Supply Abstraction Points, Report No. EPA-600/2–80–044, EPA, Washington, DC, 1980, p. 198.

117. Angelakis, A., Thairs, T., Lazarova, V., and Asano, T., Water reuse in the EU countries: Necessity of establishing EU-guidelines, Eureau Joint Water Reuse Working Group, Brussels, Belgium, 2001, p. 52.

118. The Royal Society, Endocrine Disrupting Chemicals (EDCs), London, 2000.

119. Avendt, R.J., Reuse for recharge: Cedar Creek reclamation-groundwater recharge demonstration program, in *Reuse of Sewage Effluent*, Thomas Telford, London, 1985, p. 199.

120. Lauer, W.C., Rogers, S.E., and Ray, J.M., The current status of Denvers potable water reuse project, *J. Am. Water Works Assoc.*, 77, 52, 1985.

121. Levine, B.N., Madireddi, K., Lazarova, V., Stenstrom, M.K., and Suffet, M., Treatment of trace organic compounds by membrane processes: at the Lake Arrowhead water reuse pilot plant, *Water Sci. Technol.*, 40, 293, 1999.

122. Iranpour, R., Safavi, H.R., Patel, D., and Aiyeola, T., Discussion of wastewater reclamation of Lake Arrowhead, California: an overview, *Water Environ. Res.*, 70, 1099, 1998.

123. Gibert, M., The European water industry's view of endocrine disruptors and pharmaceuticals, in *Endocrine Disrupters and Pharmaceutical Active Compounds in Drinking Water Workshop*, Center for Health Effects of Environmental Contamination, Chicago, IL, 2000.

124. Levine, B., Lazarova, V., Manem, J., and Suffet, I.H., Wastewater reuse standards: goals, status and guidelines, in *Beneficial Reuse of Water and Biosolids*, Water Environment Federation, Malaga, Spain, 1997, p. 12.

125. Arar, A., Management aspects of the use of treated sewage effluent for irrigation, in *Treatment and Reuse of Wastewater*, Biswas, A.K., and Arar, A., Eds., Butterworths, London, 1988, p. 46.

126. Muszkat, L., Degradation of organosynthetic pollutants, in *Hypertrophic Reservoirs for Wastewater Storage and Reuse: Ecology, Performance and Engineering Design*, Juanico, M. and Dor, I., Eds., Springer-Verlag Heidelberg, 1999, p. 205.

127. Harms, H.H., Bioaccumulation and metabolic fate of sewage sludge derived organic xenobiotics in plants, *Sci. Total Environ.*, 185, 83, 1986.

128. Kinlock, N.E., An Investigation into Nonylphenol Ethoxylate Surfactant Usage in the Kings Lynn Area and Their Implications for Water Reuse, unpublished MSc. thesis, Imperial College of Science, Technology and Medicine, London, 1998.

129. Shore, L.S., Gurevitz, M., and Shemesh, M., Estrogen as an environmental pollutant, *Bull. Environ. Contam. Toxicol.*, 51, 361, 1993.

130. Crook, J., Water reuse in California, *J. Am. Water Works Assoc.*, 77, 60, 1985.

131. West Basin Municipal Water District, Recycled Water: The Facility, http://www.west-basin.com/recycle_facility.html, 2000, p. 4 (accessed Mar. 21, 2002).

132. Levine, B., Reich, K., Shields, P., Suffet, I.H., and Lazarova, V., Water quality assessment for indirect potable reuse: a new methodology for controlling trace organic compounds at the West Basin Water Recycling Plant (California, USA), *Water Sci. Technol.*, 43, 249, 2001.

133. Bouwer, H., Issues in artificial recharge, *Water Sci. Technol.*, 33, 381, 1996.
134. U.S. Environmental Protection Agency, Guidelines for Water Reuse, EPA/625/R-92/004, Cincinnati, OH, 1992.
135. Kanarek, A. and Michail, M., Groundwater recharge with municipal effluent: Dan region reclamation project, Israel, *Water Sci. Technol.*, 34, 227, 1996.
136. Hammer, M.J. and Hammer, M.J.J., Water reuse, in *Water and Wastewater Technology*, 3rd ed., Francis, E., Ed., Prentice-Hall Inc., Upper Saddle River, NJ, 1996, p. 492.
137. National Research Council, Groundwater Recharge Using Waters of Impaired Quality, National Academy Press, Washington, DC, 1994, p. 283.
138. Crook, J., Water reclamation and reuse, in *Water Resources Handbook*, Mays, L.W., Ed., McGraw-Hill, New York, 1996, p. 31.
139. Montgomery, H.A.C., Water quality aspects of the recharge of sewage effluents for reuse, in *Treatment and Reuse of Wastewater*, Biswas, A.K. and Arar, A., Eds., Butterworths, London, 1988, p. 38.
140. Montgomery, H.A., Beard, M.J., and Baxter, K.M., Reuse for recharge: groundwater recharge of sewage effluents in the UK, in *Reuse of Sewage Effluent*, Thomas Telford, London, 1985, p. 219.
141. Zoller, U., Non-ionic surfactants in reused water: are activated sludge/soil aquifer treatments sufficient?, *Water Res.*, 28, 1625, 1994.
142. Buckley, C.A. and Hurt, Q.E., Membrane applications: a contaminant based perspective, in *Water Treatment: Membrane Processes*, Mallevialle, J., Odendaal, P.E. and Wiesne, M.R., Eds., McGraw-Hill, New York, 1996, pp.3.1.
143. Wehner, M., A research and demonstration project permit under the new Californian groundwater recharge regulations, *Desalination*, 87, 37, 1992.
144. Mills, W.R. and Watson, I.C., Water Factory 21: the logical sequence, *Desalination*, 98, 265, 1994.
145. Nellor, M.H., Baird, R.B., and Smyth, J.R., Health effects of indirect potable water reuse, *J. Am. Water Works Assoc.*, 77, 88, 1985.
146. Diaper, C., Jefferson, B., Parsons, S.A., and Judd, S.J., Water-recycling technologies in the UK, *J. Chart. Inst. Water Environ. Manage.*, 15, 282, 2001.
147. Drage, B.E., Upton, J. E., and Purvis, M., On-line monitoring of micropollutants in the River Trent (UK) with respect to drinking water abstraction, *Water Sci. Technol.*, 38, 123, 1998.
148. Haarhoff, J. and Van der Merwe, B., Twenty-five years of wastewater reclamation in Windhoek, Nambia, *Water Sci. Technol.*, 33, 25, 1996.
149. National Research Council, Ed., Chemical contaminants in reuse systems, in *Issues in Potable Reuse: The Viability of Augmenting Drinking Water Supplies with Reclaimed Water*, Water Science and Technology, Environment and Resources National Research Council, Eds., National Academy Press, Washington, DC, 1998, p. 45.
150. Olivieri, A.W., Eisenberg, D.M., Cooper, R.C., Tchobanoglous, G., and Gagliardo, P., Recycled water: a source of potable water: City of San Diego health effects study, *Water Sci. Technol.*, 33, 285, 1996.
151. Miller, W.H. and Sayre, I.M., Denver's reuse demonstration plant: forerunner of the future?, *J. Am. Water Works Assoc.*, 77, 13, 1985.
152. Lauer, W.C. and Rogers, S.E., The demonstration of direct potable reuse: Denver's pioneer project, in *Water Reuse Conference Proceedings*, AWWA/WEF AWWA, Denver, 1996, p. 269.
153. Lauer, W.C., Rogers, S.E., La Chance, A.M., and Nealy, M.K., Process selection for potable reuse health effect studies, *J. Am. Water Works Assoc.*, 83, 52, 1991.

154. Lester, J. N. and Birkett, J.W., *Microbiology and Chemistry for Environmental Scientists and Engineers,* 2nd ed., E & FN Spon, London, 1999.
155. Xing, C.-H., Wen, X.-H., Qian, Y., and Tardieu, E., Microfiltration-membrane-coupled bioreactor for urban wastewater reclamation, *Desalination,* 141, 63, 2001.
156. Van Kiper, C. and Geselbracht, J., Water reclamation and reuse, *Water Environ. Res.,* 70, 586, 1998.
157. EC, Council Directive 91/271/EEC of 30 May 1991 on urban wastewater treatment, *Off. J. Eur. Comm.,* L135/40, 1998.
158. Asano, T. and Levine, A.D., Wastewater reclamation, recycling and reuse: past, present, and future, *Water Sci. Technol.,* 33, 1, 1996.
159. Shackleford, W.M. and Cline, D.M., Organic compounds in water, *Environ. Sci. Technol.,* 20, 652, 1986.
160. Walker, D., The promotion of planned indirect wastewater re-use scheme in Essex, *J. Chart. Inst. Water Environ. Manage.,* 15, 271, 2001.

8 Management Strategies for Endocrine Disrupters in the Aquatic Environment

M.R. Gaterell

CONTENTS

8.1   INTRODUCTION

In developing effective management strategies intended to minimize the risks imposed by the release of compounds into the environment, the relationship between potential sources and the subsequent fate and behavior of such compounds needs to be completely understood. It is also equally important to systematically evaluate the relative costs and benefits of prevention or mitigation options designed to reduce the impact of such releases, and ensure that in alleviating one potential environmental issue the burden of so doing is not simply shifted to another.

These issues are exemplified when considered in the context of endocrine disrupting chemicals (EDCs). As outlined in the preceding chapters, there is a wide range of routes by which these compounds can both enter the aquatic environment and potentially be absorbed into the human food chain. Water treatment systems (both sewage and potable supply) offer a number of potential opportunities to control their release, within these systems EDCs are often subject to complex transformation pathways that are dependent on local environmental conditions and the presence of other compounds. Clearly, management strategies designed to control their release need to be based on a thorough understanding of the pathways and their implications for the effectiveness of other parts of the process chain. Furthermore, determining which of these opportunities represents the most efficient use of economic and environmental resources requires all potential impacts to be included in any analysis.

To ensure all the potential implications of maximizing the removal of EDCs are accounted for, a systematic assessment of the process chain needs to be undertaken. While using traditional appraisal techniques enables the economic implications to be considered readily, they rarely account for many of the environmental consequences of investment decision making. Consequently, using conventional analysis, a given option might appear to provide the most effective use of resources while actually imposing greater environmental burdens than those it seeks to address.

This chapter highlights how the systematic analysis of the process chain allows the true nature of the implications associated with prevention or mitigation strategies to be made explicit. It considers why the use of traditional economic appraisal techniques which might not be sufficient to ensure all costs and benefits associated with different strategies are included in the decision-making process. Moreover, it outlines current tools and techniques developed to address this issue and highlights their application within the aquatic environment.

8.2 ANALYZING THE PROCESS CHAIN

Achieving water quality targets, whether in terms of enhancement or preservation, often requires changes in the operation of existing processes or investment in additional treatment technologies. Identifying the economic implications of operational changes or investment is relatively straightforward and has been the subject of substantial research.[1-8] While any wider environmental impacts of such decisions are more difficult to determine, in general, it possible to distinguish two specific areas of environmental impacts that need to be considered in the decision-making process to ensure the full implications of any given option are evaluated (Figure 8.1). For the purpose of this analysis, these two areas are termed environmental costs and benefits. It is important to recognize each can be positive or negative in nature and can be measured in both monetary and nonmonetary terms.

Environmental benefits are likely to accrue to users (both direct and indirect) of a watercourse as a consequence of meeting a specified water quality target. Direct users include those that derive utility from actual consumption and include anglers, walkers, and picnickers. Indirect users are those whose utility stems from knowledge that a particular environmental resource is being enhanced or preserved and reflects a more intrinsic environmental value. For example, the willingness of individuals to financially support campaigns to protect rain forests even though they have no intention of visiting them reflects such indirect uses. Environmental costs reflect the

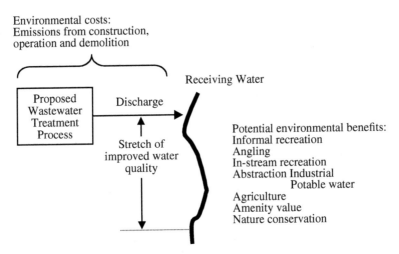

FIGURE 8.1 The focus of current benefit evaluation techniques.

potential implications of building and operating the treatment plant necessary for meeting a particular water quality target. They can be considered in terms of their contribution to different areas of environmental concern (e.g., global warming potential [GWP], acidification)(Figure 8.2).

Given the complex nature of the fate and behavior of EDCs in water treatment processes, identifying precisely how environmental costs and benefits are likely to accrue as a consequence of different management strategies is likely to be at least highly compound specific. It is also accepted that in many cases the transformation of parent compounds will be highly site or process specific and that the relative importance of the transformation compounds will be dependent on local conditions (such as the presence of other compounds or the physical/chemical characteristics of the influent).[9]

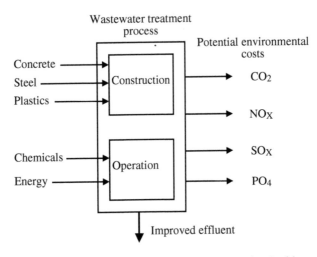

FIGURE 8.2 Example of potential environmental costs associated with constructing and operating wastewater treatment processes.

It is not intended that this section present an in-depth analysis of the costs and benefits associated with removing any single EDC. Rather the aim is to highlight how adopting the approach outlined above enables both key areas of concern and significant areas of sensitivity to be identified. For the purpose of this chapter, the analysis will focus on wastewater treatment processes. However, it is also important to recognize that many of the issues are equally applicable to the removal of EDCs in other areas of their life cycle.

In wastewater treatment processes there are potentially a number ways of removing EDCs (Figure 8.3). Each is likely to be associated with different environmental costs and benefits. EDC removal could be enhanced by altering secondary wastewater treatment process parameters. For example, for certain compounds, reducing the hydraulic loading rate could increase removal. Reducing the loading rate is likely to require additional treatment capacity, and the provision of such capacity will have a number of associated environmental costs in terms of resource and energy consumption. Similarly, reducing the influent concentration of EDCs can increase the removal efficiency. Increasing dilution, perhaps through limited recycling of effluent, will also require additional treatment plant that also has associated environmental costs. To determine whether such approaches provide an efficient means of increasing removal, these environmental costs need to be considered in terms of the benefits that accrue as a consequence of lower effluent concentrations being discharged into the environment. Such benefits could reflect improvements in the quality of the aquatic environment or reductions in the risk to human health.

Depending on the mechanism by which EDCs are being removed from the influent stream, there could be additional environmental costs that need to be considered to ensure the true environmental impacts are reflected in the analysis. For example, if EDCs are being bound into the secondary sludge, they may impose additional burdens on incumbent sludge treatment and disposal systems.

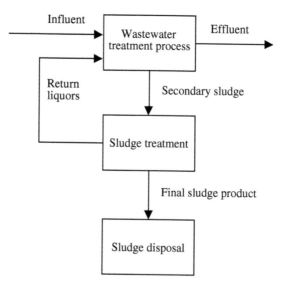

FIGURE 8.3 Potential EDC removal routes in wastewater treatment.

Again, the precise nature of any such additional burdens will be dependent on the compounds under consideration and the type of sludge treatment processes being employed. As outlined in the preceding chapters, sewage sludge treatment processes can have a significant impact on the overall concentration of EDCs in the final sludge product, and each is likely to be associated with specific impacts. For example, during sludge thickening some compounds will be released from the solid phase and returned to the head of the treatment works. As increases in the influent concentrations can decrease removal efficiencies, additional environmental costs might be incurred in dealing with return liquors as outlined above.

To enable the relative environmental costs and benefits of different prevention or mitigation strategies to be compared, it is essential they are evaluated consistently using systematic methodologies. The remainder of this chapter considers why traditional economic assessment techniques do not account for such costs and benefits adequately, and outlines how available methodologies seek to address this issue.

8.3 ECONOMIC AND ENVIRONMENTAL CONSIDERATIONS

For over 30 years it has been recognized that classical economic theory is unable to account adequately for environmental goods, or the impact economic activity has on them.[10] One reason why neoclassical economics has not addressed environmental considerations is due to the allocation of property rights. Neoclassical economics is concerned with maximizing social welfare. This is achieved by considering how supply and demand factors operate through the marketplace to ensure an efficient (Pareto optimal) allocation of resources; these markets only function properly when the exchange of property rights is involved. Environmental characteristics such as clean air and water, natural beauty, and wildlife support capacity are defined as public goods that are equally available to all members of society and have no property rights attached to them. Therefore, decisions regarding public goods in a market economy may not be consistent with Pareto optimality.[11]

A market is said to be Pareto optimal if it is not possible to make someone better off without making someone else worse off. However, it is important to note that Pareto optimality does not consider whether the initial distribution of goods is equitable or fair.[12] In basic terms, Pareto optimality is reached when supply equals demand, or using alternative terminology, where marginal cost equals marginal benefit. These ideas are presented in Figure 8.4, where supply meets demand at A, or the production of Q goods.

For market outcomes to reflect Pareto optimality, a number of ideal assumptions have to be met. Of particular importance in an environmental context, is the assumption that there are no market externalities. Externalities are the consequences of an action carried out by a firm or individual that are not included in the costs and benefits associated with the original action.[11] For example, if a firm pollutes a water course as a consequence of its production process, the quality of water available to other firms using the same water course is reduced. Consequently, lower water quality may mean additional treatment costs are imposed on all firms. The full production costs of the

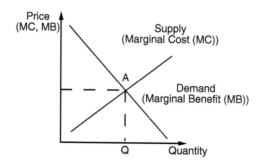

FIGURE 8.4 Pareto optimality in ideal markets.

original polluting firm should include all additional treatment costs. However, water is a public good, has no property rights associated with it and is not owned by anyone. This means that no one can charge the original firm for disposing its waste into the river. In such situations, pollution is an externality and the additional treatment costs are external costs. The impact of such externalities is presented in Figure 8.5.

If the original polluting firm were to be made to include the external costs of production, then their marginal costs would increase and hence marginal costs would now equal marginal benefits at B corresponding to the production of Q* goods. Clearly, noninclusion of this type external cost is likely to lead to over production and higher than optimal rates of consumption.[11]

The consequences of consuming public goods in this way is often referred to as the tragedy of the commons.[13] This suggests that rather than supporting the notion of the invisible hand, where decisions made by an individual to further his own gain are expected be in the best public interest, such consumption is likely to have detrimental effects on the quality or availability of public goods and hence on society. The potential impact of the tragedy of the commons is perhaps best typified by the possibility of global climate change resulting from the so-called enhanced greenhouse effect. Waste products from industrial activity, particularly carbon dioxide, are thought to be contributing to the natural greenhouse effect to such an extent that it may result in potentially catastrophic changes in the global climate.[14] In this context, "each rational man finds that his share of the cost of the wastes he discharges into the commons to be less than the cost or purifying his wastes before releasing them." [13]

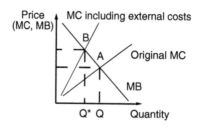

FIGURE 8.5 The impact of external costs.

Optimizing public goods consumption or availability means economic appraisals must account for all external costs associated with a particular course of action, whether it is formulation of policy or implementation of investment. As highlighted above, externalities include environmental considerations, such as the attractiveness of an area and the level of satisfaction derived from a visit to a particular resource. The way society responds to these types of environmental considerations reflects its preferences, and the welfare of society is a function of how satisfactorily their preferences are met. Determining the external costs associated with a particular course of action is dependent on identifying and measuring how it affects preferences being satisfied, or how welfare is changed. Establishing which policy or investment option is most efficient requires both changes in welfare and economic costs associated with each option to be compared directly.

As outlined above, two potential sources of environmental externality have been distinguished for the purpose of this analysis, namely environmental costs and benefits (Figure 8.3). The remainder of this chapter outlines techniques designed to enable the evaluation of such externalities to be carried out.

8.3.1 ENVIRONMENTAL BENEFIT EVALUATION

Over the last 20 years, several benefit assessment techniques have been developed for use in the aquatic environment that center around use of cost benefit analysis (CBA). CBA provides an appraisal framework that enables costs and benefits associated with a particular action to be compared systematically over time. A more in-depth review of CBA is presented in the literature.[11,15,16]

In the late 1970s, a manual for assessing the benefits of flood alleviation was developed and that intended to "simplify assessment of benefits to the community of protecting both urban areas and agricultural land from flooding" through the application of CBA.[15] However, it does not include the value of indirect or environmental costs and benefits in the analysis. Subsequently, a project appraisal guide for determining urban flood protection benefits was developed that was intended to be a companion to the earlier manual.[16] The focus of the guide is limited to urban flooding. Although it does include indirect benefits and losses in the appraisal framework, environmental consequences and the use of environmental economics are not considered.

Impacts associated with flooding together with coastal erosion were further considered in the development of an integrated approach to coastal management.[18] One significant development is the inclusion of environmental economic valuation techniques to estimate the recreational and amenity value of coastal sites and enable their inclusion in investment appraisal.

Providing justification for water quality improvement schemes is considered specifically in *Assessing the Benefits of Surface Water Quality Improvements*,[18] hereafter referred to as the Manual. The Manual is intended to "identify and value the benefits arising from proposed investments in surface water quality (rivers coasts and estuaries) in order to undertake CBA." In addition, a multi-attribute scoring and weighting system (MAT), based in part on the above manual and developed by the Environment Agency,[19] is also available. MAT uses nonmonetary measures to evaluate the magnitude of potential benefits to the same range of water quality scenarios.

Central to the operation of the Manual is the application of environmental economic valuation techniques that enable values to be ascribed to environmental benefits and allow their inclusion in policy or investment decision making.[20–22] Valuation of most goods is made by observing the markets in which they are traded. However, in the absence of markets for environmental goods, a number of potential methods of establishing value have been developed.[23] A brief review of the most important valuation techniques is presented below, and more details can be found in the literature.[11,23–26] The main techniques used for valuation in environmental economics are:

- Travel cost method (TCM)
- Contingent valuation method (CVM)
- Hedonic price method (HPM)
- Replacement cost method (RCM)

8.3.1.1 Travel Cost Method

The TCM is based on the principle that although the value of recreational use cannot be derived from actual markets, the value is implicit in the amount of time and expense an individual is willing to spend in undertaking a visit to a particular site. This method is most commonly used to value the recreational use of a resource, and can be considered to be part of either the direct or indirect use values. Initially a value is assigned to an individual's leisure time. The leisure time allows a monetary value to be calculated for the total time taken during the visit, including time taken travelling to and from the resource as well as time spent at the resource. The costs incurred in undertaking the visit, including all admission and participating fees and travel expenses, are added to this total. The summation of the two totals for each visitor to the resource reflects the opportunity cost of the visit and therefore the value placed upon the resource by each visitor. Application of this process to all visitors gives an estimation of the particular use value being estimated for that resource.

The TCM has the advantage of measuring actual expenditure by users of a resource and can be determined through on site surveys. The main disadvantage is that no consensus exists as to the value of an individual's leisure time, and determination of this value is a complex issue.[27]

8.3.1.2 Contingent Valuation Method

The CVM is most commonly used to provide estimates for values that cannot be derived from markets. This method requires surrogate valuations to be established through surveys, analogous to market surveys or political polls. The manner in which the market functions is that individuals are questioned directly about their willingness to pay for a particular resource or set of circumstances pertaining to that resource. For example, individuals may be asked about their willingness to pay to preserve a resource in its current state rather than allow it to be developed in some way. Realism of the hypothetical market is essential for several reasons:

- Respondents are being asked to "buy" public goods that they have little or no previous experience in placing a value.
- There is difficulty in knowing whether respondents have fully understood exactly what they are valuing and the consequences of their decision.
- Respondents must believe the market to be as realistic as possible to ensure that their willingness to pay is as close as possible to the amount they would actually pay in a real market.

The willingness-to-pay figures given by respondents reflect the value they place on the resource. If these individual values are aggregated across the survey sample and then applied to all members of the population considered to have an interest in the site, the full value can be estimated.

While the CVM has the advantage of simulating a market which in practice does not exist, this is also its principal disadvantage, as no actual transactions take place. Values derived by this method must be used with caution, as they can be subject to an intrinsic error.

8.3.1.3 Hedonic Price Method

The HPM is used to measure a combination of indirect and nonuse values by the observation of an actual market where values are considered to be influenced by the presence of an environmental resource or characteristic, such as a reservoir or air/water quality. In general, the property market is used, where it is assumed that by allowing for general property characteristics (e.g., size, proximity to local transport etc.), it is possible to isolate that proportion of the house price that is a direct consequence of the property being near, or subjected to, the resource or characteristic. For example, individuals may be willing to pay a premium for property that is near to, or has a view across, a reservoir. If it is possible to isolate the value of this premium and estimate the number of properties whose price includes such a premium for a particular resource, an estimation of the value of the resource may be made.

The HPM's advantage is that values are derived from actual market transactions. The disadvantage is that it is difficult to isolate the proportion of the market price that reflects the presence of an environmental resource or the characteristics of interest.

8.3.1.4 Replacement Cost Method

The RCM is used to estimate the total value of the resource in question. The principle of the method is that under certain circumstances, the costs incurred in restoring environmental damage can be taken as a measure of the benefits derived from the restoration, assuming that the costs can be readily identified. Care must be exercised in the application of this method since its use in CBA will inevitably lead to a cost-benefit ratio of unity.

8.3.2 Environmental Cost Evaluation

Considerable attention has been focused on evaluating environmental benefits. However, the environmental costs associated with water quality improvement, in terms of resources consumed and emissions released through the treatment plant's construction and operation have not been considered.

At a strategic level, the importance of considering a wide range of environmental costs and their contribution to areas of environmental concern is already recognized. Environmental reports from some United Kingdom water service companies are already defining company performance in terms of a number of key emissions (e.g., carbon dioxide, nitrogen oxides and sulphur oxides).[28] Within the U.K. water industry there is significant attention focused on the development of a range of environmental performance and sustainability indicators. These are likely to relate directly to regulatory standards and issues raised in U.K. government policy documents, and they may form the basis for setting environmental improvement targets in strategic action plans.[29]

Such balance sheets of environmental performance are essential for understanding the operational impact of a given company. In order to optimize such performance, the relative contribution of different policy/investment options needs to be established and managed at an early stage. Consideration of these factors at the investment/policy appraisal stage enables robust accounts of current and future emissions to be developed, and allows the true environmental cost of implementing legislation to be established. In addition, conflicting pressures, particularly on regulated industries, can be identified. For example, U.K. water service companies are required to comply with legislation that invariably results in the construction and operation of energy intensive treatment processes.[29] However, at the same time, they are being encouraged to reduce emissions (particularly those contributing to global climate change that to a large extent originate from energy generation and consumption) through the application of economic instruments.[30]

Increased focus on sustainability issues in forthcoming legislation, such as the Water Framework Directive,[31,32] and the necessary inclusion of environmental performance measures in sustainable development indicators mean the systematic evaluation of environmental costs will become increasingly important in investment decision making. One approach enabling the systematic evaluation of potential environmental costs associated with process construction/operation is life cycle assessment (LCA). Indeed, use of a life cycle approach is fundamental to several of the available tools, such as the sustainable process index, ecological foot-printing, and appropriate carrying capacity.[33]

8.4 DECISION SUPPORT TOOLS

In the aquatic environment, monetary and nonmonetary evaluation of environmental benefits is currently undertaken through application of the Manual or MAT, respectively. Such techniques focus on benefits associated with the impact of discharging cleaner effluents, and the subsequent effect on quality within the receiving water (Figure 8.1). Benefits are defined in terms of how they accrue to user groups whose activities are influenced directly by a given quality change.

Benefit estimates derived from application of the FWR Manual are based on changes to operational or capital expenditure (Opex and Capex), or where appropriate values obtained through environmental economic valuation techniques, such

TABLE 8.1
Example of Benefit Value Estimation Using the Manual

User Group	Quality Change	Benefit Value for Each Person Visit	Annual No. of Person Visits	Annual Benefit Stream
Informal recreation[a]	Significant reduction in sewage derived litter[a]	£1.20[a]	10,000	£12,000

[a] Defined in Foundation for Water Research, Assessing the Benefits of Surface Water Quality Improvements, Medmenham, England, 1996.

as the hedonic price method or contingent valuation.[18] In all cases, estimates are expressed in monetary terms as annual benefit streams. For example, significant reductions in sewage derived litter in a receiving water, visited by 10,000 informal recreation users each year, may realize an annual benefit stream of £12,000 (Table 8.1). Clearly, benefit estimates established in this way may be included in investment appraisal techniques, such as CBA, and consequently compared directly with financial cost data.

In contrast, estimates derived through application of the MAT are expressed in terms of a weighted score, reflecting the influence of a number of key receiving water characteristics.[19] An example application relating to the above scenario is presented in Table 8.2, where the benefit score for informal recreation is 288. This type of approach precludes direct comparison with financial cost data in CBA. However, the summation of individual scores for each affected user group may be combined with financial data to determine the cost-effectiveness of a given option. In the context of investment in the aquatic environment, where there is likely to be

TABLE 8.2
Example Application of MAT

User Group	Length (a)	Access (b)	Visitor Potential (c)	Water Quality (d)	Aesthetics (e)	Score (a * b * c)*(d + e)
Informal recreation	3 (13 km affected)	4 (fair proportion of stretch accessible: some car parking)	3 (>35,000 population within 3 km with few alternative sites)	3 (River ecosystem class 4 to class 3)	5	288

Note: Individual scores in bold; definitions and scores from Environment Agency, Multi-Attribute Scoring and Weighting System, Bristol, England, 1999.

a potentially large number of competing schemes due to legislative pressures, investment prioritization through cost-effectiveness scores can be a powerful tool.

The above techniques account for benefits associated with a particular change in quality (Figure 8.1). However, no account is taken of the potential environmental costs incurred in constructing and operating a wastewater treatment process required to actually deliver that change. An outline of such potential environmental costs is presented in Figure 8.2.

8.4.1 Adopting a Life Cycle Assessment Framework

One technique enabling the systematic evaluation of potential environmental costs associated with process construction/operation is LCA. LCA is a technique for assessing the environmental aspects and potential impacts of a product or process throughout its life, from raw material extraction through production, operation, and demolition (disposal).[34] The key stages of LCA are presented in Figure 8.6:

- Goal and scope definition: States the indented application and reasons behind an assessment. Also includes a definition of the process system to be studied and its boundaries, data requirements, assumptions, etc.[34]
- Inventory analysis: Involves the collection of necessary data defining all resource, materials, and energy inputs. Also includes outputs associated with the process system, and the application of calculation procedures to determine all releases to air, water, and land arising from these inputs and outputs.[34]
- Impact assessment: Results from the life cycle inventory analysis are used to evaluate the significance of potential environmental impacts.[34]

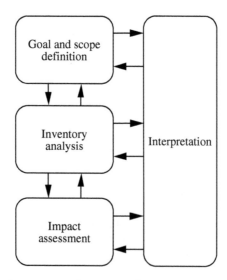

FIGURE 8.6 Life cycle assessment framework. (From International Organization for Standardization, International Standard 14040, Environmental Management: Life Cycle Assessment: Principles and Framework, Switzerland, 1997.)

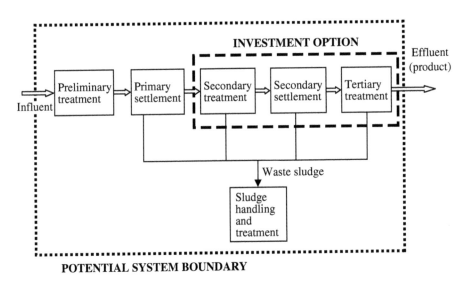

FIGURE 8.7 Example of potential process units to be included in a system boundary.

The selection of an appropriate system boundary is central to the LCA's success. This boundary defines those process units or stages to be included in the assessment. This is particularly important where LCA is used to make comparative assertions regarding the relative performance of a number of options, as any comparison must be executed on the basis of equivalent function. For example, in the context of wastewater treatment, one of the options outlined above could produce significantly different volumes of waste sludge as a consequence of delivering a given quality change. If this is the case, then processes involved in subsequent sludge handling and treatment should be included in the system boundary. Such considerations must extend to differences between options at each stage of the process chain, and may result in a significantly more process units being included in the system boundary than were in the original investment option (Figure 8.7).

Clearly, the need to establish an inventory of all material, resource, and energy flows for each process within a system boundary is likely to make a comprehensive LCA of every potential investment or policy option uneconomic. However, research suggests a majority of the environmental burdens associated with a range of wastewater treatment scenarios are due to a limited number of key system inputs and outputs.[6,35–37] Of these perhaps the most important is the consumption of energy during the operational phase of the life cycle. Even when a relatively conservative asset life of 20 years is used (some civil engineering assets may have an expected life in excess of 40 years), environmental impacts associated with the operational phase still dominate. In particular, this phase may contribute over 90% of global warming potential, acidification, and ozone depletion (Figures 8.8a to 8.8c), virtually all of which originates from energy consumption.[36,37]

Emissions relating to the construction phase of wastewater treatment processes also have a distinct impact on overall environmental performance (Figures 8.8a to 8.8c). However, of the wide range of resources and materials that may be required during

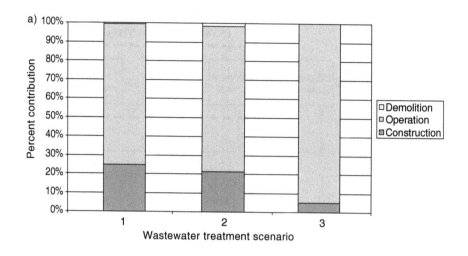

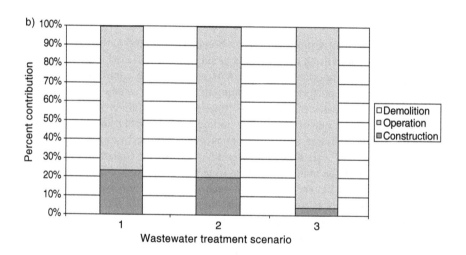

FIGURE 8.8 (a) The contribution of different life cycle stages to GWP. (1) distributed works; (2) centralized works (From Gay, R.J., A Comparison of the Advantages and Disadvantages of Regional versus Distributed Wastewater Treatment Plant, MSc. thesis, Imperial College, London, 1999); (3) UV disinfection (From Barrett, J.L., An Examination of the Whole Costs of Introducing Advanced Wastewater Treatment at Cromer Sewage Treatment Works on the North Norfolk coast, MSc. thesis, Imperial College, London, 1998). (b) The contribution of different life cycle stages to acidification. (1) distributed works; (2) centralized works (From Gay, R.J., A Comparison of the Advantages and Disadvantages of Regional versus Distributed Wastewater Treatment Plant, MSc. thesis, Imperial College, London, 1999); (3) UV disinfection (From Barrett, J.L., An Examination of the Whole Costs of Introducing Advanced Wastewater Treatment at Cromer Sewage Treatment Works on the North Norfolk coast, MSc. thesis, Imperial College, London, 1998).

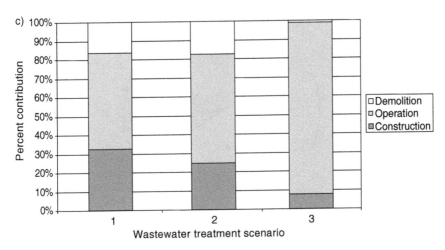

FIGURE 8.8c The contribution of different life cycle stages to ozone depletion. (1) distributed works; (2) centralized works (From Gay, R.J., A Comparison of the Advantages and Disadvantages of Regional versus Distributed Wastewater Treatment Plant, MSc. thesis, Imperial College, London, 1999); (3) UV disinfection (From Barrett, J.L., An Examination of the Whole Costs of Introducing Advanced Wastewater Treatment at Cromer Sewage Treatment Works on the North Norfolk coast, MSc. thesis, Imperial College, London, 1998).

manufacture and construction, only a limited number make any significant contribution, such as concrete, steel, and iron (Figure 8.9). Furthermore, environmental impacts associated with the demolition and disposal life stage may only contribute up to 20% of the total; in many cases, this contribution may be significantly less (Figures 8.8a to 8.8c).

The nature of different treatment processes is likely to not only change the relative contribution of individual life stages, but also the absolute magnitude of their impact. For example, research suggests carbon dioxide emissions associated with activated sludge plant are dominated by operational energy requirements, with manufacture and construction having only a slight impact. However, not only are operational emissions significantly lower for biological filters, but also manufacture and construction have a considerably greater impact on the total.[35] To account for how such differences contribute to environmental performance indicators and to include them as part of the decision making process in investment appraisal, they need to be quantified in a consistent and transparent manner.

A number of LCA software tools are available currently that enable the impact of a wide range of material and resource flows to be evaluated, and related directly to areas of environmental concern such as those employed for environmental performance measurement (Figure 8.10).[38,39] Clearly, such tools allow large amounts of impact and emissions data to be considered systematically and in a cost-effective way. Furthermore, effects related to variations in manufacture, operation, or disposal techniques, as well as data quality, can be modeled rapidly through sensitivity analysis. For example, in terms of GWP, each GW.hr of electricity consumed in the United Kingdom is likely to generate between approximately 0.75 and 1 tonne of carbon dioxide equivalent, depending on the generation mix (gas, coal, nuclear, etc.).[39]

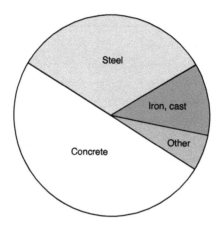

FIGURE 8.9 The contribution of different construction phase material flows to GWP. From (A) distributed works (From Gay, R.J., A Comparison of the Advantages and Disadvantages of Regional versus Distributed Wastewater Treatment Plant, MSc. thesis, Imperial College, London, 1999).

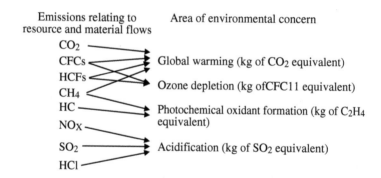

FIGURE 8.10 Example of the relationship between emissions and environmental impacts. (Adapted from PIRA International, PEMS Life Cycle Assessment Computer Model: User Manual, Surrey, London, 1997.)

Defining process environmental performance in terms of their quantitative contribution to areas of particular concern (Table 8.3), based on data outlined above, enables such considerations to be included in investment appraisal. Where investment prioritization dominates, absolute measures of individual contributions can be combined with existing nonmonetary measures of environmental costs and benefits (MAT technique) to form part of a cost-effectiveness appraisal.

8.5 SUMMARY

Before detailed strategies for managing the potential implications of releasing EDCs into the aquatic environment can be developed, it is essential that the fate and behavior of different compounds (both parent and transformation compounds) is

TABLE 8.3
Magnitude of Contributions to Areas of Environmental Concern

Investment Option	GWP (tons CO_2 equiv.)	Acidification (tons SO_2 equiv.)	Ozone Depletion (kg CFC11 equiv.)
A	2940	15.6	1.5
B	3130	16.2	0.85

understood fully. This necessarily includes consideration of how local conditions are likely to influence their impact. Notwithstanding these requirements, strategies that are developed need to be evaluated in terms of their overall effectiveness, including environmental and economic considerations. While the economic implications are likely to be readily identified, the environmental impacts have historically been far more difficult to determine. In determining likely environmental impacts, it is important not only to consider the potential benefits of improvements in environmental quality (an area which has received considerable recent attention), but also the environmental costs associated with such improvement. Tools and techniques outlined in this chapter are designed to address these issues. Their application could help to ensure management strategies designed to mitigate one problem do not simply contribute to another.

However, external barriers, such as the role of public perception and attitudes surrounding these issues, should also be addressed to complement the success of the existing tools and techniques. Influencing factors that sway public opinion involve the role of the media, the level of information received or accessed within the public domain, and more important, public confidence in decision makers.[40,41]

REFERENCES

1. Morse, G.K., Brett, S., Guy, J.G., and Lester, J.N. Phosphorus sustainability: a review of phosphorus removal and recovery technologies, *Sci. Total Environ.*, 212, 69–81, 1998.
2. Brett S.W., Morse, G.K., and Lester, J.N., Operational expenditure in the water industry. I. A methodology for estimating variable costs at an advanced water treatment works, *Euro. Water Manage.*, 1, 31–38, 1998.
3. Guy, J.A., Morse, G.K., and Lester, J.N., Operational expenditure in the water industry II: a methodology for determining operational expenditure in wastewater treatment, *Euro. Water Manage.*, 1, 47–50, 1998.
4. Butt, E.P., Morse, G.K., Guy, J.A., and Lester, J.N., Co-recycling of sludge and municipal solid waste: a cost benefit analysis, *Environ. Technol.*, 19, 1163–1175, 1998.
5. Brett, S.W., Gaterell M.R., Morse, G.K., and Lester, J.N., A cost comparison of potable water softening technologies, *Environ. Technol.*, 20, 1021–1031, 1999.
6. Gaterell, M.R., Gay, R., Wilson, R., Gochin, R.J. and Lester, J.N.L., An economic and environmental evaluation of the opportunities for substituting phosphorus recovered from wastewater treatment works in existing UK fertiliser markets, *Environ. Technol.*, 21, 1067–1084, 2000.

7. Zakkour, P.D., Gaterell, M.R, Griffin, P., Gochin, R.J. and Lester, J.N., Anaerobic treatment of domestic wastewater in temperate climates: treatment plant modeling and costing, *Water Res.*, 35, 4137–4149, 2001.
8. Zakkour, P.D., Gaterell, M.R, Griffin, P., Gochin R.J. and Lester J. N., An economic evaluation of anaerobic domestic wastewater pre-treatment for temperate climates, *Euro. Water Manage.*, 4, 35–46, 2001.
9. Perry, R., Kirk, P.W.W., Stephenson, T., and Lester, J.N., Environmental aspects of the use of NTA as a detergent builder, *Water Res.*, 18, 255–267, 1984.
10. Boulding, K., The coming spaceship Earth, in *Environmental Economics*, Markandya A. and Richardson J., Eds., Earthscan, London, 1992.
11. Common, M., *Environmental and Resource Economics: An Introduction*, Longman, New York, 1988.
12. Sen, A., *On Ethics and Economics*, Basil Blackwell, Oxford, 1987.
13. Hardin, G., The tragedy of the commons, in *Environmental Economics*, Markandya A. and Richardson J., Eds., Earthscan, London, 1992.
14. Malnes, R., *Valuing the Environment*, Manchester University Press, Manchester, England, 1995.
15. Penning-Rowsell, E.C. and Chatterton, J.B., *The Benefits of Flood Alleviation: A Manual of Assessment Techniques*, Saxon House, Farnborough, England, 1977.
16. Penning-Rowsell, E.C., Green, C.H., Thompson, P.M., Coker, A.M., Tunstall, S.M., Richards, C., and Parker, D.J., *The Economics of Coastal Management: A Manual of Benefit Assessment Techniques*, Belhaven Press, London, 1992.
17. Parker, D.J., Green, C.H., and Thompson, P.M., *Urban Flood Protection Benefits: A Project Appraisal Guide*, Ashgate Publishing Company, Burlington, VT, 1987.
18. Foundation for Water Research, *Assessing the Benefits of Surface Water Quality Improvements,* Medmenham, England, 1996.
19. Environment Agency, Multi-Attribute Scoring and Weighting System, Bristol, England, 1999.
20. Gaterell, M.R., Morse, G.K., and Lester, J.N., Economic assessment of water quality improvements in the Orwell Estuary using the Foundation for Water Research benefit assessment procedure, *Euro. Water Manage.*, 1, 31–36, 1998.
21. Gaterell, M.R., Morse, G.K., and Lester, J.N., Investment in the aquatic environment. I. A marketing based approach, *J. Environ. Manage.*, 56, 1–10 1999.
22. Gaterell, M.R., Morse, G.K., and Lester, J.N., Investment in the aquatic environment. II. Comparison of two techniques for evaluating environmental benefits, *J. Environ. Manage.*, 56, 11–24, 1999.
23. Pearce, D., Markandya, A., and Barbier, E., *Blueprint for a Green Economy*, Earthscan, London, 1989.
24. Brookshire, D.S., Eubanks, L.S., and Randall, A., Estimating option prices and existence value for wildlife resources, *Land Economics*, 59, 1–15, 1983.
25. Cesario, F.J., Value of time in recreation benefit studies, *Land Economics*, 52, 32–41, 1976.
26. Faber S., The value of coastal wetlands for recreation: an application of travel cost and contingent valuation methodologies, *J. Environ. Manage.*, 26, 299–312, 1988.
27. McConnell, K.E., Measuring the cost of time in recreation demand analysis: an application to sports fishing, *Am. J. Agr. Econ.*, 63, 153–156, 1981.
28. Severn Trent Water, Stewardship 1998: Environmental Report, Coventry, England, 1998.
29. Anon., Energy and the environment, *e.water*, 140, 1999.

30. UK Government, Economic Instruments and the Business Use of Energy: A Report by Lord Marshall, HM Treasury, London, 1998.
31. World Wildlife Fund, EU Water Framework Directive, available at: http://www.panda.org/europe/freshwater/initiatives/wfd-hist.html, accessed Sep. 15, 2001.
32. Europa, Community legislation in force, available at http://europa.eu.int/eur-lex/en/lif/dat/2000/en_300L0060.html, accessed Mar 1, 2001.
33. Krotscheck, C., Measuring eco-sustainability: comparison of mass and/or energy flow based highly aggregated indicators, *Environemetrics*, 8, 661–681, 1997.
34. International Organization for Standardization, International Standard 14040, Environmental Management: Life Cycle Assessment: Principles and Framework, Switzerland, 1997.
35. Emmerson, R.H.C, Morse, G.K., and Lester, J.N., The life-cycle analysis of small-scale sewage treatment processes, *Water Environ. Manage. J.*, 9, 317–326, 1995.
36. Barrett, J.L., An Examination of the Whole Costs of Introducing Advanced Wastewater Treatment at Cromer Sewage Treatment Works on the North Norfolk coast, MSc. thesis, Imperial College, London, 1998.
37. Gay, R.J., A Comparison of the Advantages and Disadvantages of Regional versus Distributed Wastewater Treatment Plant, MSc thesis, Imperial College, London, 1999.
38. Rice, G., Clift, R., and Burns, R., LCA software review, *Int. J. Life Cycle Assess.*, 2, 53–59, 1997.
39. PIRA International, PEMS Life Cycle Assessment Computer Model: User Manual, Surrey, London, 1997.
40. Myatt-Bell, L.B., Scrimshaw, M.D., Lester J.N., and Potts, J., Public perception of managed realignment: Brancaster West Marsh, North Norfolk, UK, *Mar. Policy*, 26, 45, 2002.
41. Myatt, L.B., Scrimshaw, M.D., and Lester, J.N., Public perceptions and attitudes towards a current managed realignment scheme: Brancaster West Marsh, North Norfolk, UK., *J. Coastal Res.*, submitted, 2002.

Index

A

Acid dissociation constant, 104
Acinetobacter spp., in polyaromatic compound degradation, 126
Adsorption, of endocrine disrupting chemicals
  in drinking water treatment, 223, 228, 229
  in receiving waters, 180–181
  in sewage treatment, 108–113
Adsorption potential, 113, 224
Aerobic degradation of EDCs, in sludge, 152, 163
Agonistic effects, 5–6
Agricultural runoff
  estrogens in, 41
  organotins in, 45
  sludge disposal and, 166–167
Alcaligenes spp., in polyaromatic compound degradation, 126
Algae (*Chlorella vulgaris*), 200
Alkylphenolic compounds, 10, 16–17
  anaerobic degradation of, in sludge, 151, 158–159
  analysis of, in wastewater, 78–79
  chromatographic analysis of, 75, 78–80
  degradation of, in receiving waters, 200, 202–204
  in drinking water sources, 224
  as leachates in drinking water, 226–227
  primary removal of, in sewage treatment, 119–122
  removal of, in drinking water treatment, 230
  in sludge, 147
  sources of, 45–46
Amended soils, 166–167
Anaerobic degradation of EDCs, in sludge, 151–154
  mechanisms of, 153–154, 156
Analytical methods, 59–60
  chemical, 72–85
  immunological, 85, 88
  in screening systems, 70–72
  in vitro, 61–67
  in vivo, 67, 69–70
Androgens, definition of, 8
Antagonistic effects, 5–6
Antifouling paint, 2–3, 44, 45, *see also* Organotins
Aquatic environment, see Receiving waters
Aqueous solubility, 104, 113, 187–188, 224

Arthrobacter spp., in polyaromatic compound degradation, 126, 191
Assessing the Benefits of Surface Water Quality Improvements, 273–274, *see also* Manual
Atmospheric pressure chemical ionization (APCI), 73, 75
Atrazine
  chemical and physical properties of, 110
  degradation of, in receiving waters, 195
  removal of, in drinking water treatment, 231

B

Bacillus spp., in polyaromatic compound degradation, 198
Barnacle (*Elminius modestus*), 183
Benz(a)anthracene, 19
  chemical and physical properties of, 111
Benzo(k)fluoroanthene, chemical and physical properties of, 111
Benzo(a)pyrene, 19
  chemical and physical properties of, 111
Bioaccumulation of EDCs, in receiving waters, 187–188
Bioassays, 60
Bioavailability of EDCs, in receiving waters, 181, 187
Biochemical oxygen demand (BOD), 106
Biocides, see Organotins
Bioconcentration factor (BCF), 187–188, 224
Biologic degradation/transformation of EDCs, in sewage treatment, 114–115
Biophore, 21
Bioreactor, 157
Bisphenol A, 15
  anaerobic degradation of, in sludge, 162
  bacterial degradation of, 191
  chemical and physical properties of, 112
  degradation of, in receiving waters, 190–191
  as leachate in drinking water, 226–227
  primary removal of, in sewage treatment, 125
  removal of, in drinking water treatment, 230
  sources of, 42–43
Bisphenols, 15
  sources of, 42–43
Breast cancer cell lines, human, 63–65